Learni

We work with leading authors to develop the strongest
educational materials in Building and Construction,
bringing cutting-edge thinking and best learning
practice to a global market.

Under a range of well-known imprints, including
Prentice Hall, we craft high quality print and electronic
publications which help readers to understand and
apply their content, whether studying or at work.

To find out more about the complete range of our
publishing, please visit us on the World Wide Web at:
www.pearsoned.co.uk

CONSTRUCTION TECHNOLOGY

FOURTH EDITION

ROY CHUDLEY   *MCIOB*

Revised by
ROGER GREENO   *BA (HONS)*, *FCIOB, FIPHE, FRSA*

Harlow, England • London • New York • Boston • San Francisco • Toronto • Sydney • Singapore • Hong Kong
Tokyo • Seoul • Taipei • New Delhi • Cape Town • Madrid • Mexico City • Amsterdam • Munich • Paris • Milan

Pearson Education Limited
Edinburgh Gate
Harlow
Essex CM20 2JE
England

and Associated Companies throughout the world

Visit us on the World Wide Web at:
www.pearsoned.co.uk

First published (as *Construction Technology*) 1973 (Volume 1), 1974 (Volume 2)
Second edition 1987
Third edition (published as a single volume, with revisions by Roger Greeno) 1999
Reprinted 2002 (revised edition), 2003 (twice) (third edition update)

British Library Cataloguing in Publication Data
A catalogue entry for this title is available from the British Library

Library of Congress Cataloging-in-Publication Data
Chudley, R.
    Construction technology / Roy Chudley ; revised by Roger Greeno.—4th ed.
      p. cm.
    Includes bibliographical references and index.
    ISBN 0-13-128642-0 (alk. paper)
    1. Building.   I. Greeno, Roger.   II. Title
  TH145.C49 2005
  690—dc22

                                           2005045899

ISBN 0 131 28642 0

10 9 8 7 6 5 4 3 2 1
10   09   08   07   06   05

Set by 35 in 10/12pt Ehrhardt
Printed in Great Britain by Henry Ling Ltd., at the Dorset Press, Dorchester, Dorset

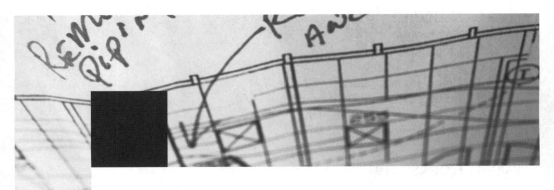

CONTENTS

PREFACE

Since the previous edition, reprint opportunities have permitted some amendments to include new procedures resulting from legislative and practice changes. This edition develops these further, with greater attention to information and detail. It also addresses many more recent issues, especially aspects of the Building Regulations that now require housing to be designed and built to more environmentally responsible and thermally efficient standards. These include reducing fuel energy consumption of heating and hot water equipment and the establishment of continuous insulation about the building envelope. Reference is also provided for carbon emission assessment relative to the fuel efficiency construction of dwellings.

Notwithstanding contemporary requirements, some well established building practices and techniques described in previous editions are purposely retained. These remain a valid reference to existing building stock.

Human rights issues are considered with regard to building designers and owners responsibilities for providing accessibility for the disabled. The implications of the Disability Discrimination Act and the associated Building Regulations for new dwellings are outlined in a new chapter containing basic provisions.

As with previous editions the content is neither extensive nor prescriptive, space would never permit inclusion of every possible means for constructing buildings. However, the content is generally representative and details and explanations typical of adopted procedures. Allowance should be made for regional traditions, material resources and local standards.

The original concept of providing supplementary lecture support material for students of construction is maintained. This book should be read in conjunction with experiential learning in the work place or by observation. Further study of associated legislation, practice guidance papers, product manufacturer's literature and specialised text is encouraged.

In conjunction with this edition's companion volume, *Advanced Construction Technology*, the reader should gain an appreciation of the subject material to support progression through technical, academic and professional qualifications.

Roger Greeno
Guildford 2005

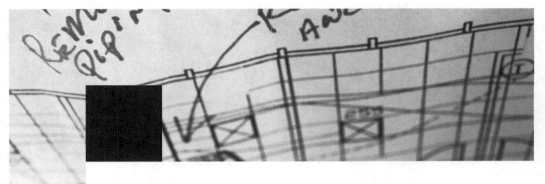

PREFACE TO THIRD EDITION

Roy Chudley's *Construction Technology* was first published in four volumes, between 1973 and 1977. The material has since been continuously updated through numerous reprints and full second editions in 1987. The books have gained a world-wide readership, and their success – and their impact on construction education – is a tribute to Roy Chudley's experience in further and higher education and his talents as a skilled technologist, illustrator and writer.

As a former colleague, it has been a privilege to once again work with Roy, on this occasion revising his original work, and compiling the material into two books: *Construction Technology* and *Advanced Construction Technology*. The content forms a thorough study for all students of building, construction management, architecture, surveying and the many other related disciplines within the diverse construction profession.

The original presentation of comprehensive text matched by extensive illustration is retained. Changes in legislation, such as the Building and Construction Regulations, have been fully incorporated into the text; however, as much of the original work as possible has been purposely retained as it contains many relevant examples of existing construction. Additional material discusses the new developments and concepts of contemporary practice.

The two new volumes are complementary, as many of the topics introduced here are further developed in *Advanced Construction Technology*. Together the books provide essential reading for all students aspiring to management, technologist and professional qualifications. They should be read alongside the current local building regulations and national standards, and where possible supplemented by direct experience in the workplace.

Roger Greeno
Guildford 1998

PREFACE TO SECOND EDITION

Note: This book is a combined and updated edition of Volumes 1 and 2 of Roy Chudley's original *Construction Technology*.

Volume 1

In writing this book it is not my intention to present a comprehensive reference book on elementary building technology, since there are many excellent textbooks of this nature already in existence, which I urge all students to study.

My purpose, therefore, has been to prepare in concise note form, with ample illustrations, the basic knowledge the student should acquire in the first year of any building technology course of study.

To keep the book within a reasonable cost limit I have deliberately refrained from describing in depth what has been detailed in the drawings.

Building technology is an extensive but not necessarily exact subject. There are many ways of obtaining a satisfactory construction in building, but whichever method is used they are all based upon the same basic principles and it is these which are learnt in building technology.

The object of any building technology course is to give a good theoretical background to what is essentially a practical subject. With the knowledge acquired from such a course, coupled with observations of works in progress and any practical experience gained, a problem should be able to be tackled with confidence by the time the course of study has reached an advanced level.

Another aspect of a course of this nature is to give sufficient basic knowledge, over the whole field of building activities, to enable the technologist to hold and understand discussions with other related specialists.

Volume 2

The aim of this second volume is to provide a continuity of study with the contents of Volume 1, which dealt mainly with domestic building. The second year of a typical construction technology course will study further aspects of domestic construction and introduce the student to framing techniques and materials. The presentation is in the form of notes accompanied by ample illustrations with as little repetition between notes and drawings as possible. It is also written with the assumption that the reader has covered, understood and retained the technical knowledge contained in a basic first-year course.

The depth of study presented in this volume is limited to the essential basic knowledge required for a second-year construction technology course. The student is therefore urged to consult all other possible sources of reference to obtain a full and thorough understanding of the subject of construction technology.

Roy Chudley

ACKNOWLEDGEMENTS

This book originated over 30 years ago as two separate publications. The two were combined to create a single volume at the last edition. The continuing success and popularity of this study resource can be attributed to contributions from numerous sources, but unfortunately space does not permit credit to them all. Much can be attributed to the observations and suggestions of some of my former colleagues, professional associates and not least the positive response of so many of my past students.

The book's agreeable presentation of comprehensive text and simple illustration is attributed to the late Colin Bassett as General Editor and of course, Roy Chudley as founding author. I am especially grateful to Roy for his cooperation, trust and permission to work on his original manuscript and hope that my attempts to emulate his illustrative style bear some comparison with his original work.

Without the publisher, the book would not exist. The enthusiasm and support of the staff at Pearson Education is appreciated; in particular that of Pauline Gillett for her direction and patience throughout the preparation of this latest edition.

Roger Greeno
Guildford 2005

Extracts from British Standards are reproduced with the permission of BSI. Complete copies can be obtained by post from BSI Customer Services, 389 Chiswick High Road, London, W4 4AL.

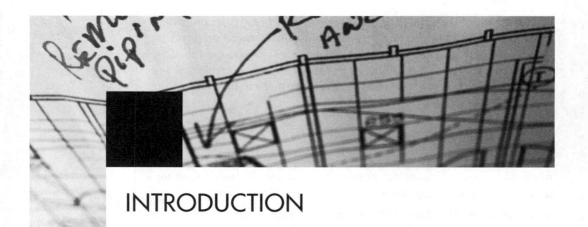

INTRODUCTION

There are two general aspects to the construction of buildings:

- conventional or traditional methods;
- modern or industrialised methods.

Conventional or traditional methods are studied in the first two years of most construction courses, with the intention of forming a sound knowledge base before proceeding to studies of advanced techniques in the final years. There is, nevertheless, an element of continuity and overlap between traditional and contemporary, and both are frequently deployed on the same building, e.g. traditional brick facing to a prefabricated steel-framed commercial building or to a factory-made timber-framed house.

Initial studies of building construction concentrate on the smaller type of structure, such as a domestic dwelling of one or two storeys built by labour-intensive traditional methods. Generally it is more economic to construct this type of building by these methods, unless large numbers of similar units are required on the same site. In these circumstances, economies of scale may justify factory-manufactured, prefabricated elements of structure. These industrialised methods are usually a rationalised manufacturing process used to produce complete elements, i.e. floors, walls, roof frames, etc. in modules or standardised dimensional increments of 300 mm.

Very few building contractors in the UK and other developed countries employ many staff directly. They are therefore relatively small companies when compared with the capital value of the work they undertake. This is partly due to the variable economic fortunes of the construction industry and the need for flexibility. Hence most practical aspects of building are contracted out to specialist subcontracting organisations, e.g. bricklayers, electricians, carpenters, in response to the main contractor's work load. The main contractor is effectively a building management company, which could be engaged on a variety of work, including major serial developments for the same client, maintenance work or aftercare programmes, extensions to existing structures, or possibly just small one-off projects.

It is essential that all students of building have an awareness of the variable methods of construction and application of materials in both traditional and industrial practice, in order to adapt their career pattern to the diverse expectations of the industry.

■■■ THE BUILDING TEAM

Building is essentially a team process in which each member has an important role to play. Figure 1 shows the organisation structure of a typical team for a large project, and the function of each member is outlined below.

- **Building owner**   The client; the person or organisation who finances and commissions the work. They directly or indirectly employ all other personnel, with particular responsibility for appointing the planning supervisor (usually the architect) and nominating the principal contractor – see Construction (Design and Management) Regulations 1994.
- **Architect**   Engaged by the building owner as agent to design, advise and ensure that the project is kept within cost and complies with the design.
- **Clerk of works**   Employed on large contracts as the architect's on-site representative. The main function is to liaise between architect and main contractor and to ensure that construction proceeds in accordance with the design. They can offer advice, but directives must be through the architect.
- **Quantity surveyor**   Engaged to prepare cost evaluations and bills of quantities, check tenders, prepare interim valuations, effect cost controls, and advise the architect on the cost of variations.
- **Consulting engineers**   Engaged to advise and design on a variety of specialist installations, e.g. structural, services, security. They are employed to develop that particular aspect of the design within the cost and physical parameters of the architect's brief.
- **Principal or main contractor**   Employed by the client on the advice of the architect, by nomination or competitive tendering. They are required to administer the construction programme within the architect's direction.
- **Contract's manager or site agent**   On large projects, the main contractor's representative on site, with overall responsibility for ensuring that work proceeds effectively and efficiently, i.e. in accordance with the design specification and to time. Sometimes known as the **general foreman**, but this title is more appropriate on small to modest-size contracts.
- **Surveyor**   Employed by the main contractor to check work progress and assist the quantity surveyor in the preparation of interim valuations for stage payments and final accounts. May also be required to measure work done for bonus and subcontractor payments.
- **Estimator**   Prepares unit rates for the pricing of tenders, and carries out pre-tender investigations into the cost aspects of the proposed contract.
- **Buyer**   Orders materials, obtains quotations for the supply of materials and services.

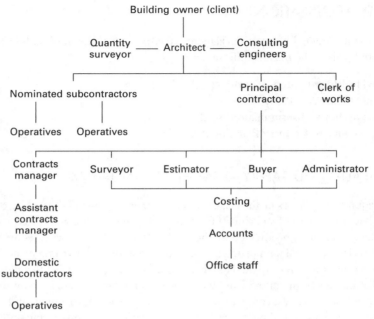

Figure 1 The building team.

- **Accountant** Prepares and submits accounts to clients and makes payments to suppliers and subcontractors. May also have a costing department that would allocate the labour and material costs to each contract to assist with the preparation of accounts.
- **Administrator** Organises the general clerical duties of the contractor's office for the preparation of contract documents and payment of salaries, subcontractors' and suppliers' invoices, insurances and all necessary correspondence.
- **Assistant contract manager** Often a trainee, in the process of completing professional examinations. Assists with the general responsibility for administering site proceedings.
- **Nominated subcontractor** Engaged by the client or architect for specialist construction or installation work, e.g. lifts, air conditioning.
- **Domestic subcontractor** Employed by the principal contractor to assist with the general construction, e.g. ground workers, bricklayers.
- **Operatives** The main workforce on-site; includes craftsmen, apprentices and labourers.

The size of the building firm or the size of the contract will determine the composition of the construction team. For medium-sized contracts some of the above functions may be combined, e.g. that of the surveyor and estimator. Furthermore, many design-and-build practices have been created by combining the professional expertise of architect, builder and consultants. The objective is to improve communications and create better working relationships to provide the client with a more efficient and cost-effective service.

■■■ LEGISLATION

Many statutes made by Act of Parliament affect the construction of buildings and associated work. The most significant are:

- The Health and Safety at Work etc. Act 1974;
- The Building Act 1984;
- The Disability Discrimination Act 1995–2004;
- The Town and Country Planning Act 1990.

THE HEALTH AND SAFETY AT WORK ETC. ACT 1974

This requires employers or their agents to implement an overall duty of care through a safe system of working. This applies to people in the workplace and others who could be affected by the work activity. The Act provides an all-embracing standard supporting a framework of statutory instruments or regulations administered by the Health and Safety Executive (HSE) through specialist inspectors operating from local offices. The inspectorate has powers to access premises, review a company's safety records, issue improvement and prohibition notices and, if necessary, effect prosecutions for non-compliance.

The following are the principal statutory instruments made under the Health and Safety at Work etc. Act. They are legally binding on the client, architect and builder:

The Construction (Design and Management) Regulations 1994

Establish mutual responsibility between client, designer and builder for matters pertaining to the health and safety of site personnel throughout the duration of a construction project. The client has prime responsibility for appointment of a 'planning supervisor' (usually the architect) and a 'principal contractor' (usually the main contractor), and for ensuring that both are adequately resourced, competent, and sufficiently informed of issues relating to the development.

The planning supervisor
Main responsibilities are to notify the HSE of the project details, to ensure cooperation between designers, to coordinate the design with regard to avoidance of undue risks, to prepare a pre-tender health and safety plan and manual for the client's use for advice, and to provide updates on health and safety issues.

The principal contractor
Responsibilities are mainly to ensure site personnel comply with the health and safety plan, to develop and update it throughout the duration of the work, and to exclude from site unauthorised and uninsured persons.

The Construction (Health, Safety and Welfare) Regulations 1996

Establish objectives measured against assessment of risk, for the well-being of personnel on a building site throughout the duration of work. The main areas for assessment include: temporary timbering and other non-permanent support facilities; safety barriers to excavations; air quality in the workplace; safe use of doors, gates and other possible means of entrapment; defined traffic and pedestrian routes; safe means of access to and egress from all workplaces, including special considerations for scaffolds; and emergency lighting and power. Also, provision must be made for welfare facilities, to include sanitation, hot and cold water supply, first aid equipment/personnel, protective clothing, facilities to dry clothes, and appropriate accommodation for meals. Implicit is good site management with regard for organisation and planning.

Control of Substances Hazardous to Health Regulations 1994

Redress the balance in favour of health, since the introduction of the Health and Safety at Work etc. Act has greater emphasis on safety. Manufacturing companies are obliged to monitor and declare health risks of their products, and building contractors must provide operatives with protective clothing and/or a well-ventilated environment if required to use them. COSHH has promoted the removal of harmful substances from materials, e.g. toxins, irritants and solvents, but finding substitutes is not always possible. Timber preservatives, welding fumes, dust from cement and plaster and insulating fibres are, as yet, a few of the unavoidable harmful constituents in building materials. Where these are applied, employers are obliged to monitor exposure levels, retain records, identify personnel who could be at risk, and document the facilities provided for their protection.

The Manual Handling Operations Regulations 1992

Determine the employer's responsibility to ensure that employees under their control are not expected to undertake manual tasks that impose an undue risk of injury. The client, or their agent the main contractor, must assess all manual handling operations and reduce risk or injury or safety to the lowest practicable level. This will be manifest in the provision of work systems, such as cranes and hoists, that the employee is obliged to make full use of. Due regard must be applied to:

- **tasks** space availability, manipulating distance, body movement, excess pulling and pushing and prolonged physical effort;
- **load** unwieldy or bulky, eccentric, excessive, liability to shift, temperature and sharpness of finish;
- **work environment** surface finish (slippery, uneven or variable), lighting, ventilation and temperature variation;
- **individual capacity** unusual weight or dimension, health problems/ limitations, special training and provision of personal protective clothing.

THE BUILDING ACT 1984

This is a consolidating Act of primary legislation relating to building work. It contains enabling powers and a means for the Office of the Deputy Prime Minister to produce building regulations for the outline purpose of:

- maintaining the health, well-being and convenience of people using and relating to buildings;
- promoting the comfort of building occupants, with due regard for the conservation of fuel and efficient use of energy within the structure;
- preventing the misuse of water, excessive consumption and contamination of supply.

The building regulations or statutory instruments currently in place under the Building Act are:

- The Building Regulations applicable to England and Wales;
- The Building (Approved Inspectors, etc.) Regulations;
- The Building (Prescribed Fees) Regulations;
- The Building (Inner London) Regulations;
- The Building (Disabled People) Regulations.

The Building Regulations contain minimum performance standards expected of contemporary buildings. They are supported by a series of Approved Documents that are not mandatory, but which give practical guidance on compliance with the requirements of the regulations. This guidance often incorporates British Standard, Building Research Establishment, British Board of Agrément and other authoritative references.

Control of the Building Regulations is vested in the local council authority. However, a developer/builder can opt for private certification whereby the developer and an approved inspector jointly serve an initial notice on the local authority. This describes the proposed works, which the local authority can reject (if they have justification) within 10 days of application. With this option, the responsibility for inspecting plans, quality of work, site supervision and certification of satisfactory completion rests with the approved inspector as defined in The Building (Approved Inspectors, etc.) Regulations.

Building legislation is regionally divided throughout the UK, each area having its own regulations:

- The Building Regulations applicable to England and Wales;
- The Building Standards (Scotland) Regulations;
- The Building Regulations (Northern Ireland);
- The Building (Inner London) Regulations.

There are also statutory provisions for the Channel Islands, Isle of Man and the Republic of Ireland.

Approved Documents

The current Building Regulations came into effect in 2000. They are constantly under review in response to new and changing technologies, public demands and environmental directives. Therefore, since 2000 there have been numerous amendments to the support series of Approved Documents. Those made under the Regulations applicable to England and Wales are currently listed from A to T (I, O and R omitted):

- A – *Structure*;
- B – *Fire safety*;
- C – *Site preparation and resistance to contaminants and moisture*;
- D – *Toxic substances*;
- E – *Resistance to the passage of sound*;
- F – *Ventilation*;
- G – *Hygiene*;
- H – *Drainage and waste disposal*;
- J – *Combustion appliances and fuel storage systems*;
- K – *Protection from falling, collision and impact*;
- L1 – *Conservation of fuel and power in dwellings*;
- L2 – *Conservation of fuel and power in buildings other than dwellings*;
- M – *Access to and use of buildings*;
- N – *Glazing – safety in relation to impact, opening and cleaning*;
- P – *Electrical safety*;
- Q – *Electronics communications services*;
- S – *Security systems*;
- T – *Telecommunications systems*.

There is another Approved Document made under Regulation 7 – Materials and workmanship. This requires that any building subject to the Building Regulations be carried out with proper materials and in a workmanlike manner.

The Approved Documents provide practical and technical guidance for satisfying the Building Regulations. There is no obligation to adopt any of these, provided the Building Regulations' performance requirements are shown to be satisfied in some other way. This may include European Technical Approvals, British Board of Agrément certification, CE marking of products, and calculations in accordance with acceptable structural standards for selection of components.

THE DISABILITY DISCRIMINATION ACT 1995–2004

This Act was introduced in three stages between 1995 and 2004. It is designed to assist and benefit an estimated 8 million UK residents who suffer some sort of disability. In principle, the Act requires that all new, adapted and refurbished buildings be constructed with unobstructed access and facilities that can be used by wheelchair occupants. This requirement extends to service providers and owners of public buildings.

Building Regulations Part M – 2004

Part M and Approved Document M: *Access to and use of buildings* has been revised in support of the Act. The Regulations provide design and practical guidance for disabled user convenience in new and refurbished buildings. All new homes (houses and flats) are required to be constructed with sufficient accessibility and use of facilities for disabled people. The objective is to allow disabled and elderly people greater freedom of use and independence in their own homes for a longer period of time. The Regulations also allow for internal and external accessibility for wheelchair users in buildings other than dwellings. Movement in and around all buildings is to enable the disabled the same means of access that ambulant people enjoy when visiting friends, relatives, shops, entertainments and other conveniences without being impeded.

The main features of the Approved Document are as follows:

- Access approach from car parking area to the main entrance of a building to be level or ramped.
- Main entrance threshold to be level, not stepped.
- Main entrance door wide enough for a wheelchair.
- WC facilities at ground or entrance floor.
- WC compartment with sufficient manoeuvrability space for a wheelchair user.
- WC facilities at accessible levels for ambulant and disabled people.
- Switches for lighting, power sockets, heating control etc. at convenient heights above floor level.
- In flats, lifts provided to access all floors.
- In houses, the structure about a stair to be strong enough to support a stair lift.

BS 8300: *Design of buildings and their approaches to meet the needs of disabled people. Code of Practice*

This British Standard complements the Building Regulations and the Disability Discrimination Act. It was produced from government-commissioned research into the ergonomics of modern buildings. The Standard presents the built environment from the perspective of the disabled user, with particular regard to residential buildings as occupier and visitor, and public buildings as spectator, customer and employee. The Standard also considers the disabled as participants in sports events, conferences and performances.

THE TOWN AND COUNTRY PLANNING ACT 1990

This Act establishes the procedures for development of land and construction of buildings. It is administered through the hierarchy of government, regional and local offices:

- central government;
- county planning departments;
- local borough planning departments.

Central government is represented by the Office of the Deputy Prime Minister (ODPM). They issue departmental circulars or policy planning guidance (PPG) to county planning departments. PPGs contain directives for implementing government policy for overall development on a large scale, e.g. Thames Gateway.

Included are objectives and a perspective for housing needs, communications systems, transport, social facilities, green belts, redevelopment, retail and commercial plans.

County planning departments formulate development policy within the framework determined by the ODPM. This is known as the **Structure Plan**. Structure plans are prepared with regard for 'brownfield' sites (usually redundant industrial buildings) and 'greenfield' sites (undeveloped land), in accordance with anticipated future needs. These include social and economic demands for housing, and commercial, social and recreational facilities within a viable communications and transport infrastructure. Structure plans are subject to public consultation, and once established they normally remain in place for 15 years.

Local planning departments or local authorities establish a **Local Plan** for their borough or region. This is produced within the framework of the Structure Plan with regard for an economic, social and practical balance of facilities for the various communities under their administration. To maintain a fair and equitable interest, local plans are subject to public and ODPM consultation. Local authorities are also responsible for processing applications for development in their area. Applications range from a nominal addition to an existing building to substantial estate development. Procedures for seeking planning consent vary depending on the scale of construction. All require the deposit of area and site plans, building elevations, forms declaring ownership or nature of interest in the proposal, and a fee for administration. If an application is refused, the applicant has a right of appeal to the ODPM.

■■■ BRITISH STANDARDS

These function as support documents to building design and practice. Many are endorsed by the Building Regulations and provide recommendations for minimum material and practice specifications. They are published in four possible formats:

- **Codes of practice (BS Code of practice for . . .)**   These are guides to good practice in particular areas of construction activity, e.g. scaffolding and structural steelwork.
- **Standard specifications (BS)**   These are specifications appropriate to materials and components such as bricks and windows. Products complying may be endorsed with the Institution's kitemark.
- **Draft for development (DD)**   These are issued instead of a Code of practice or Standard specification where there is insufficient data or information to make a firm or positive recommendation. They have a maximum life of five years, during which time sufficient data may be accumulated to upgrade the draft to a Standard specification.
- **Published document (PD)**   These are publications that are difficult to locate in any of the preceding categories.

Codes of practice and standard specifications are compiled by specialist committees of professional interest in the particular subject. It must be remembered that these codes and specifications are only recommendations. However, many are used in the

Building Regulations, Approved Documents as a minimum acceptable standard for specific applications. Copies of the codes and specifications can be obtained from the British Standards Institution (Customer Services), 389 Chiswick High Road, London, W4 4AL, or online from http://bsonline.techindex.co.uk

■■■ AGRÉMENT CERTIFICATES

These are administered by the British Board of Agrément, Bucknalls Lane, Garston, Watford WD25 9DA. Their purpose is to establish the quality and suitability of new products and innovations not covered by established performance documents such as British (BSI) or European Standards (CEN). The Board assesses, examines and tests materials and products with the aid of the Building Research Establishment and other research centres, to produce reports. These reports are known by the acronym MOATS (Methods Of Assessment and Test). Products satisfying critical examination relative to their declared performance are issued with an Agrément Certificate.

■■■ INTERNATIONAL STANDARDS

Many British Standards have been harmonised with the dimensional and material specifications of European continental products and vice versa. For example, BS EN 196-2: 1995: *Chemical analysis of cement* has replaced BS 4550: Part 2: 1970: *Methods of testing cement; chemical tests*. European standards are administered by the Comité Européen de Normalisation (CEN), which incorporates bodies such as the British Standards Institution (BSI) from member states.

The International Standards Organisation also produces documents compatible with British Standards. These are prefixed ISO: for example, BS EN ISO 9004: *Quality management systems* complements BS 5750-8: *Quality systems*.

■■■ METRICATION

Since the building industry adopted the metric system of measurement during the early 1970s there has been a great deal of modification and rationalisation as designers and manufacturers have adapted themselves to the new units.

Most notable has been product developments coordinated about the preferred dimension of 300 mm. The second preference is 100 mm, with subdivisions of 50 mm and 25 mm as third and fourth preferences respectively. Many manufacturers now base their products on the 100 mm module, and coordinate their components within a three-dimensional framework of 100 mm cubes. This provides product compatibility and harmonisation with international markets that are also metric and dimensionally coordinated. Kitchen fitments provide a good example of successful development, with unit depths of 600 mm and widths in 300, 600, 1200 mm and so on. Less successful has been the move to metric bricks in lengths of 200 or 300 mm × 100 mm wide × 75 mm high. They are just not compatible with existing buildings, and bricklayers have found them less suited to the hand.

Although the metric system is now well established, it should be remembered that a large proportion of the nation's buildings were designed and constructed using imperial units of measurement, which are incongruous with their metric replacements. Many building materials and components will continue to be produced in imperial dimensions for convenience, maintenance, replacement and/or refurbishment, and where appropriate some manufacturers provide imperial–metric adaptors, e.g. plumbing fittings. The prohibitive capital cost of replacing manufacturing plant such as brick presses and timber converters will preserve traditional imperial dimensions for some time.

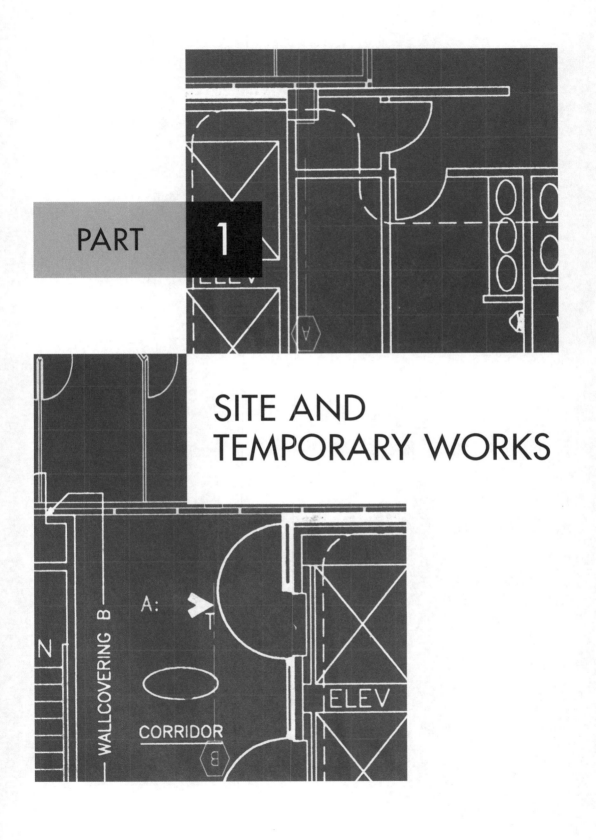

PART 1

SITE AND
TEMPORARY WORKS

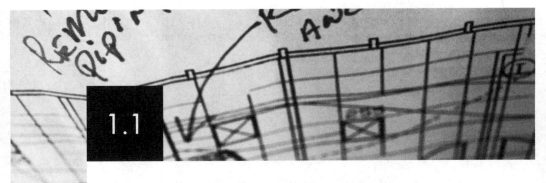

SITE WORKS AND SETTING OUT

When a builder is given possession of a building site the contractor will have been provided with the site layout plan and the detail drawings necessary for him to construct the building(s). Under most forms of building contract it is the builder's responsibility to see that the setting out is accurate.

The site having been taken over, the task of preparing for and setting out the building can be commenced. These operations can be grouped under three headings:

- clearing the site;
- setting out the building;
- establishing a datum level.

■■■ CLEARING THE SITE

This may involve the demolition of existing buildings, the grubbing out of bushes and trees, and the removal of soil to reduce levels. Demolition is a skilled occupation and should be tackled only by an experienced demolition contractor. The removal of trees can be carried out by manual or mechanical means. The removal of large trees should be left to the specialist contractors.

Building Regulation C1, 'The ground to be covered by the building shall be reasonably free from vegetable matter.' This is in effect to sterilise the ground, because the top 300 mm or so will contain plant life and decaying vegetation. This means that the topsoil is easily compressed and would be unsuitable for foundations. Topsoil is valuable as a dressing for gardens, and will be retained for reinstatement when the site is landscaped. The method chosen for conducting the site clearance work will be determined by the scale of development, and by consideration for any adjacent buildings.

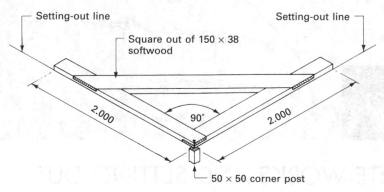

Setting-out line

Setting-out line

Square out of 150 × 38 softwood

2.000

90°

2.000

50 × 50 corner post

Typical builder's square

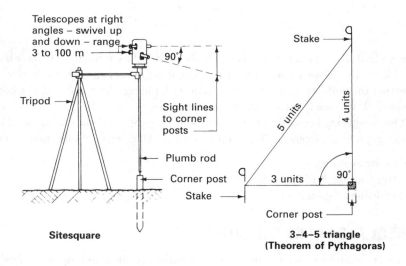

Telescopes at right angles – swivel up and down – range 3 to 100 m

90°

Tripod

Sight lines to corner posts

Plumb rod

Corner post

Stake

Sitesquare

Stake

5 units

4 units

3 units

90°

Corner post

Stake

3–4–5 triangle (Theorem of Pythagoras)

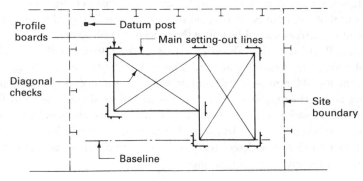

Profile boards

Datum post

Main setting-out lines

Diagonal checks

Site boundary

Baseline

Figure 1.1.1   Setting out and checking methods.

■ ■ ■ SETTING OUT THE SITE

The first task is to establish a baseline from which the whole of the building can be set out. The position of this line must be clearly marked on-site so that it can be re-established at any time. For on-site measuring a steel tape should be used (30 m would be a suitable length). Linen and plastic-coated tapes are also available. The disadvantage with linen tapes is that they are liable to stretch.

After the baseline has been set out, marked and checked, the main lines of the building can be set out, each corner being marked with a stout peg. A check should now be made of the setting-out lines for right angles and correct lengths. There are several methods of checking whether a right angle has been established, and in fact the setting out would have been carried out by one of these methods. A check must still be made, and it is advisable to check by a different method to that used for the setting out. The setting-out procedure and the methods of checking the right angles are illustrated in Fig. 1.1.1.

After the setting out of the main building lines has been completed and checked, profile boards are set up as shown in Fig. 1.1.2. These are set up clear of the foundation trench positions to locate the trench, foundations and walls. Profile boards are required at all trench and wall intersections.

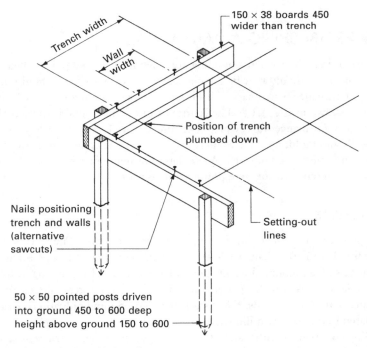

Figure 1.1.2   Typical profile board.

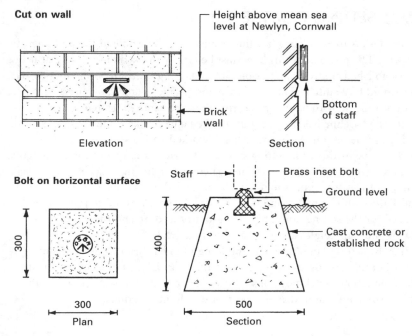

Figure 1.1.3 Common types of benchmark.

■ ■ ■ ESTABLISHING A DATUM LEVEL

It is important that all levels in a building are taken from a fixed point called a
datum. This point should now be established; wherever possible it should be
related to an ordnance benchmark. This is an arrow with a horizontal mark above
the arrow as shown in Fig. 1.1.3. The centreline of the horizontal is the actual level
indicated on an Ordnance Survey map. Benchmarks are found cut or let into the
sides of walls and buildings. Where there are no benchmarks on or near the site, a
suitable datum must be established. A site datum or temporary benchmark could be
a post set in concrete or a concrete plinth set up on site.

TAKING LEVELS

The equipment used is an engineer's level and a levelling staff. The level is simply a
telescope fitted with cross-hairs to determine alignment. The telescope rotates on a
horizontal axis plate, mounted on a tripod. The staff is usually 4 m long in folding
or extendable sections. The 'E' pattern shown in Fig. 1.1.4 is generally used, with
graduations at 10 mm intervals. Some staffs may have 5 mm graduations. Readings
are estimated to the nearest millimetre.

   Levelling commences with a sight to a benchmark from the instrument stationed
on firm ground. Staff stations are located at measured intervals such as a 10 m grid.
From these, instrument readings are taken as shown in Fig. 1.1.5. The level

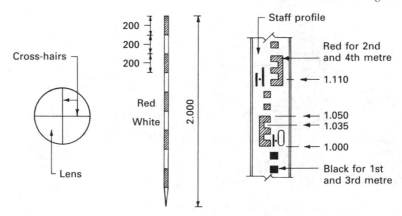

Figure 1.1.4   Levelling 'scope, ranging rod and 'E' pattern staff.

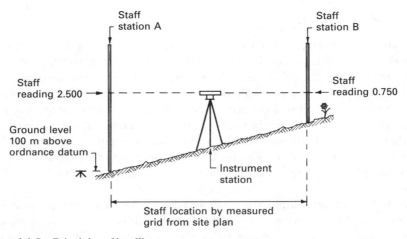

Figure 1.1.5   Principles of levelling.

differentials can then be combined with plan area calculations to determine the volume of site excavation or cut and fill required to level the site.

From Fig. 1.1.5:

Rise and fall method:
   Staff reading at A = 2.500 m
   Staff reading at B = 0.750 m
   Ground level at A = 100 m above ordnance datum (AOD)
   Level at B = 100 m + rise (− fall if declining)
   Level at B = 100 m + (2.500 − 0.750) = 101.750 m.

Alternative height of collimation (HC) method:
   HC at A = Reduced level (RL) + staff reading
         = 100 m + 2.500 = 102.500 (AOD)
   Level at B = HC at A − staff reading at B
         = 102.500 − 0.750 = 101.750 m

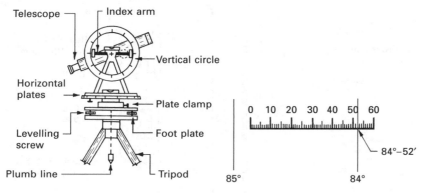

Figure 1.1.6   Traditional theodolite with direct reading.

MEASURING ANGLES

The sitesquare shown in Fig. 1.1.1 is accurate for determining right angles in the horizontal plane. Where acute or obtuse angles occur in the horizontal, or vertical angles are to be established or checked, a **theodolite** is used. This instrument is basically a focusing telescope with cross-hairs, mounted on horizontal index plates over a tripod. A vertical measurement circle with index is attached to one side of the telescope. Figure 1.1.6 shows the outline features of a traditional vernier theodolite. Traditional theodolites require visual or manual measurement of angles on the micrometer index or scale. In contrast, contemporary instruments are far more sophisticated, with automatic settings, liquid crystal displays, and facilities for data transfer to computers.

With the instrument firmly stationed, the telescope and horizontal (vertical if appropriate) plate are rotated from an initial sighting through the required angle. A pole-mounted target may be used for location. A check can be made by rotating the telescope through 180° vertically and the index through 180° horizontally for a second reading. Angles are recorded in degrees, minutes and seconds, the extent of accuracy determined by the quality of instrument and the skill of the user.

SLOPING SITES

Very few sites are level, and therefore before any building work can be commenced the area covered by the building must be levelled. In building terms this operation is called **reducing levels**. Three methods can be used, and it is the most economical that is usually employed.

- **Cut and fill**   The usual method because, if properly carried out, the amount of cut will equal the amount of fill.
- **Cut**   This method has the advantage of giving undisturbed soil over the whole of the site, but has the disadvantage of the cost of removing the spoil from the site.

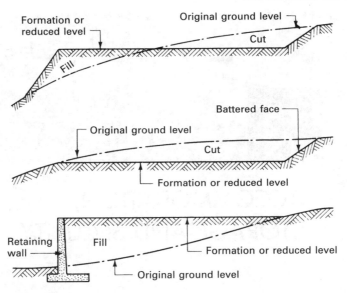

Figure 1.1.7 Sloping sites.

- **Fill** A method not to be recommended because, if the building is sited on the filled area, either deep foundations would be needed or the risk of settlement at a later stage would have to be accepted. The amount of fill should never exceed a depth of 600 mm.

The principles of the above methods are shown in Fig. 1.1.7.

1.2

ACCOMMODATION, STORAGE AND SECURITY

The activities and the temporary nature of a building site do not generally justify the provision of permanent buildings for staff accommodation or for the storage of materials. It is, however, within the builder's interest to provide the best facilities that are economically possible for any particular contract. This should promote good relationships between management and staff; it should also reduce the loss of materials due to theft, accidental damage or vandalism. The better the facilities and amenities provided on a building site, the greater will be the contentment of the site staff, which will ultimately lead to higher productivity.

◼◼◼ ACCOMMODATION

The Construction (Health, Safety and Welfare) Regulations 1996 is a statutory instrument that establishes objectives for accommodation and facilities for staff to be provided on sites throughout the construction industry. Requirements will vary with regard to the number of personnel on site and in some cases the anticipated duration of the contract. An extract from the forerunner to the current regulations is shown in Table 1.2.1, which indicates absolute minimum standards for guidance only.

Units of staff accommodation usually come in one of two forms:

- semi-portable units;
- mobile caravans or cabins.

Preliminary planning is necessary to anticipate the amount and type of temporary accommodation, site space/location and facilities required for material storage and use by site personnel. Offices need to be weatherproof, heated, insulated, lit and furnished with desks, work-surfaces, plan chests and chairs to suit the office activity. A typical semi-portable site office is shown in Fig. 1.2.1. The same basic units can be used for all accommodation including meal rooms and toilets equipped

Table 1.2.1 The Construction (Health and Welfare) Regulations

No. of persons employed by contractor on site	Requirement	0 5 10 20 25	40 50	100
First aid	Box to be clearly marked and in charge of named person.	First-aid boxes	First-aid boxes and person trained in first aid →	→
Stretcher ambulance			Stretcher provided. Local health authority informed of site, work and completion date. If no 'phone or radio, ambulance kept ready →	→
First-aid room	To be used only for treatment and in charge of trained person.		Where number of persons on site exceeds 250 each employer of more than 40 persons must provide first-aid room →	→
Shelter and clothing	All persons to have shelter and place for depositing clothing.	Where possible, means of warming themselves and drying wet clothing → / Adequate means of warming themselves and drying wet clothing →	→	→
Meals room	All persons to have drinking water provided and facilities for boiling water and eating meals.		Facilities for heating food if hot meals are not available on site →	→
Washing facilities	All persons on site for more than 4 hours to have washing facilities with ventilation and light.	Where work is likely to last 6 weeks / hot and cold or warm water, soap and towel provided →	Where work is likely to last 12 months →	4 washplaces plus 1 for every 35 persons more than 100
Sanitary facilities	To be maintained and kept clean – provision to be made for lighting and ventilation.	1 convenience for every 25 persons →	→	1 convenience for every 35 persons

Notes

- Washing facilities to be close to meals room, sanitary conveniences and changing rooms. Separate facilities for men and women.
- Protective clothing to be provided where person is required to work in inclement weather.
- Subcontractors may use the facilities provided by another contractor, and for the purpose of these regulations their workforce on site is included in the total workforce on site.

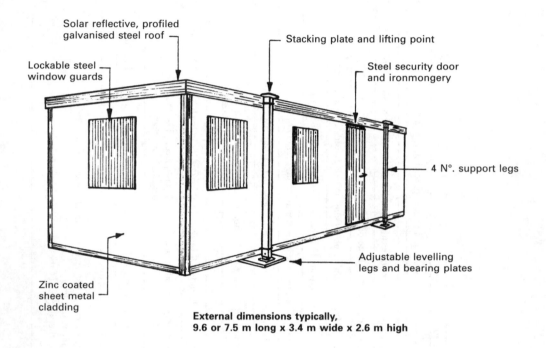

Solar reflective, profiled galvanised steel roof

Stacking plate and lifting point

Lockable steel window guards

Steel security door and ironmongery

4 N°. support legs

Zinc coated sheet metal cladding

Adjustable levelling legs and bearing plates

External dimensions typically, 9.6 or 7.5 m long x 3.4 m wide x 2.6 m high

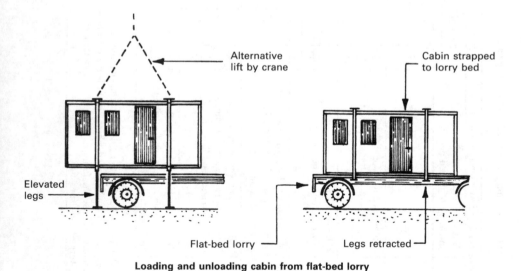

Alternative lift by crane

Cabin strapped to lorry bed

Elevated legs

Flat-bed lorry

Legs retracted

Loading and unloading cabin from flat-bed lorry

Figure 1.2.1   Semi–mobile portable cabin site accommodation.

as indicated in Table 1.2.1. Where site space is limited, most semi-portable units are designed to be stacked.

Semi-mobile cabins are available in a wide variety of sizes, styles and applications. The outer construction is generally of galvanised sheet steel over a structural steel frame, suitably insulated and finished internally with plasterboard walls and ceiling. Floor finishes vary from chequerplate to carpeted boards. To prevent over-heating, the roof is covered with a solar reflective material on profiled galvanised steel. Units may be hired or purchased, usually pre-wired and plumbed as appropriate for connection to mains supplies and drains. Inclusion of furniture is also an option. Cabins are transported to site by flat bed lorry as shown in Fig. 1.2.1 and craned into position.

Subject to the limitations below, mobile caravans as shown in Fig. 1.2.2 can be used as offices on small sites. Timber sectional huts are of very limited use and prohibited where reference to the "Joint Fire Code" is included in the building contract. Most contracts now incorporate this code and quote Clause 12 for fire rated temporary buildings and temporary accommodation.

Ref. *Fire Prevention On Construction Sites – The Joint Code of Practice on the Protection From Fire of Construction Sites and Buildings Undergoing Renovation*; published by Construction Confederation, Loss Prevention Council and National Contractors Group.

■■■ STORAGE

The type of storage facilities required of any particular material will depend upon the following factors:

■ durability – will it need protection from the elements?
■ vulnerability to damage;
■ vulnerability to theft.

Cement, plaster and lime supplied in bag form require a dry store free from draughts, which can bring in moist air and may cause an air set of material. These materials should not be stored for long periods on site: therefore provision should be made for rotational use so that the material being used comes from the older stock.

Aggregates such as sand and ballast require a clean firm base to ensure that foreign matter is not included when extracting materials from the base of the stockpile. Different materials and grades must be kept separated so that the ultimate mix batches are consistent in quality and texture. Care must be taken, by careful supervision, to ensure that the stockpiles are not used as a rubbish tip. If the storage piles are exposed to the elements a careful watch should be kept on the moisture content; if this rises it must be allowed to drain after heavy rain, or alternatively the water/cement ratio of the mix can be adjusted.

Bricks and blocks should be stacked in stable piles on a level and well-drained surface in a position where double handling is reduced to a minimum. Facing bricks and light-coloured bricks can become discoloured by atmospheric pollution and/or adverse weather conditions; in these situations the brick stacks should be covered

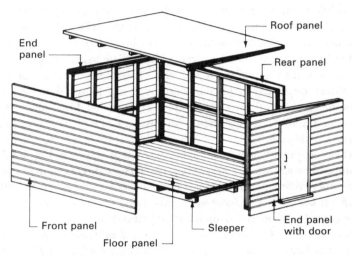

Sectional store hut (not suitable where Joint Fire Code applies)

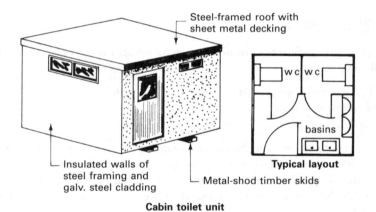

Cabin toilet unit

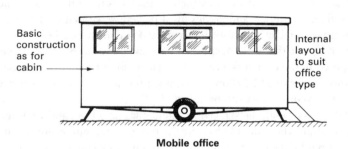

Mobile office

Figure 1.2.2   Typical site accommodation units.

with tarpaulin or polythene sheeting, adequately secured to prevent dislodgement. Blocks, being less dense than bricks, should be stacked to allow air movement around them, and should always be covered with a suitable sheet material.

Roof tiles have a greater resistance to load when it is imposed on the edge: for this reason tiles should be stacked on edge and in pairs, head to tail, to give protection to the nibs. An ideal tile stack would be five to seven rows high, with end tiles laid flat to provide an abutment. Tile fittings such as ridge and hip tiles should be kept separate and if possible placed on end.

Drainage goods, like tiles, may be stored in an open compound; they should be stacked with their barrels horizontal and laid with spigots and sockets alternately reversed, or placed in layers with the spigots and sockets reversed in alternate layers. Plain-ended clay pipes are delivered to site on pallets with pipes separated by timber battens recessed to suit the pipe profile. Pipes should be stored in this way until they are required. Fittings should be kept separate, and those such as gullies, which can hold water, should be placed upside down.

Timber is a hygroscopic material, and therefore to prevent undue moisture movement it should be stored in such a manner that its moisture content remains fairly constant. A rack of scaffold tubulars with a sheet roof covering makes an ideal timber store: the various section sizes allow good air flow around the timber, and the roof provides protection from the rain and snow.

Ironmongery, hand tools and paint are some of the most vulnerable items on a building site. Small items such as locks, power drills and cans of paint should be kept in a locked hut and issued only against an authorised stores requisition. Large items such as baths can be kept in the compound and suitably protected; it is also good practice only to issue materials from the compound against a requisition order.

▪▪▪ SECURITY AND PROTECTION

FENCING

A building site and the compound can be given a degree of protection by surrounding with a fence. The fence fulfils two functions:

- it defines the limit of the site or compound;
- it acts as a deterrent to the would-be trespasser or thief.

A fence can be constructed to provide a physical barrier of solid construction or a visual barrier of open-work construction. If the site is to be fenced as part of the contract it may be advantageous to carry out this work at the beginning of the site operations. The type of fencing chosen will depend upon the degree of security required, cost implications, type of neighbourhood and duration of contract.

A security fence around the site or compound should be at least 1.800 m high above the ground and include the minimum number of access points, which should each have a lockable barrier or gate. Standard fences are made in accordance with the recommendations of BS 1722, which covers 13 forms of fencing giving suitable methods for both visual and physical barriers: typical examples are shown in Fig. 1.2.3.

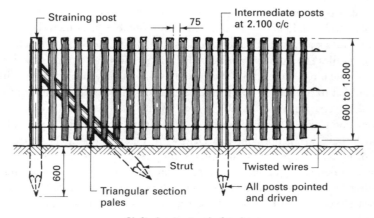

Straining post

75

Intermediate posts
at 2.100 c/c

600 to 1.800

600

Strut

Triangular section
pales

Twisted wires

All posts pointed
and driven

Cleft chestnut pale fencing

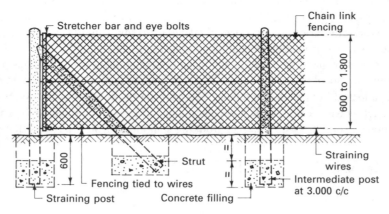

Stretcher bar and eye bolts

Chain link
fencing

600 to 1.800

600

Strut

Fencing tied to wires

Straining post

Concrete filling

Straining
wires

Intermediate post
at 3.000 c/c

Chain link fence with concrete posts

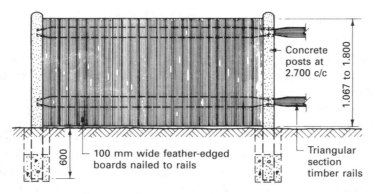

Concrete
posts at
2.700 c/c

1.067 to 1.800

600

100 mm wide feather-edged
boards nailed to rails

Triangular
section
timber rails

Close boarded fence with concrete posts

Figure 1.2.3 Typical fencing details.

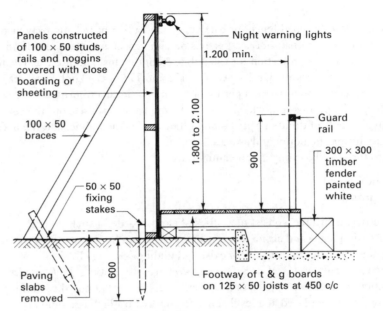

Panels constructed
of 100 × 50 studs,
rails and noggins
covered with close
boarding or
sheeting

Night warning lights

1.200 min.

1.800 to 2.100

900

Guard
rail

300 × 300
timber
fender
painted
white

100 × 50
braces

50 × 50
fixing
stakes

600

Paving
slabs
removed

Footway of t & g boards
on 125 × 50 joists at 450 c/c

Typical free-standing vertical hoarding

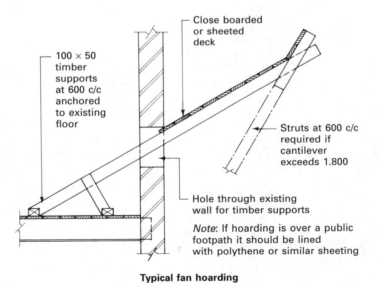

Close boarded
or sheeted
deck

100 × 50
timber
supports
at 600 c/c
anchored
to existing
floor

Struts at 600 c/c
required if
cantilever
exceeds 1.800

Hole through existing
wall for timber supports

Note: If hoarding is over a public
footpath it should be lined
with polythene or similar sheeting

Typical fan hoarding

Figure 1.2.4 Timber hoardings.

HOARDINGS

These are close-boarded fences or barriers erected adjacent to a highway or public footpath to prevent unauthorised persons obtaining access to the site, and to provide a degree of protection for the public from the dust and noise associated with building operations. Under Sections 172 and 173 of the Highways Act 1980 it is necessary to obtain written permission from the local authority to erect a hoarding. The permission, which is in the form of a licence, sets out the conditions and gives details of duration, provision of footway for the public, and the need for lighting during the hours of darkness.

Two forms of hoarding are in common use:

- vertical hoardings;
- fan hoardings.

The vertical hoardings consist of a series of close-boarded panels securely fixed to resist wind loads and accidental impact loads. It can be free standing, or fixed by stays to the external walls of an existing building (see Fig. 1.2.4).

The Construction (Health, Safety and Welfare) Regulations 1996 require protection from falling objects to be given to persons. A fan hoarding fulfils this function by being placed at a level above the normal traffic height and arranged in such a manner that any falling debris is directed back towards the building or scaffold (see Fig. 1.2.4).

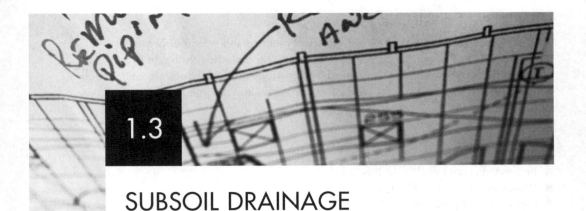

1.3

SUBSOIL DRAINAGE

■ ■ ■ BUILDING REGULATION C2 Resistance to moisture

Subsoil drainage shall be provided if it is needed to avoid:

■ the passage of ground moisture to the interior of the building;
■ damage to the fabric of the building.

The ideal site (see Fig. 1.3.1) will not require any treatment, but sites with a high water table will require some form of subsoil drainage. The water table is the level at which water occurs naturally below the ground, and this level will vary with the seasonal changes.

The object of subsoil drainage is to lower the water table to a level such that it will comply with the above Building Regulation, i.e. not rise to within 0.25 m of the lowest floor of a building. It also has the advantage of improving the stability of the ground, lowering the humidity of the site, and improving its horticultural properties.

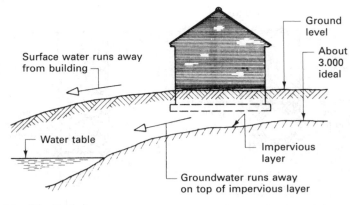

Figure 1.3.1 The ideal site.

MATERIALS

The pipes used in subsoil drainage are usually dry jointed and are either porous or perforated pipes. The porous pipes absorb the water through their walls and thus keep out the fine particles of soil or silt, whereas perforated pipes, which are laid with the perforations at the base, allow the water to rise into the pipe, leaving any silt behind.

Suitable pipes

- Perforated clayware: BS EN 295-5.
- Porous concrete: BS 5911-114.*
- Clayware field pipes: BS 1196.
- Profiled and slotted polypropylene or uPVC: BS 4962.
- Perforated uPVC: BS 4660.

Note: Porous concrete is rarely manufactured now, therefore the BS has been withdrawn.

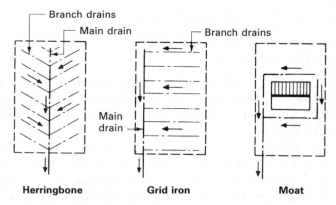

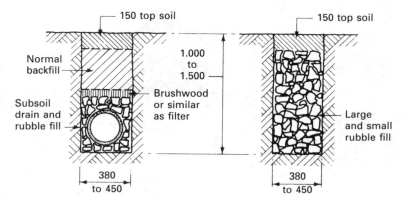

Figure 1.3.2   Subsoil drainage systems and drains.

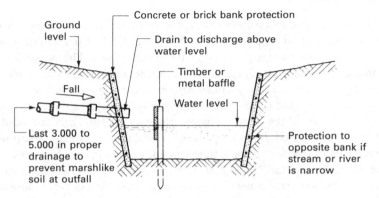

Figure 1.3.3   Outfall to stream or river.

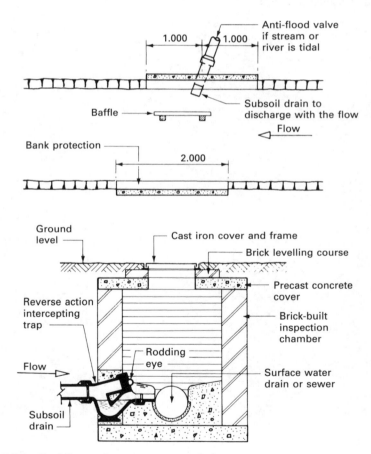

Figure 1.3.4   Outfall to surface water sewer or drain.

DRAINAGE LAYOUTS

The pipes are arranged in a pattern to cover as much of the site as is necessary.
Typical arrangements are shown on the plans in Fig. 1.3.2. Water will naturally
flow towards the easy passage provided by the drainage runs. The system is
terminated at a suitable outfall such as a river, stream or surface water sewer.
In all cases permission must be obtained before discharging a subsoil system.
The banks of streams and rivers will need protection against the turbulence set up
by the discharge, and if the stream is narrow the opposite bank may also need
protection (see Fig. 1.3.3). If discharge is into a tidal river or stream, precautions
should be taken to ensure that the system will not work in reverse by providing
an outlet for the rising tide. On large schemes sediment chambers or catch pits are
sometimes included to trap some of the silt which is the chief cause of blockages
in subsoil drainage work. The construction of a catch pit is similar to the manhole
shown in Fig. 1.3.4 except that in a catch pit the inlet and outlet are at a high
level; this interrupts the flow of subsoil water in the drains and enables some of the
silt to settle on the base of the catch pit. The collected silt in the catch pit must be
removed at regular intervals.

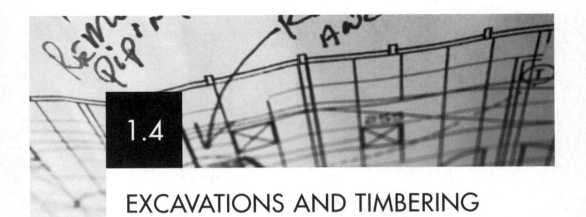

1.4

EXCAVATIONS AND TIMBERING

Before a foundation can be laid it is necessary to excavate a trench of the required depth and width. On small contracts such as house extensions this is effectively carried out by hand, but on large works it will be more economic to use some form of mechanical excavator. The general procedure for the excavation of foundation trenches is illustrated in Fig. 1.4.1.

■■■ TIMBERING

This is a term used to cover temporary supports to the sides of excavations and is sometimes called **planking and strutting**. The sides of some excavations will need support to:

- protect the operatives while working in the excavation;
- keep the excavation open by acting as a retaining wall to the sides of the trench.

The type and amount of timbering required will depend upon the depth and nature of the subsoil. Over a short period many soils may not require any timbering, but weather conditions, depth, type of soil and duration of the operations must all be taken into account, and each excavation must be assessed separately.

Suitable timbers for this work are:

- Scots pine;
- Baltic redwood;
- Baltic whitewood;
- Douglas fir;
- larch;
- hemlock.

Typical details of timbering to trenches are shown in Figs 1.4.2–1.4.6.

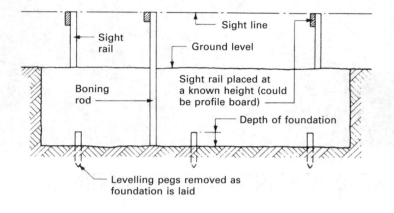

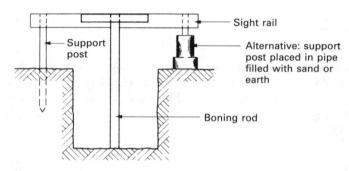

Figure 1.4.1   Trench excavations.

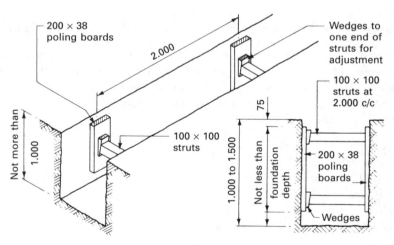

Figure 1.4.2   Typical timbering in hard soils.

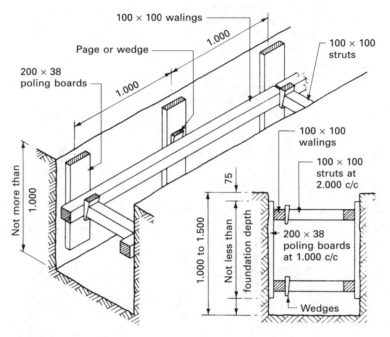

Figure 1.4.3 Typical timbering in firm soils.

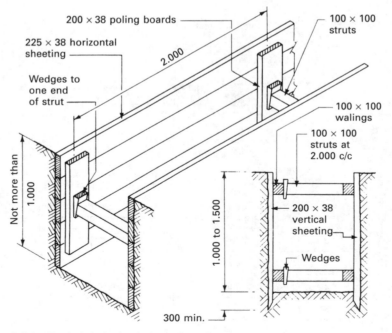

Figure 1.4.4 Typical timbering in loose dry soils.

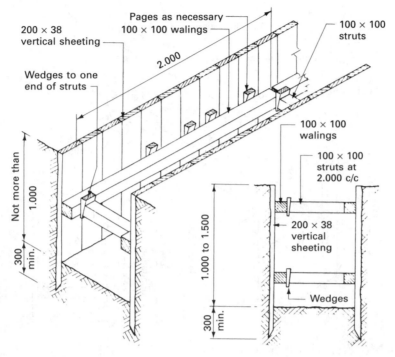

Figure 1.4.5   Typical timbering in loose wet soils.

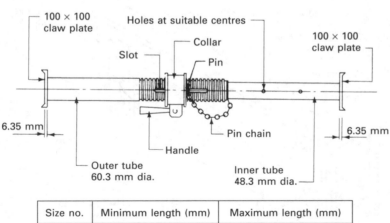

Size no.	Minimum length (mm)	Maximum length (mm)
0	300	480
1	480	680
2	680	1060
3	1060	1670

Figure 1.4.6   Adjustable metal status – BS 4074.

THE CONSTRUCTION (HEALTH, SAFETY AND WELFARE) REGULATIONS 1996

This document establishes objectives for employers, the self-employed and employees, to ensure safe working and support in excavations.

Regulation 5 – Safe places of work

This is a general requirement applying not least to work in the ground. It places an obligation on all involved to assess the risks and to ensure reasonably practicable means to safeguard work in excavations.

Regulations 12 and 13 – Excavations

Timber or other suitable material must be provided and used to prevent danger from a fall or dislodgement of materials forming the sides of an excavation.

Underground services must be foreseen (if possible), located, identified and assessed for risk to operatives working in excavations. Positive action is necessary to eliminate any risk and to prevent injury.

Deep excavations, tunnels, cofferdams and caissons require special consideration and must be correctly designed, constructed and maintained with regard for prevailing conditions. (See Chapter 3 of *Advanced Construction Technology*.)

Regulations 28, 29 and 30 – Training, inspection and reports

These require excavation support to be installed or supervised by adequately trained, suitably knowledgeable and experienced personnel. Prior to general access being given, an inspection of excavations by a competent person (usually the main contractor's safety supervisor) must be made. That person must ensure that work can proceed safely. Following inspection, a written report must be filed for presentation if requested by the Health and Safety Executive. Further documented inspections should be undertaken daily or at the beginning of each shift or after severe weather conditions. Changes to the trench format and unexpected falls of earth should also be monitored.

BUILDING REGULATIONS

Notice of commencement and completion of certain stages of work.

Building Regulation 14 requires that the building control office of the local authority is notified by a person carrying out building work prior to commencement and at specific stages during construction work. The notice should be given in writing or by such means as may be agreed with the local authority.

Notices of the stages when statutory inspections are required under this regulation occur as follows:

- Commencement   48 h
- Foundation excavation   24 h
- Foundation concrete   24 h
- Oversite preparation, before concreting   24 h
- Damp-proof course   24 h
- Drains before backfilling (foul and rainwater)   24 h
- Drain test after covering   5 days after
- Occupation before completion   5 days before
- Completion   within 5 days

With the exception of notice of commencement, drain testing and completion, the amount of time required before progressing with other parts of the work is one day. Weekends and bank holidays are excluded.

1.5

SCAFFOLDING

A scaffold is a temporary structure from which persons can gain access to a place of work in order to carry out building operations. It includes any working platforms, ladders and guard rails. Basically there are two forms of scaffolding:

- putlog scaffolds;
- independent scaffolds.

■ ■ ■ PUTLOG SCAFFOLDS

This form of scaffolding consists of a single row of uprights or standards set away from the wall at a distance that will accommodate the required width of the working platform. The standards are joined together with horizontal members called **ledgers** and are tied to the building with cross-members called **putlogs**. The scaffold is erected as the building rises, and is used mostly for buildings of traditional brick construction (see Fig. 1.5.1).

■ ■ ■ INDEPENDENT SCAFFOLDS

An independent scaffold has two rows of standards, which are tied by cross-members called **transoms**. This form of scaffold does not rely upon the building for support and is therefore suitable for use in conjunction with framed structures (see Fig. 1.5.2).

Every scaffold should be securely tied to the building at intervals of approximately 3.600 m vertically and 6.000 m horizontally. This can be achieved by using a horizontal tube called a **bridle** bearing on the inside of the wall and across a window opening with cross-members connected to it (see Fig. 1.5.1); alternatively a tube with a reveal pin in the opening can provide a connection point for the cross-members (see Fig. 1.5.2). If suitable openings are not available then the scaffold should be strutted from the ground using raking tubes inclined towards the building.

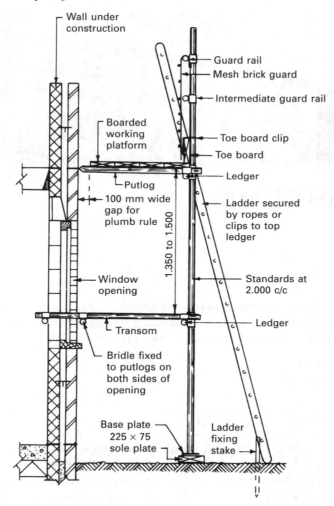

Figure 1.5.1   Typical tubular steel putlog scaffold.

◼◼◼ MATERIALS

Scaffolding can be of:

- tubular steel;
- tubular aluminium alloy;
- timber.

TUBULAR STEEL

British Standard 1139 gives recommendations for both welded and seamless steel
tubes of 48 mm outside diameter with a nominal 38 mm bore diameter. Steel tubes
can be obtained galvanised (to guard against corrosion); ungalvanised tubes will

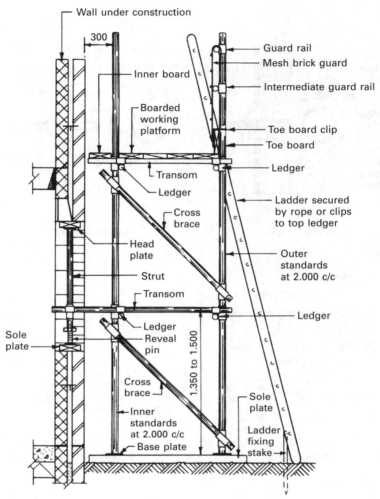

Figure 1.5.2 Typical tubular steel independent scaffold.

require special care such as painting, varnishing or an oil bath after use. Steel tubes are nearly three times heavier than comparable aluminium alloy tubes, but are far stronger, and as their deflection is approximately one-third that of aluminium alloy tubes, longer spans can be used.

ALUMINIUM ALLOY

Seamless tubes of aluminium alloy with a 48 mm outside diameter are specified in BS 1139 for metal scaffolding. No protective treatment is required unless they are to be used in contact with materials such as damp lime, wet cement or seawater, which can cause corrosion of the aluminium alloy tubes. A suitable protective treatment would be to coat the tubes with bitumastic paint before use.

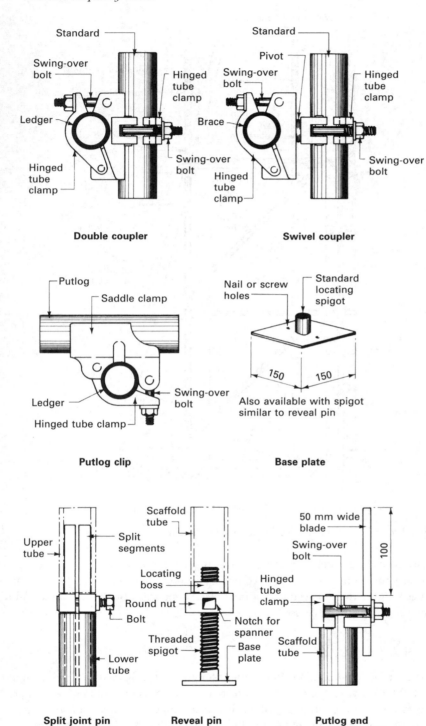

Figure 1.5.3 Typical steel scaffold fittings.

TIMBER

The use of timber as a temporary structure in the form of a scaffold is now rarely encountered in the UK, although it is extensively used in the developing world. The timber used is structural-quality softwood in either putlog or independent format. The members are lashed together with wire or rope instead of the coupling fittings used with metal scaffolds.

SCAFFOLD BOARDS

These are usually boards of softwood timber, complying with the recommendations of BS 2482, used to form the working platform at the required level. They should be formed out of specified softwoods of 225 mm × 38 mm section and not exceeding 4.800 m in length. To prevent the ends from splitting they should be end bound with not less than 25 mm wide × 0.9 mm galvanised hoop iron extending at least 150 mm along each edge and fixed with a minimum of two fixings to each end. The strength of the boards should be such that they can support a uniformly distributed load of 6.7 kN/m^2 when supported at 1.200 m centres.

SCAFFOLD FITTINGS

Fittings of either steel or aluminium alloy are covered by the same British Standard as quoted above for the tubes. They can usually be used in conjunction with either tubular metal unless specified differently by the manufacturer. The major fittings used in metal scaffolding are:

- **Double coupler**   The only real loadbearing fitting used in scaffolding; used to join ledgers to standards.
- **Swivel coupler**   Composed of two single couplers riveted together so that it is possible to rotate them and use them for connecting two scaffold tubes at any angle.
- **Putlog coupler**   Used solely for fixing putlogs or transoms to the horizontal ledgers.
- **Base plate**   A square plate with a central locating spigot, used to distribute the load from the foot of a standard on to a sole plate or firm ground. Base plates can also be obtained with a threaded spigot and nut for use on sloping sites to make up variations in levels.
- **Split joint pin**   A connection fitting used to joint scaffold tubes end to end. A centre bolt expands the two segments, which grip on the bore of the tubes.
- **Reveal pin**   Fits into the end of a tube to form an adjustable strut.
- **Putlog end**   A flat plate that fits on the end of a scaffold tube to convert it into a putlog.

Typical examples of the above fittings are shown in Fig. 1.5.3.

■ ■ ■ MOBILE ACCESS TOWER

Mobile access towers are temporary structures used for gaining access to buildings for maintenance and repair. They are preferred to conventional scaffolding for work

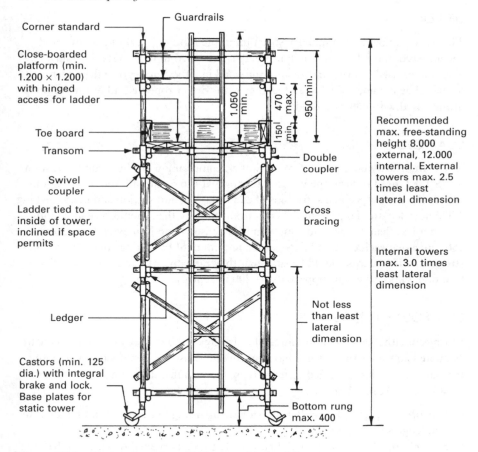

Corner standard

Guardrails

Close-boarded
platform (min.
1.200 × 1.200)
with hinged
access for ladder

Toe board

Transom

Swivel
coupler

Ladder tied to
inside of tower,
inclined if space
permits

Ledger

Castors (min. 125
dia.) with integral
brake and lock.
Base plates for
static tower

1.050 min.

470 max.

150 min.

950 min.

Double
coupler

Cross
bracing

Not less
than least
lateral
dimension

Bottom rung
max. 400

Recommended
max. free-standing
height 8.000
external, 12.000
internal. External
towers max. 2.5
times least
lateral dimension

Internal towers
max. 3.0 times
least lateral
dimension

Figure 1.5.4   Section through scaffold – assembled mobile access tower.

of a relatively short duration and where a ladder would be inadequate. Access
towers are also easily moved where work is continuous, e.g. painting building
exteriors, or gutter renewal.

Mobile access and working towers comply with the following:

- They are assembled from prefabricated components, either standard scaffold
  tube and fittings as shown in Fig. 1.5.4, or interconnecting 'H' shaped steel or
  aluminium tubular frames.
- They have a facility to be moved manually on firm, level ground.
- They have dimensions to a predetermined design.
- They are freestanding, with supplementary support optional.
- They have at least one platform to work from.
- They have at least four legs; normally each leg is fitted with castors. A base plate
  can be used at the bottom of each leg where mobility is not required.
- The platform is accessed by a ladder or steps contained within the base
  dimensions of the tower. A ladder must be firmly attached to the tower;

inclined is preferred. A ladder should not touch the ground; the first rung is located not more than 400 mm above the ground.

■ The working platform should have a hinged opening for ladder access, adequate guard rails, and toeboards.

STABILISERS, OUTRIGGERS OR DIAGONAL BRACING

An optional attachment that can be adjusted to ensure ground contact where the surface is uneven. They should be attached securely to enable direct transfer of loads without slipping or rotating.

ADDITIONAL SAFETY GUIDANCE

■ Constructed by competent persons.
■ Height to base ratio:
  ■ 3:1 maximum indoors.
  ■ 2.5:1 maximum outdoors.
■ Maximum height 12 m indoors, 8 m outdoors.
■ Never moved with persons, equipment or materials on the platform or frame.
■ Access ladders fitted within the frame.
■ Castors fitted with a locking device and secured before access is permitted.
■ Stable ground essential.
■ Components visually inspected for damage before assembly.
■ Inspected by a competent person before use and every 7 days if it remains in the same place. Inspected after any substantial alteration or period of exposure to bad weather.
■ Local authority Highways Department approval required prior to use on a public footpath or road. Licence or permit to be obtained.
■ Barriers or warning tape used to prevent people walking into the tower. Where appropriate, illuminated.
■ No work to be undertaken below a platform in use or within the tower.
■ Maximum of two persons working from the platform at any one time.
■ Material storage on the platform to be minimal.
■ Ladders or other means of additional access must not be used from the platform or any other part of the structure.

Further details and additional reading on this topic can be obtained from, BS EN 1298: *Mobile access and working towers* and BS EN 12811-1: *Temporary works equipment. Scaffolds. Performance requirements and general design.*

■■■ THE WORK AT HEIGHT REGULATIONS 2005

This statutory instrument is designed to ensure that suitable and sufficient safe access to and egress from every place at which any person at any time works are provided and properly maintained. Scaffolds and ladders are covered by this document, which sets objective requirements for materials, maintenance, inspection

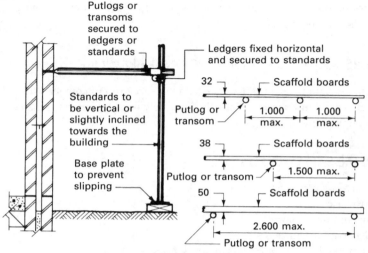

Putlogs or transoms secured to ledgers or standards

Ledgers fixed horizontal and secured to standards

Standards to be vertical or slightly inclined towards the building

Base plate to prevent slipping

32 — Scaffold boards

Putlog or transom

1.000 max.   1.000 max.

38 — Scaffold boards

Putlog or transom — 1.500 max.

50 — Scaffold boards

2.600 max.

Putlog or transom

Standards, putlogs and transoms

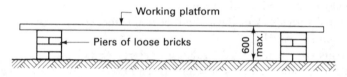

Working platform

Piers of loose bricks

600 max.

Stability of scaffolds

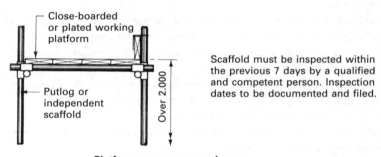

Close-boarded or plated working platform

Putlog or independent scaffold

Over 2.000

Scaffold must be inspected within the previous 7 days by a qualified and competent person. Inspection dates to be documented and filed.

Platforms, gangways and runs

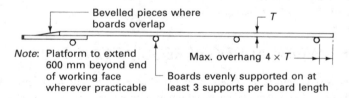

Bevelled pieces where boards overlap

T

Note: Platform to extend 600 mm beyond end of working face wherever practicable

Max. overhang 4 × *T*

Boards evenly supported on at least 3 supports per board length

Boards in working platforms

Figure 1.5.5   Scaffolds regulations – 1.

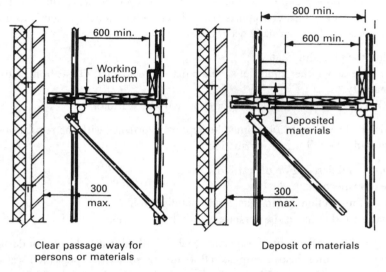

Clear passage way for persons or materials

Deposit of materials

Widths of working platforms for putlog and independent scaffolds

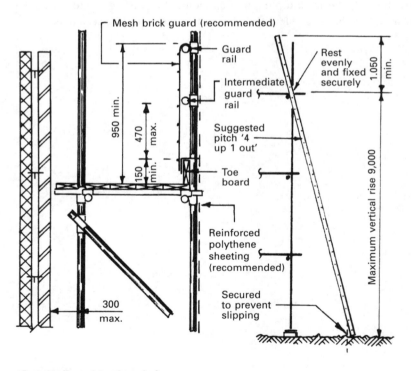

Guard rails and toe boards for putlog and independent scaffolds

Use of ladders

Figure 1.5.6 Scaffold regulations – 2.

and construction of these working places. The main constructional requirements of these regulations are illustrated in Figs 1.5.5 and 1.5.6. Supervision of scaffold erection and progress/safety reports are undertaken by a suitably experienced and qualified person. This normally occurs:

- within the preceding 7 days;
- after adverse weather conditions, which may have affected the scaffold's strength or stability;
- whenever alterations or additions are made to the scaffold.

Records of all such inspections must be kept in accordance with the regulations, and must provide the following information:

- location and description of scaffold;
- date of inspection;
- result of inspection, stating the condition of scaffold;
- signature and office of the person making the inspection.

The importance of providing a safe and reliable scaffold from which to undertake building work cannot be overemphasised. Badly assembled and neglected scaffolds have been a significant contributory factor to the high accident rate associated with the construction industry. The Health and Safety Executive (HSE) are promoting instructional literature and training programmes for scaffold erectors, as lack of formal training has been identified as the source of many site accidents.

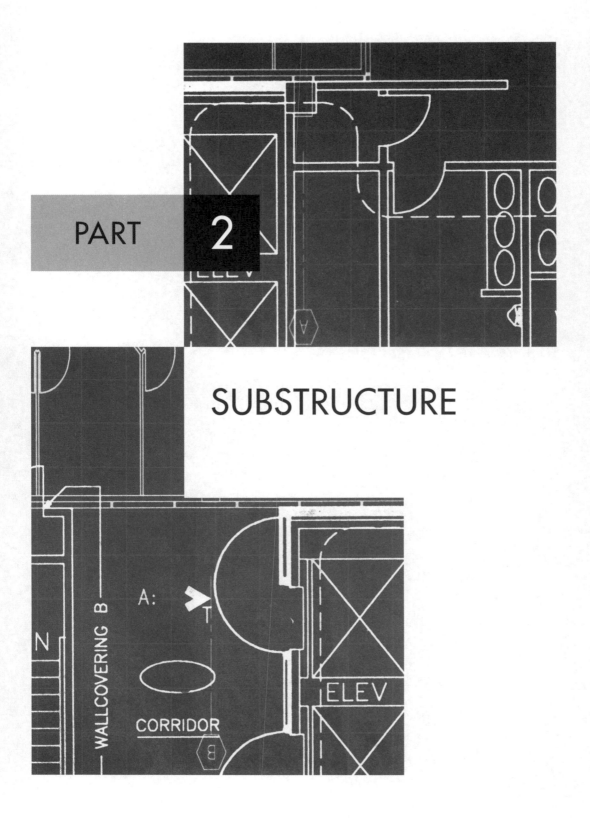

PART 2

SUBSTRUCTURE

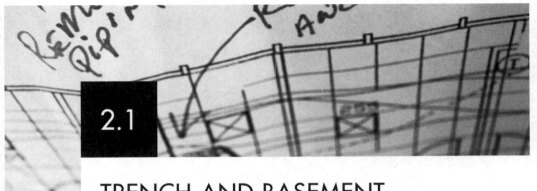

TRENCH AND BASEMENT EXCAVATION

Excavations may be classified as shallow, medium or deep as follows:

- shallow – up to 1.500 m deep;
- medium – 1.500–3.000 m deep;
- deep – over 3.000 m deep.

The method of excavation and timbering to be used in any particular case will depend upon a number of factors:

- The nature of the subsoil can determine the type of plant or hand tools required and the amount of timbering necessary.
- The purpose of the excavation can determine minimum widths, minimum depths and the placing of support members to give a reasonable working space within the excavation.
- The presence of groundwater may necessitate the need for interlocking timbering, sump pits and pumps; large quantities of groundwater may prompt the use of de-watering techniques.
- The position of the excavation may impose certain restrictions such as the need for a licence or wayleave, highway authority or police requirements when excavating in a public road.
- Non-availability of the right type of plant for bulk excavation may mean that a different method must be used.
- The presence of a large number of services may restrict the use of machinery to such an extent that it becomes uneconomic.
- The disposal of the excavated spoil may restrict the choice of plant because the load and unload cycle does not keep pace with the machine output.

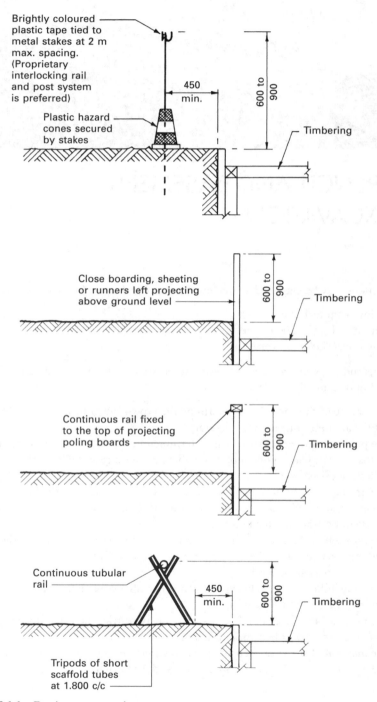

Figure 2.1.1   Barriers to excavations.

■■■ SAFETY

As previously described in Chapter 1.4, safety requirements in subsurface excavations, including basements, are strictly enforced through the Construction (Health, Safety and Welfare) Regulations 1996. Additionally, a suitable fence or barrier must be provided to the sides of excavations. For small pits, such as manholes and other limited excavation, the void may be covered. Methods of providing a suitable barrier are shown in Fig. 2.1.1. Materials must not be placed near the edge of any excavation, nor must plant be placed or moved near excavations so that persons working in the excavation are endangered.

■■■ TRENCH EXCAVATIONS

Long, narrow trenches in firm soil may be excavated to the full depth by mechanical excavators, enabling the support timbering to be placed in one continuous operation. Weak and waterlogged ground must be supported before excavation commences by driving timber runners or steel trench sheeting to a position below the formation level, or by a drive and dig procedure. In the latter method the runners can be driven to a reasonable depth of approximately 1.500 m followed by an excavation cut of 1.200 m and then the operation repeated until the required level has been reached; this will make the driving of the runners easier and enable a smaller driving appliance to be used.

In medium–depth trenches different soil conditions are very often encountered throughout the depth of the excavation, and therefore the method of timbering must be changed to suit the new soil conditions; a typical example of trench timbering in these circumstances is shown in Fig. 2.1.2.

Hand trimming is usually required in the trench bottom to form an accurate line and level; this process is called **bottoming of trenches**. Approximately 150 mm should be allowed for trimming by hand, and it is advisable to cover the trimmed surface with hardcore to protect the soil at formation level from being disturbed or drying out and shrinking.

■■■ BASEMENT EXCAVATIONS

There are three methods that can be used for excavating a large pit or basement:

- complete excavation with sloping sides;
- complete excavation with timbered sides;
- perimeter trench method.

Excavation for a basement on an open site can be carried out by cutting the perimeter back to the natural angle of repose of the soil. This method requires sufficient site space around the intended structure for the over-excavation. No timbering is required, but the savings on the temporary support work must pay for the over-excavation and consequent increase in volume of backfilling to be an economic method.

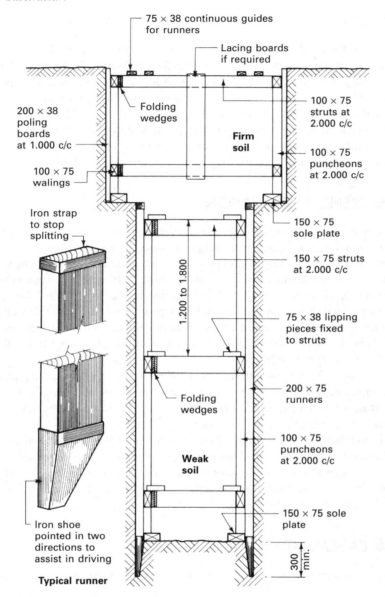

Figure 2.1.2   Trench timbering in different soils.

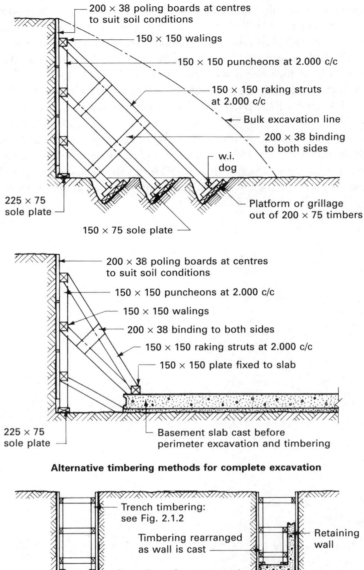

200 × 38 poling boards at centres to suit soil conditions

150 × 150 walings

150 × 150 puncheons at 2.000 c/c

150 × 150 raking struts at 2.000 c/c

Bulk excavation line

200 × 38 binding to both sides

w.i. dog

Platform or grillage out of 200 × 75 timbers

225 × 75 sole plate

150 × 75 sole plate

200 × 38 poling boards at centres to suit soil conditions

150 × 150 puncheons at 2.000 c/c

150 × 150 walings

200 × 38 binding to both sides

150 × 150 raking struts at 2.000 c/c

150 × 150 plate fixed to slab

225 × 75 sole plate

Basement slab cast before perimeter excavation and timbering

Alternative timbering methods for complete excavation

Trench timbering: see Fig. 2.1.2

Timbering rearranged as wall is cast

Retaining wall

Dumpling of unexcavated soil between trenches

Perimeter trench method

Figure 2.1.3 Basement excavations and timbering.

In firm soils where poling boards can be placed after excavation an economic method is to excavate the bulk of the pit and then trim the perimeter, placing the poling boards with their raking struts in position as the work proceeds. Alternatively the base slab could be cast before the perimeter trimming takes place and the rakers anchored to its edge or side (see Fig. 2.1.3).

The perimeter trench method is used where weak soils are encountered; a trench wide enough to enable the retaining walls to be constructed is excavated around the perimeter of the site, and timbered according to the soil conditions. The permanent retaining walls are constructed within the trench excavation and the timbering is removed; the dumpling or middle can then be excavated and the base cast and joined to the retaining walls (see Fig. 2.1.3). This method could also be used in firm soils when the mechanical excavators required for bulk excavation are not available.

FOUNDATIONS

A foundation is the base on which a building rests, and its purpose is to safely transfer the load of a building to a suitable subsoil.

The Building Regulations require all foundations of buildings to:

- safely sustain and transmit to the ground the combined dead and imposed loads so as not to cause any settlement or other movement in any part of the building or of any adjoining building or works;
- be of such a depth, or be so constructed, as to avoid damage by swelling, shrinkage or freezing of the subsoil;
- be capable of resisting attack by deleterious material, such as sulphates, in the subsoil.

Subsoils are the soils below the topsoil, the topsoil being about 300 mm deep. Typical bearing capacities of subsoils are given in Table 2.2.1.

Table 2.2.1 Typical subsoil bearing capacities

Type	Bearing capacity (kN/m^2)
Rocks, granites and chalks	600–10 000
Non-cohesive soils; compact sands; loose uniform sands	100–600
Cohesive soils; hard clays; soft clays and silts	< 600
Peats and made ground	To be determined by investigation

Terminology

- **Backfill**   Materials excavated from site and if suitable used to fill in around the walls and foundations.
- **Bearing capacity**   Safe load per unit area that the ground can carry.

- **Bearing pressure**   The pressure produced on the ground by the loads.
- **Made ground**   Refuse, excavated rock or soil deposited for the purpose of filling in a depression or for raising the site above its natural level.
- **Settlement**   Ground movement, which may be caused by:
  - deformation of the soil due to imposed loads;
  - volume changes of the soil as a result of seasonal conditions;
  - mass movement of the ground in unstable areas.

■■■ CHOICE OF FOUNDATION TYPE

The choice and design of foundations for domestic and small types of buildings depends mainly on two factors:

- the total loads of the building;
- the nature and bearing capacity of the subsoil.

The total loads of a building are taken per metre run and calculated for the worst case. The data required is:

1. Roof material dead load on the wall plus imposed load from snow, i.e. $1.5 \ kN/m^2$ $< 30°$ pitch or $0.75 \ kN/m^2 > 30°$ pitch (1 m wide strip from ridge to eaves).
2. Floor material dead load on the wall, plus an imposed loading allowance of $< 1.5 \ kN/m^2$ for people and furniture (1 m wide strip from centre of the floor to the wall).
3. Wall load on the foundations (1 m wide strip of wall from top to foundation).
4. Total load on the foundations (summation of 1, 2 and 3, plus any additional allowances for wind loading that may be necessary in exposed situations).

[*Note*: Material loading due to the materials can be found in BS 648: *Schedule of weights of building materials*.]

The average total loading for a two-storey domestic dwelling of traditional construction is 30–50 kN/m.
   The nature and bearing capacity of the subsoil can be determined by:

- trial holes and subsequent investigation;
- boreholes and core analysis;
- local knowledge.

   Clay is the most difficult of all subsoils with which to deal. Down to a depth of about 1 m clays are subject to seasonal movement, which occurs when the clay dries and shrinks in the summer and conversely swells in the winter with heavier rainfall. This movement occurs whenever a clay soil is exposed to the atmosphere, and special foundations may be necessary.
   Subsoils that readily absorb and hold water are subject, in cold weather, to **frost heave**. This is a swelling of the subsoil due to the expansion of freezing water held in the soil; like the movement of clay soils, it is unlikely to be even, and special foundations may be needed to overcome the problem.

■■■ TYPES OF FOUNDATION

Having ascertained the nature and bearing capacity of the subsoil, the width of the foundation can be determined by one of the following methods:

1. Calculating the total (dead + imposed) load per metre run of foundation and relating this to the analysed safe bearing capacity of the subsoil, i.e.

$$\frac{\text{total load of building per metre}}{\text{safe bearing capacity of subsoil}} = \text{minimum foundation width}$$

For example, if the total load is 40 kN/m and the subsoil safe bearing capacity is 80 kN/m², then the foundation width is:

$$\frac{40}{80} = 0.5 \text{ m} \quad \text{or} \quad 500 \text{ mm}$$

[*Note*: Safe bearing capacity is determined by obtaining the actual bearing capacity by laboratory analysis and applying a factor of safety.]

2. The minimum guidance as given in design tables, such as Table 2.2.2, where size of foundations are related to subsoil type, wall loading and field tests on the soil.

Foundations are usually made of either mass or reinforced concrete and can be considered under two headings:

- **Shallow foundations**   Those that transfer the loads to subsoil at a point near to the ground floor of the building such as strips and rafts.
- **Deep foundations**   Those that transfer the loads to a subsoil some distance below the ground floor of the building such as a pile.

Raft foundations are often used on poor soils for lightly loaded buildings, and are considered capable of accommodating small settlements of the soil. In poor soils the upper crust of soil (450–600 mm) is often stiffer than the lower subsoil, and to build a light raft on this crust is usually better than penetrating it with a strip foundation.

Typical details of the types of foundation suitable for domestic and similar buildings are shown in Figs 2.2.1–2.2.3.

Table 2.2.2 Guide to strip foundation width relative to subsoil type

Subsoil type	Subsoil condition	Field test	Bearing capacity (kN/m²)	Minimum width of strip (mm) for a total load (kN/m) of:					
				20	30	40	50	60	70
Rock	Stronger than sandstone limestone or chalk	Requires mechanical device to break up	> 600	Wall thickness at least					
Chalk	Solid	Requires a pick to remove	600	Wall thickness at least					
Gravel	Medium density	Requires a pick to excavate	> 600	250	300	400	500	600	650
Sand	Compact	Breaks down when dry	> 300	250	300	400	500	600	650
Clay Sandy clay	Stiff	Pick or mechanical device required to remove	150 to 300	250	300	400	500	600	650
Clay Sandy clay	Firm	Manually excavated and can be moulded by hand	75 to 150	300	350	450	600	750	850
Sand or Gravel Silty sand Clayey Sand	Loose	Manually excavated with a spade, 50 mm square peg easily driven in	> 200 Lab. test needed	400	600	Traditional strip foundation not suitable for loading > 30 kN/m, consider reinforced strip, deep strip or short bored piling			
Silt Clay Sandy or silty clay	Soft	Easily excavated manually, can be moulded by hand	Lab. test needed	450	650	As above			
Silt, clay, sandy or silty clay	Very soft	Samples exude water when squeezed	Lab. test needed	Subsoil of this condition is not suitable to receive unreinforced strip foundations. Consider reinforced strip, deep strip and end bearing piled foundations to solid strata					

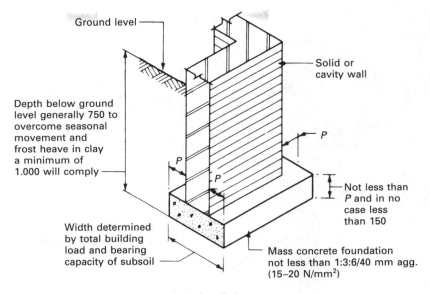

Ground level

Solid or cavity wall

Depth below ground level generally 750 to overcome seasonal movement and frost heave in clay a minimum of 1.000 will comply

P

P

P

Not less than *P* and in no case less than 150

Width determined by total building load and bearing capacity of subsoil

Mass concrete foundation not less than 1:3:6/40 mm agg. (15–20 N/mm^2)

Strip foundations

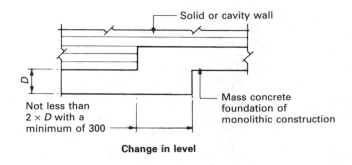

Solid or cavity wall

D

Not less than 2 × *D* with a minimum of 300

Mass concrete foundation of monolithic construction

Change in level

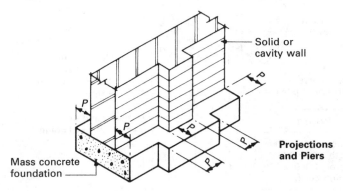

Solid or cavity wall

P

P

P

P

P

Projections and Piers

Mass concrete foundation

Figure 2.2.1   Strip foundations and building regulations.

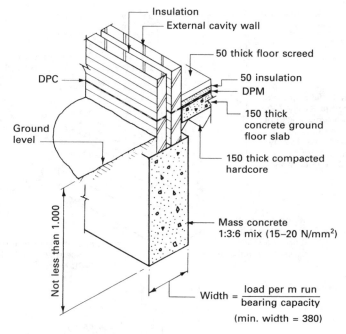

Insulation
External cavity wall
50 thick floor screed
DPC
50 insulation
DPM
Ground level
150 thick concrete ground floor slab
150 thick compacted hardcore
Not less than 1.000
Mass concrete 1:3:6 mix (15–20 N/mm²)

$$\text{Width} = \frac{\text{load per m run}}{\text{bearing capacity}}$$

(min. width = 380)

Deep strip foundation

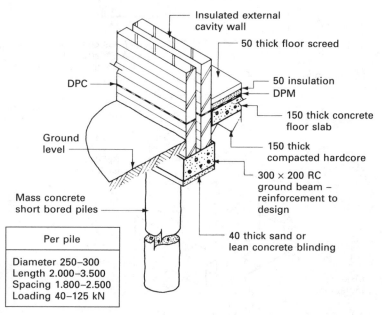

Insulated external cavity wall
50 thick floor screed
DPC
50 insulation
DPM
Ground level
150 thick concrete floor slab
150 thick compacted hardcore
300 × 200 RC ground beam – reinforcement to design
Mass concrete short bored piles
40 thick sand or lean concrete blinding

Per pile
Diameter 250–300
Length 2.000–3.500
Spacing 1.800–2.500
Loading 40–125 kN

Short bored pile foundation

Figure 2.2.2   Alternative foundations for clay soils.

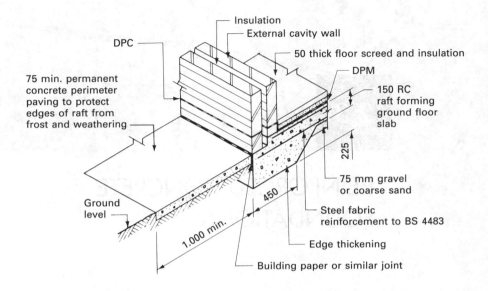

DPC

Insulation

External cavity wall

50 thick floor screed and insulation

DPM

150 RC raft forming ground floor slab

75 min. permanent concrete perimeter paving to protect edges of raft from frost and weathering

225

450

75 mm gravel or coarse sand

Steel fabric reinforcement to BS 4483

Edge thickening

Building paper or similar joint

Ground level

1.000 min.

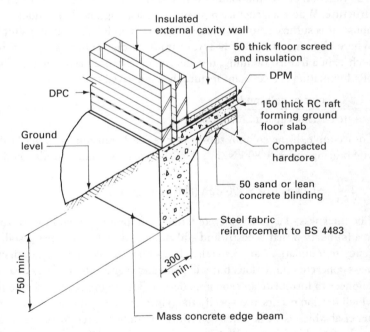

Insulated external cavity wall

50 thick floor screed and insulation

DPM

DPC

150 thick RC raft forming ground floor slab

Ground level

Compacted hardcore

50 sand or lean concrete blinding

Steel fabric reinforcement to BS 4483

750 min.

300 min.

Mass concrete edge beam

Figure 2.2.3   Typical raft foundations.

2.3

REINFORCED CONCRETE
FOUNDATIONS

The function of any foundation is to transmit to the subsoil the loads of the structure. Where a structure has only light loadings, such as a domestic dwelling house, it is sufficient to use a mass concrete strip foundation or a simple raft. Where buildings are either heavy, or transmit the loadings at a series of points, such as in a framed building, it is uneconomic to use mass concrete. The plan size of a foundation is a constant feature, being derived from:

$$\frac{\text{point or column load}}{\text{safe bearing capacity of subsoil}}$$

For example, if the column shown in Fig. 2.3.1 transmits a 50 kN load to subsoil of safe bearing capacity 80 kN/m², then the square column foundation dimensions are:

$$\sqrt{\frac{50}{80}} = 0.79 \text{ m} \quad \text{or} \quad 790 \text{ mm (800 mm square)}$$

The thickness of a mass concrete foundation for a heavy point load would result in a foundation that is costly and adds unnecessary load to the subsoil. Reinforced concrete foundations are generally cheaper and easier to construct than equivalent mass concrete foundations but will generally require the services of a structural engineer to formulate an economic design. The engineer must define the areas in which tension occurs and specify the reinforcement required, as concrete is a material which is weak in tension. Further studies of reinforced concrete principles can be found in Chapter 10.2.

■ ■ ■ TYPES OF FOUNDATION

The principal types of reinforced concrete foundation for buildings are:

1. strip foundations;
2. isolated or pad foundations;

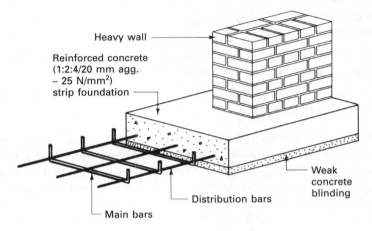

Heavy wall

Reinforced concrete
(1:2:4/20 mm agg.
– 25 N/mm²)
strip foundation

Weak
concrete
blinding

Distribution bars

Main bars

RC strip foundation

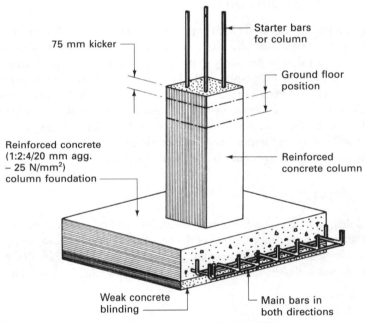

75 mm kicker

Starter bars
for column

Ground floor
position

Reinforced concrete
(1:2:4/20 mm agg.
– 25 N/mm²)
column foundation

Reinforced
concrete column

Weak concrete
blinding

Main bars in
both directions

RC isolated or pad foundation

Figure 2.3.1   Reinforced concrete strip and pad foundations.

3. raft foundations;
4. combinations of 1, 2 and 3;
5. piled foundations.

   The foundations listed in 4 and 5 above are included in *Advanced Construction Technology* and are therefore not considered in this volume.

STRIP FOUNDATIONS

Reinforced concrete strip foundations are used to support and transmit the loads from heavy walls. The effect of the wall on the relatively thin foundation is to act as a point load, and the resultant ground pressure will induce tension on the underside across the width of the strip. Tensile reinforcement is therefore required in the lower face of the strip, with distribution bars in the second layer running longitudinally (see Fig. 2.3.1). The reinforcement will also assist the strip in spanning any weak pockets of soil encountered in the excavations.

ISOLATED OR PAD FOUNDATIONS

This type of foundation is used to support and transmit the loads from piers and columns. The most economic plan shape is a square, but if the columns are close to the site boundary it may be necessary to use a rectangular plan shape of equivalent area. The reaction of the foundation to the load and ground pressures is to cup, similar to a saucer, and therefore main steel is required in both directions. The depth of the base will be governed by the anticipated moments and shear forces, the calculations involved being beyond the scope of this volume. Incorporated in the base will also be the starter bars for a reinforced concrete column or the holding-down bolts for a structural steel column (see Fig. 2.3.1).

RAFT FOUNDATIONS

The principle of any raft foundation is to spread the load over the entire area of the site. This method is particularly useful where the column loads are heavy and thus require large bases, or where the bearing capacity is low, again resulting in the need for large bases. Raft foundations can be considered under three headings:

- solid slab rafts;
- beam and slab rafts;
- cellular rafts.

   Solid slab rafts are constructed of uniform thickness over the whole raft area, which can be wasteful because the design must be based on the situation existing where the heaviest load occurs. The effect of the load from columns and the ground pressure is to create areas of tension under the columns and areas of tension in the upper part of the raft between the columns. Very often a nominal mesh of reinforcement is provided in the faces where tension does not occur to control shrinkage cracking of the concrete (see Fig. 2.3.2).

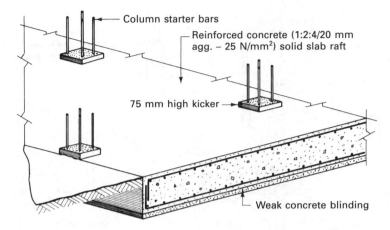

— Column starter bars

— Reinforced concrete (1:2:4/20 mm agg. – 25 N/mm^2) solid slab raft

75 mm high kicker —

— Weak concrete blinding

RC solid slab raft foundation

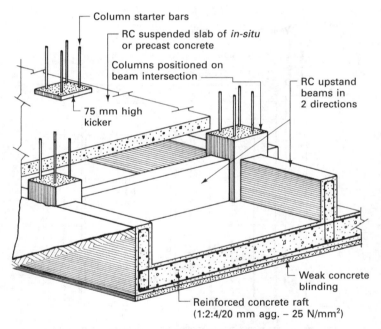

— Column starter bars

— RC suspended slab of *in-situ* or precast concrete

Columns positioned on beam intersection —

— RC upstand beams in 2 directions

— 75 mm high kicker

— Weak concrete blinding

— Reinforced concrete raft (1:2:4/20 mm agg. – 25 N/mm^2)

RC beam and slab raft foundation

Figure 2.3.2   Reinforced concrete raft foundations.

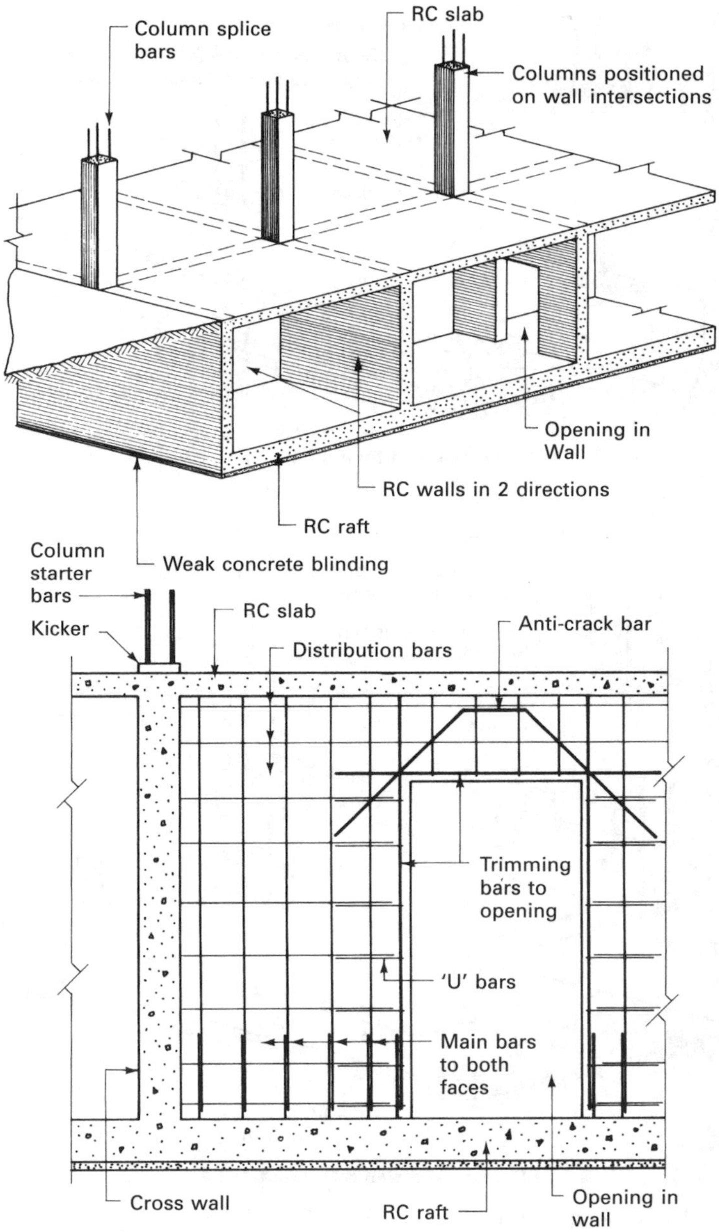

Figure 2.3.3   Typical cellular raft details.

Beam and slab rafts are an alternative to the solid slab raft and are used where poor soils are encountered. The beams are used to distribute the column loads over the area of the raft, which usually results in a reduction of the slab thickness. The beams can be upstand or downstand depending upon the bearing capacity of the soil near the surface. Downstand beams will give a saving on excavation costs, whereas upstand beams create a usable void below the ground floor if a suspended slab is used (see Fig. 2.3.2).

CELLULAR RAFTS

This form of foundation can be used where a reasonable bearing capacity subsoil can be found only at depths where beam and slab techniques become uneconomic. The construction is similar to reinforced concrete basements except that internal walls are used to spread the load over the raft and divide the void into cells. Openings can be formed in the cell walls, allowing the voids to be utilised for the housing of services, store rooms or general accommodation (see Fig. 2.3.3).

■■■ BLINDING

A blinding layer 50 to 75 mm thick of weak concrete or coarse sand should be placed under all reinforced concrete foundations. The functions of the blinding are to fill in any weak pockets encountered during excavations and to provide a true level surface from which the reinforcement can be positioned. If formwork is required for the foundation some contractors prefer to lay the blinding before assembling the formwork; the alternative is to place the blinding within the formwork and allow this to set before positioning the reinforcement and placing the concrete.

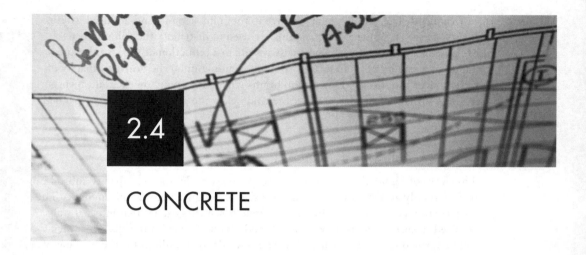

CONCRETE

Concrete is a mixture of cement, fine aggregate, coarse aggregate and water. The proportions of each material control the strength and quality of the resultant concrete.

■■■■ PORTLAND CEMENT

Cement is the setting agent of concrete, and the bulk of cement used in this country is Portland cement. This is made from chalk or limestone and clay, and is generally produced by the **wet process**.

In this process the two raw materials are washed, broken up and mixed with water to form a slurry. This slurry is then pumped into a steel rotary kiln, which is from 3 to 4 m in diameter and up to 150 m long and lined with refractory bricks. While the slurry is fed into the top end of the kiln a pulverised coal is blown in at the bottom end and fired. This raises the temperature at the lower end of the kiln to about 1400 °C. First the slurry passing down the kiln gives up its moisture; then the chalk or limestone is broken down into carbon dioxide and lime, and finally it forms a white-hot clinker, which is transferred to a cooler before being ground. The grinding is carried out in a ball mill, which is a cylinder some 15 m long and up to 4.5 m in diameter containing a large number of steel balls of various sizes, which grind the clinker into a fine powder. As the clinker is being fed into the ball mill, gypsum (about 5%) is added to prevent a flash setting off the cement.

The alternative method for the preparation of Portland cement is the **dry process**. The main difference between this and the wet process is the reduction in the amount of water that has to be driven off in the kiln. A mixture of limestone and shale is used, which is proportioned, ground and blended to form a raw meal of low moisture content. The meal is granulated in rotating pans with a small amount of water before being passed to a grate for preheating prior to entering the kiln. The kiln is smaller than that used in the wet process but its function is the same – that is, to form a clinker, which is then cooled, ground and mixed with a little gypsum as described for the previous process.

Rapid-hardening Portland cement is more finely ground than ordinary Portland cement. Its main advantage is that it gains its working strength earlier than ordinary cement. The requirements for both ordinary Portland and rapid-hardening Portland cement are given in BS EN 197-1.

▨■■ HIGH-ALUMINA CEMENT – BS 915-2

This is made by firing limestone and bauxite (aluminium ore) to a molten state, casting it into pigs, and finally grinding it into a fine powder. Its rate of hardening is very rapid, and produces a concrete that is resistant to the natural sulphates found in some subsoils. It can, however, cost up to two and a half times as much as ordinary Portland cement. However, high-alumina cement has its limitations. When hydrated, the cement converts or undergoes a chemical change that can reduce the concrete's strength and resistance to aggressive chemicals. The manner and extent of conversion vary with the amount of free water/cement ratio, the temperature in which the concrete is located, and the amount of moisture or humidity in the surrounding environment. All these factors are applicable during the cement's manufacture, subsequent curing and design life. Because of a number of structural failures, notably in the humid atmospheres of swimming pools, high-alumina cement is not now normally permitted in exposed concrete structures.

Other forms of cement available are as follows:

- Portland blastfurnace – BS 146;
- low-heat Portland – BS 1370;
- low-alkali sulphate-resisting – BS 4027;
- high-slag blastfurnace – BS 146;
- supersulphated – BS 4248;
- masonry – BS EN 413-1;
- Portland pulverised fuel ash – BS 6588;
- pozzolanic pulverised fuel ash – BS 6610;
- Portland limestone – BS 7583.

Cement should be stored on a damp-proof floor in the dry and kept for short periods only, because eventually it will harden as a result of the action of moisture in the air. This is known as **air hardening**, and any hardened cement should be discarded.

▨■■ AGGREGATES

These are the materials that are mixed with the cement to form concrete, and are classed as a fine or coarse aggregate. Fine aggregates are those that will pass a standard 5 mm sieve, and coarse aggregates are those that are retained on a standard 5 mm sieve. All-in aggregate is a material composed of both fine and coarse aggregates.

A wide variety of materials (for example, gravel, crushed stone, brick, furnace slag and lightweight substances, such as foamed slag, expanded clay and vermiculite) are available as aggregates for the making of concrete.

In making concrete, aggregates must be graded so that the smaller particles of the fine aggregate fill the voids created by the coarse aggregate. The cement paste fills the voids in the fine aggregate thus forming a dense mix.

Aggregates from natural sources and synthetic aggregates are defined in BS EN 12620: *Aggregates for concrete.*

■■■ WATER

The water used in the making of concrete must be clean and free from impurities that could affect the concrete. It is usually specified as being of a quality fit for drinking. A proportion of the water will set up a chemical reaction, which will harden the cement. The remainder is required to give the mix workability, and will evaporate from the mix while it is curing, leaving minute voids. An excess of water will give a porous concrete of reduced durability and strength.

The quantity of water to be used in the mix is usually expressed in terms of the **water/cement ratio,** which is:

$$\frac{\text{total weight of water in the concrete}}{\text{weight of cement}}$$

For most mixes the ratio is between 0.4 and 0.7.

Concrete mixes can be expressed as volume ratios thus:

- 1:2:4 = 1 part cement, 2 parts fine aggregate and 4 parts coarse aggregate;
- 1:5   = 1 part cement and 5 parts all-in aggregate.

SOME COMMON MIXES

- 1:10 – not a strong mix but suitable for filling weak pockets in excavations and for blinding layers.
- 1:8 – slightly better than the last, suitable for paths and pavings.
- 1:6 – a strong mix suitable for mass concrete foundations, paths and pavings.
- 1:3:6 – the weakest mix equivalent to that quoted in Approved Document A as deemed to satisfy the requirements of Building Regulation A1.
- 1:2:4 – a strong mix that is practically impervious to water; in common use especially for reinforced concrete.

MIXING CONCRETE

Concrete can be mixed or batched by two methods:

- by volume;
- by mass.

A 25 kg bag of cement has a volume of approximately 0.02 m^3.

Batching by volume

This method is usually carried out using an open bottom box (of such dimensions as to make manual handling possible) called a **gauge box**. For a 1:2:4 mix a gauge box is filled once with cement, twice with fine aggregate and four times with coarse aggregate, the top of the gauge box being struck off level each time.

If the fine aggregate is damp or wet its volume will increase by up to 25%, and therefore the amount of fine aggregate should be increased by this amount. This increase in volume is called **bulking**.

Batching by mass

This method involves the use of a balance that is linked to a dial giving the exact mass of the materials as they are placed in the scales. This is the best method, because it has a greater accuracy, and the balance can be attached to the mixing machine.

Hand mixing

This should be carried out on a clean, hard surface. The materials should be thoroughly mixed in the dry state before the water is added. The water should be added slowly, preferably using a rose head, until a uniform colour is obtained.

Machine mixing

The mix should be turned over in the mixer for at least 2 minutes after adding the water. The first batch from the mixer tends to be harsh, because some of the mix will adhere to the sides of the drum. This batch should be used for some less important work such as filling in weak pockets in the bottom of the excavation.

Ready mixed

This is used for large batches with lorry transporters up to 6 m^3 capacity. It has the advantage of eliminating site storage of materials and mixing plant, with the guarantee of concrete manufactured to quality-controlled standards. Placement is usually direct from the lorry: therefore site-handling facilities must be coordinated with deliveries.

Handling

If concrete is to be transported for some distance over rough ground the runs should be kept as short as possible, because vibrations of this nature can cause segregation of the materials in the mix. For the same reason concrete should not be dropped from a height of more than 1 m. If this is unavoidable a chute should be used.

Placing

If the concrete is to be placed in a foundation trench it will be levelled from peg to peg (see Fig. 1.4.2), or if it is to be used as an oversite bed the external walls could act as a levelling guide. The levelling is carried out by tamping with a straight-edge board; this tamping serves the dual purpose of both compacting the concrete and bringing the excess water to the surface so that it can evaporate. Concrete must not be over-tamped, as this will bring not only the water to the surface but also the cement paste that is required to act as the matrix. Concrete should be placed as soon as possible after mixing to ensure that the setting action has not commenced. Concrete that dries out too quickly will not develop its full strength: therefore new concrete should be protected from the drying winds and sun by being covered with canvas, straw, polythene sheeting or damp sawdust. This protection should be continued for at least 3 days, because concrete takes about 28 days to obtain its working strength.

Specifying concrete

Any of four methods may be used to specify concrete:

- designed mix;
- prescribed mix;
- standard mix;
- designated mix.

Designed mix

The mix is specified by a grade corresponding to the required characteristic compressive strength at 28 days. There are 12 grades, from C7.5 to C60, the C indicating compressive and the number the strength in N/mm^2 or MPa. Flexural (F) strength grades may also be specified as F3, F4 or F5, e.g. 3, 4 or 5 N/mm^2. In addition to stating the strength grades the purchaser must also specify any particular requirements for cement and aggregate content and maximum free water/cement ratio.

Prescribed mix

This is a recipe of constituents with their properties and quantities used to manufacture the concrete. The specifier must state:

- the type of cement;
- type of aggregates and their maximum size;
- mix proportions by weight;
- degree of workability (slump and/or water/cement ratio) and the application.

Prescribed mixes are based on established data indicating conformity to strength, durability and other characteristics. Basic examples could include 1:3:6/40 mm aggregate and 1:2:4/20 mm aggregate.

Standard mix

Mixes are produced from one of five grades, ranging from ST1 to ST5, with corresponding 28-day strength characteristics of 7.5 to a limit of only 25 N/mm^2. Mix composition and details are specified by:

- cement to aggregate by weight;
- type of cement;
- aggregate type and maximum size;
- workability;
- use or omission of reinforcement.

These mixes are most suited to site production, where the scale of operations is relatively small. Alternatively, they may be used where mix design procedures would be too time consuming, inappropriate or uneconomic.

Designated mix

These mixes are selected relative to particular applications and site conditions, in place of generalisations or use of alternative design criteria that may not be entirely appropriate. Grading and strength characteristics are extensive, and vary with application as indicated:

- General (GEN), graded 0–4, ranging from 7.5 to 25 N/mm^2 characteristic strength. For foundations, floors and external works.
- Foundations (FND), graded 2, 3, 4A and 4B with characteristic strength of 35 N/mm^2. These are particularly appropriate for resisting the effects of sulphates in the ground.
- Paving (PAV), graded 1 or 2 in 35 or 45 N/mm^2 strengths respectively. A strong concrete for use in driveways and heavy-duty pavings.
- Reinforced (RC) and prestressed concrete graded 30, 35, 40, 45 and 50 corresponding with characteristic strength and exposures ranging from mild to most severe.

In addition to application, the purchaser needs to specify:

- reinforced or unreinforced;
- prestressed;
- heated or ambient temperature;
- maximum aggregate size (if not 20 mm);
- exposure to chemicals (chlorides and sulphates);
- exposure to subzero temperatures (placing and curing).

Because of the precise nature of these designated mix concretes, quality control is of paramount importance: therefore producers are required to have quality assurance product conformity accreditation to BS EN ISO 9001.

[*Note*: Further details of concrete grades and analysis are found in BS EN 206-1: *Concrete. Specification, performance, production and conformity.*]

■ ■ ■ TESTING OF CONCRETE

There are many different test procedures for determining the properties of concrete. These are detailed in BS 1881: *Testing concrete*, BS EN 12350: *Testing fresh concrete* and BS EN 12390: *Testing hardened concrete*. The two most common tests are the slump test, applied to wet or fresh concrete, and the compression test, applied to hardened concrete.

SLUMP TEST – BS EN 12350-2

The slump test is suitable for establishing uniformity of mixes in subsequent batches or deliveries. It is not strictly a test for workability, but it can be used as a guide when constituents are known to be constant. Mixes of the same slump can vary with regard to their cement content and grade of aggregates.

Equipment

Shown in Fig. 2.4.1 comprises an open-ended steel frustum of a cone, a tamping rod and a rule.

   Procedure – the cone is one-quarter filled with concrete and tamped 25 times. A further three layers are applied, each layer tamped as described. Surplus concrete is struck from the surface, and the cone is raised immediately.

Typical slump

Mass concrete/thick sections of reinforced concrete,	50 mm.
Roads, hand tamped and general use,	100 mm.
Thin sections of reinforced concrete,	150 mm.
Maximum aggregate size, 40 mm.	

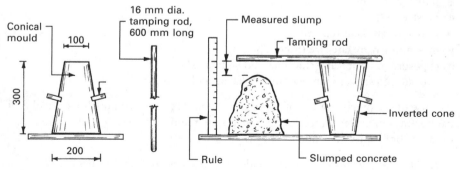

Figure 2.4.1   Slump test.

COMPRESSION TEST – BS EN 12390-1

Concrete test cubes are made from samples taken from site. Specimens are taken before and during the placing of concrete.

Equipment

Shown in Fig. 2.4.2. Standard machined steel mould, 150 mm × 150 mm × 150 mm for aggregate size up to 40 mm. 100 mm cube moulds may be used for aggregates up to 20 mm. Internal faces of the mould are lightly oiled prior to receiving concrete. A 25 mm square steel tamping rod, 380 mm long.

Procedure

Concrete is placed in the mould in 50 mm layers; each layer is tamped 35 times for 150 mm cubes or 25 times for 100 mm cubes. Alternatively, the concrete may be compacted by vibration. Surplus concrete is struck off. Samples remain in the mould for 24 hours ± 30 minutes, covered with a damp sack or similar. After this time specimens are marked, removed from the mould and submersed in water at a temperature between 10 and 21 °C until required for testing.

   The cube strength is the stress failure after 7 days. If the strength specification is not achieved at 7 days, a further test is undertaken at 28 days. If the specification is not achieved at 28 days, specimen cores may be taken from the placed concrete for laboratory analysis.

   Typical 28-day characteristic crushing strengths (that below which not more than 5 per cent of the test results are allowed to fall) are graded 7.5, 10, 15, 20, 25, 30, 35 and 40 N/mm^2.

   As a guide, shear stress of the concrete is taken at approximately one-tenth of the stress under compression.

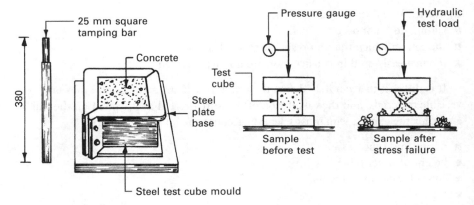

Figure 2.4.2   Compression test.

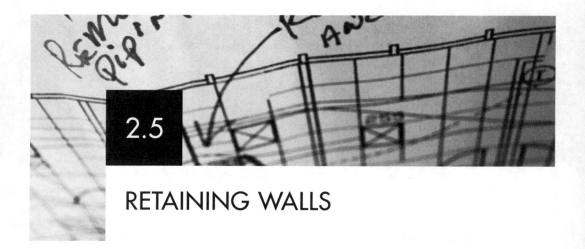

RETAINING WALLS

The basic function of a retaining wall is to retain soil at a slope that is greater than it would naturally assume, usually at a vertical or near-vertical position. The natural slope taken up by any soil is called its **angle of repose** and is measured in relationship to the horizontal. Angles of repose for different soils range from 45° to near 0° for wet clays, but for most soils an average angle of 30° is usually taken. It is the wedge of soil resting on this upper plane of the angle of repose that a retaining wall has to support. The walls are designed to offer the necessary resistance by using their own mass to resist the thrust or relying upon the principles of leverage. The terminology used in retaining wall construction is shown in Fig. 2.5.1.

■■■ DESIGN PRINCIPLES

The design of any retaining wall is basically concerned with the lateral pressures of the retained soil and any subsoil water. The wall must be designed to ensure that:

- overturning does not occur;
- sliding does not occur;
- the soil on which the wall rests is not overloaded;
- the materials used in construction are not overstressed.

It is difficult to accurately define the properties of any soil, because they are variable materials, and the calculation of pressure exerted at any point on the wall is a task for the expert, who must take into account the following factors:

- nature and type of soil;
- height of water table;
- subsoil water movements;
- type of wall;
- materials used in the construction of the wall.

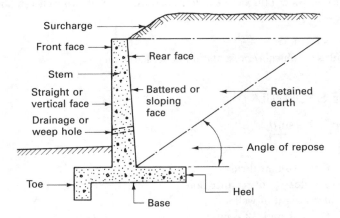

Retaining wall terminology

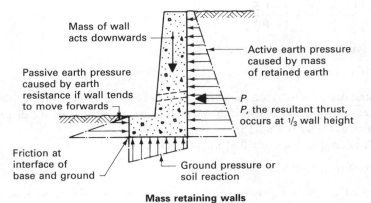

Mass retaining walls

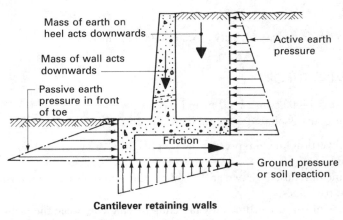

Cantilever retaining walls

Figure 2.5.1   Retaining wall terminology and pressures.

Design calculations relate to the resultant thrust of retained material behind a 1 m length of wall. This can be determined by either of two well-established methods:

- Rankine's formula;
- Coulomb's graphical representation or wedge theory.

RANKINE'S FORMULA

$$P = \frac{wh^2}{2} \times \frac{1 - \sin\theta}{1 + \sin\theta}$$

where P = resultant thrust
w = density of retained soil
h = height of wall
θ = soil angle of repose

For example, if $w = 1400$ kg/m^3, $h = 2$ m and $\theta = 30°$, then:

$$P = \frac{1400 \times (2)^2}{2} \times \frac{1 - 0.5}{1 + 0.5} = 934 \text{ kg}$$

kg $\times$ gravity = newtons, therefore

$934 \times 9.81 = 9.16$ kN

Structural design is based on the overturning or bending moment (BM), which occurs at one-third wall height if the earth pressure is uniform and the diagram triangular, as shown in Fig. 2.5.1.

In this example, $\frac{1}{3} \times 2$ m = 0.67 m. Therefore:

BM = 9.16 kN $\times$ 0.67 m = 6.14 kN m

P, the resultant thrust for water, can be calculated from the simple formula

$$P = \frac{wh^2}{2}$$

where w in this instance = density of water (1000 kg/m^3).

COULOMB'S THEORY

As before, $w = 1400$ kg/m^3, $h = 2$ m and $\theta = 30°$.

The procedure is plotted graphically in Fig. 2.5.2.

1. A scaled vertical line to represent the wall height is drawn from a horizontal baseline.
2. The plane of repose is drawn at the soil angle of repose from the intersection of wall and base lines.
3. The plane of rupture is drawn by bisecting the angle between the plane of repose and the wall line.

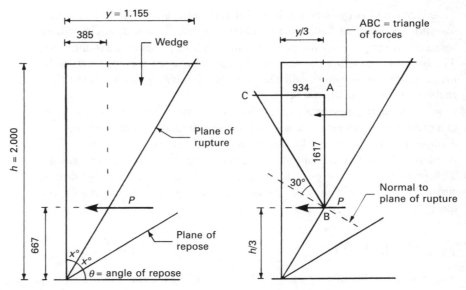

Figure 2.5.2   Coulomb's theory.

4. Coulomb's wedge is completed by drawing a horizontal line (y) from the top of the wall line until it intersects with the plane of rupture. Dimension y is scaled or calculated:

> Tangent $30° = y/2$
> Tangent $30° = 0.5774$, therefore $y = 0.5774 \times 2 = 1.155$ m.
> Wedge area $= (2 \times 1.155)/2 = 1.155$ m^2.
> Wedge volume per metre run of wall $= 1.155 \times 1 = 1.155$ m^3.
> Weight of wedge per metre run of wall $= 1.155 \times 1400 = 1617$ kg.

5. A vertical vector line AB is drawn through the wedge section centre of gravity at $y/3$.
6. AB is drawn to a scale to represent the wedge weight (1617), from where it intersects the line of thrust ($h/3$) and the plane of rupture.
7. Vector line BC is drawn at the angle of repose ($30°$) to the normal to the plane of rupture, until it intersects horizontal vector CA.
8. ABC is the triangle of forces for the wedge section of retained soil. CA is scaled (934) to represent P, the resultant thrust of 934 kg.

■■■ EARTH PRESSURES

The designer is mainly concerned with the effect of two forms of earth pressure:

■ active earth pressure;
■ passive earth pressure.

Active earth pressures are those that at all times are tending to move or overturn the retaining wall, and are composed of the earth wedge being retained together with any hydrostatic pressure caused by the presence of groundwater. The latter can be reduced by the use of subsoil drainage behind the wall, or by inserting drainage openings called **weep holes** through the thickness of the stem, enabling the water to drain away.

Passive earth pressures are reactionary pressures that will react in the form of a resistance to movement of the wall. If the wall tends to move forward, the earth in front of the toe will be compressed, and a reaction in the form of passive pressure will build up in front of the toe to counteract the forward movement. This pressure can be increased by enlarging the depth of the toe or by forming a rib on the underside of the base. Typical examples of these pressures are shown in Fig. 2.5.1.

▧■■ STABILITY

The overall stability of a retaining wall is governed by the result of the action and reaction of a number of loads:

■ **Applied loads**   For example, soil and water pressure on the back of the wall; the mass of the wall; and, in certain forms of cantilever wall, the mass of the soil acting with the mass of the wall.
■ **Induced loads**   For example, the ground pressure under the base, the passive pressure at the toe, and the friction between the underside of the base and the soil.

EFFECTS OF WATER

Groundwater behind a retaining wall, whether static or percolating through a subsoil, can have adverse effects upon the design and stability. It will increase the pressure on the back of the wall, and by reducing the soil shear strength it can reduce the bearing capacity of the soil; it can reduce the frictional resistance between the base and the soil and reduce the possible passive pressure in front of the wall. It follows therefore that the question of drainage of the water behind the retaining wall is of the utmost importance in the design.

SLIP CIRCLE FAILURE

This type of failure, shown in Fig. 2.5.3, is sometimes encountered with retaining walls in clay soils, particularly where there is a heavy surcharge of retained material. It takes the form of a rotational movement of the soil and wall along a circular arc. The arc commences behind the wall and passes under the base, resulting in a tilting and forward movement of the wall. Further movement can be prevented by driving sheet piles into the ground in front of the toe, to a depth that will cut the slip circle arc.

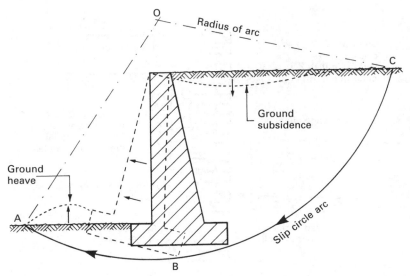

Moment due to weight of retained earth and wall above slip circle arc about O
is greater than restoring moment RM.
RM = permissible shear stress × length of arc ABC × arc radius OC.
Result: Mass above ABC rotates about O. Wall tilts forward and earth
heaves in front.

Figure 2.5.3   Retaining wall failure due to rotational movement.

■ ■ ■ TYPES OF WALL

MASS RETAINING WALLS

These are sometimes called **gravity walls** and rely upon their own mass together
with the friction on the underside of the base to overcome the tendency to slide or
overturn. They are generally economic only up to a height of 1.800 m. Mass walls
can be constructed of semi-engineering quality bricks bedded in a 1:3 cement
mortar or of mass concrete. The latter could have some light fabric reinforcement
to control surface cracking. Natural stone is suitable for small walls up to 1.000 m
high, but generally it is used as a facing material for walls over 1.000 m. Typical
examples of this are shown in Fig. 2.5.4.

CANTILEVER WALLS

These are usually of reinforced concrete, and work on the principles of leverage.
Two basic forms can be considered: a base with a large heel so that the mass
of earth above can be added to the mass of the wall for design purposes, or, if
this form is not practicable, a cantilever wall with a large toe (see Fig. 2.5.5).
The drawings show typical sections and patterns of reinforcement encountered
with these basic forms of cantilever retaining wall. The main steel occurs on the
tension face of the wall, and nominal steel (0.15% of the cross-sectional area
of the wall) is very often included in the opposite face to control the shrinkage

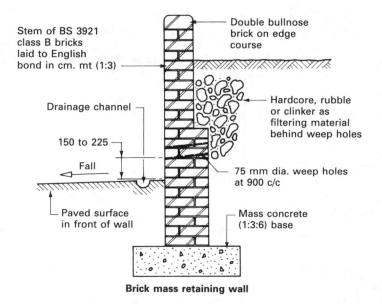

Brick mass retaining wall

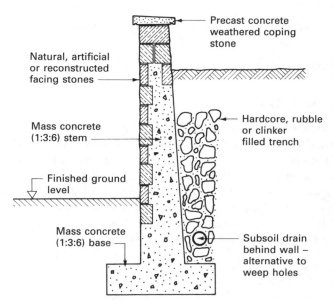

Mass concrete retaining wall with stone facings

Figure 2.5.4 Typical mass retaining walls.

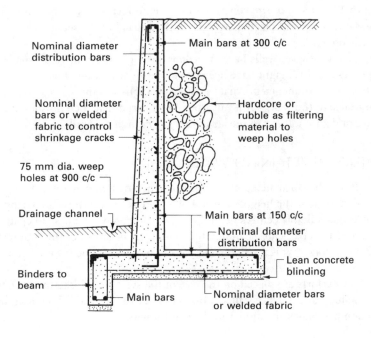

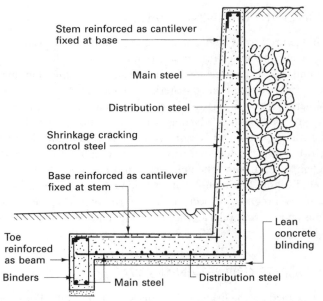

Figure 2.5.5 Typical reinforced concrete cantilever retaining walls.

cracking that occurs in *in-situ* concrete work. Reinforcement requirements, bending, fabricating and placing are dealt with in detail in the section on reinforced concrete.

Reinforced cantilever walls have an economic height range of 1.200–6.000 m; walls in excess of this height have been economically constructed using prestressing techniques. Any durable facing material may be applied to the surface to improve the appearance of the wall, but it must be remembered that such finishes are decorative and add nothing to the structural strength of the wall.

COUNTERFORT RETAINING WALLS

These walls can be constructed of reinforced or prestressed concrete, and are considered suitable if the height is over 4.500 m. The counterforts are triangular beams placed at suitable centres behind the stem and above the base to enable the stem and base to act as slabs spanning horizontally over or under the counterforts. Figure 2.5.6 shows a typical section and pattern of reinforcement for a counterfort retaining wall.

If the counterforts are placed on the face of the stem they are termed **buttresses**, and the whole arrangement is called a **buttress-retaining wall**. The design and construction principles are similar in the two formats.

PRECAST CONCRETE RETAINING WALLS

These are manufactured from high-grade precast concrete on the cantilever principle, usually to a 600 mm wide module (see Fig. 2.5.7). They can be erected on a foundation as a permanent retaining wall, or be free standing to act as a dividing wall between heaped materials such as aggregates for concrete. In the latter situation they can increase by approximately three times the storage volume for any given area. Other advantages are a reduction in time by eliminating the curing period that is required for *in-situ* walls and eliminating the need for costly formwork together with the time required to erect and dismantle the temporary forms. The units are reinforced on both faces to meet all forms of stem loading. Lifting holes are provided, which can be utilised as strap-fixing holes if required. Special units to form internal angles, external angles, junctions and curved walls are also available to provide flexible layout arrangements.

PRECAST CONCRETE CRIB-RETAINING WALLS

Crib walls are designed on the principle of a mass retaining wall. They consist of a framework or crib of precast concrete or timber units within which the soil is retained. They are constructed with a face batter of between 1:6 and 1:8 unless the height is less than the width of the crib ties, in which case the face can be constructed vertical. Subsoil drainage is not required, because the open face provides for adequate drainage (see Fig. 2.5.7).

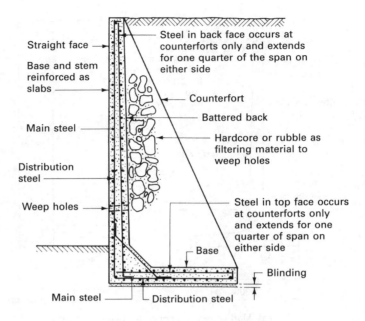

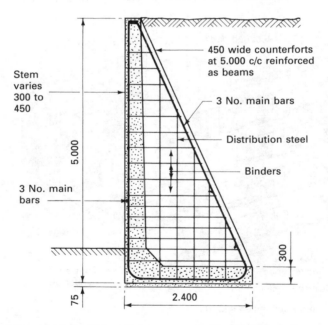

Figure 2.5.6   Typical reinforced concrete counterfort retaining wall.

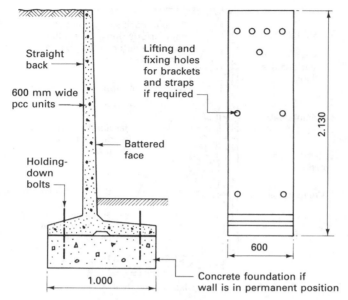

Straight back

600 mm wide pcc units

Holding-down bolts

Battered face

Lifting and fixing holes for brackets and straps if required

2.130

600

1.000

Concrete foundation if wall is in permanent position

Typical 'Marley' precast concrete retaining wall

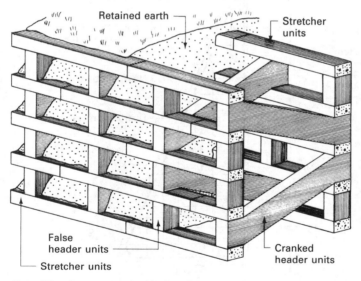

Retained earth

Stretcher units

False header units

Stretcher units

Cranked header units

Note: All units connected with dowels

'Anda-Crib' precast concrete retaining wall

Figure 2.5.7   Precast concrete retaining walls.

REINFORCED MASONRY RETAINING WALLS

Steel reinforcement may be used in brick retaining walls to resist tensile forces and to prevent the effects of shear. A brick bonding arrangement known as **quetta bond** is used to create a uniform distribution of vertical voids: see Part 3 – Superstructure, Fig. 3.2.9. Vertical steel reinforcement is tied to the foundation reinforcement and spaced to coincide with the purpose-made voids. When the brickwork is completed, the voids are filled with concrete to produce a series of reinforced concrete mini-columns within the wall.

Where appearance is not important, or the wall is to receive a surface treatment, reinforcement and *in-situ* concrete within hollow concrete blockwork provide for economical and functional construction. Figure 2.5.8 shows the application of standard-profile, hollow, dense concrete blocks laid in stretcher bond as permanent formwork to continuous vertical columns.

The height potential and slenderness ratio (effective height to width) for reinforced masonry walls can be enhanced by post-tensioning the structure. The principle is explained in Part 10 of *Advanced Construction Technology*. For purposes of brick walls there are a number of construction options, including:

- quetta bond with steel bars and concrete in the voids;
- stretcher-bonded wide cavity with reinforced steel bars coated for corrosion protection;
- solid wall of perforated bricks with continuous voids containing grouted steel reinforcement bars.

Figure 2.5.9 shows some examples.

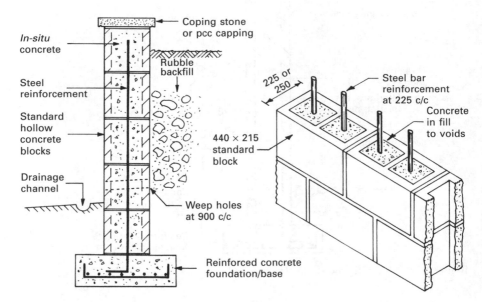

Figure 2.5.8   Reinforced concrete block retaining wall.

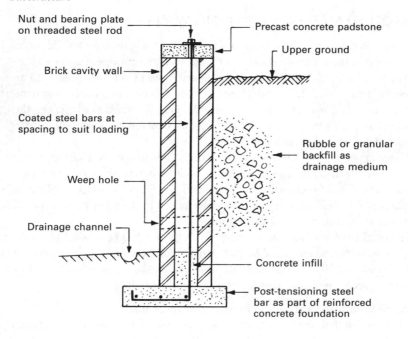

Nut and bearing plate on threaded steel rod

Precast concrete padstone

Upper ground

Brick cavity wall

Coated steel bars at spacing to suit loading

Rubble or granular backfill as drainage medium

Weep hole

Drainage channel

Concrete infill

Post-tensioning steel bar as part of reinforced concrete foundation

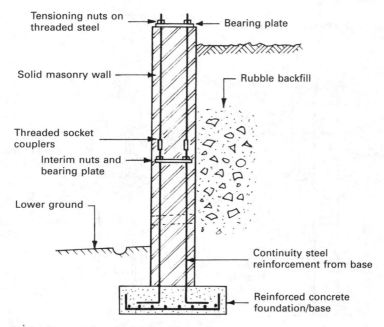

Tensioning nuts on threaded steel

Bearing plate

Solid masonry wall

Rubble backfill

Threaded socket couplers

Interim nuts and bearing plate

Lower ground

Continuity steel reinforcement from base

Reinforced concrete foundation/base

Figure 2.5.9   Post–tensioned brick retaining walls.

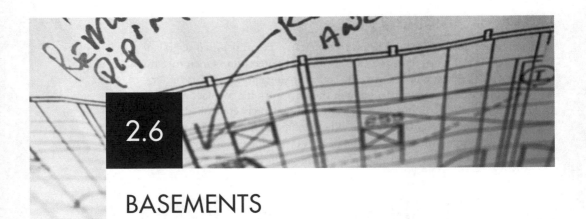

2.6

BASEMENTS

Appendix E in Approved Document B of the Building Regulations defines a basement storey as a storey with a floor that at some point is more than 1.200 m below the highest level of ground adjacent to the outside walls. This definition is given in the context of inhibiting the spread of fire within a building, and generally the fire resistance requirements for basements are more onerous than for the ground or upper storeys in the same building. This section on basements is concerned only with basement storeys that are below ground level.

The structural walls of a basement below ground level are in fact retaining walls, which have to offer resistance to the soil and groundwater pressures as well as assisting to transmit the superstructure loads to the foundations. It is possible to construct a basement free of superstructural loadings, but these techniques are beyond the scope of this book.

■ ■ ■ WATERPROOFING

Apart from the structural design of the basement walls and floor, waterproofing presents the greatest problem in basement construction. Building Regulation C2 requires such walls to be constructed so that they will not transmit moisture from the ground to the inside of the building or to any material used in the construction that would adversely be affected by moisture. Building Regulation C2 also imposes similar conditions on the construction of floors. Basement structures can be waterproofed by one of three basic methods:

■ monolithic structures;
■ drained cavities;
■ membranes.

MONOLITHIC STRUCTURES

These are basements of dense reinforced concrete using impervious aggregates
for the walls and floor to form the barrier to water penetration. Great care must
be taken with the design of the mix, the actual mixing and placing, together
with careful selection and construction of the formwork if a satisfactory water
barrier is to be achieved. Shrinkage cracking can be largely controlled by forming
construction joints at regular intervals. These joints should provide continuity of
reinforcement and by the incorporation of a PVC or rubber water bar a barrier
to the passage of water: typical examples are shown in Fig. 2.6.1. Monolithic

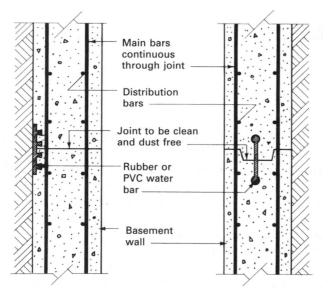

Main bars continuous through joint

Distribution bars

Joint to be clean and dust free

Rubber or PVC water bar

Basement wall

Note: Horizontal joints positioned at 12 to 15 times wall thickness

Construction joints

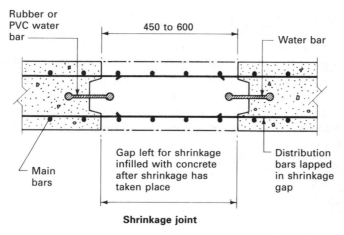

Rubber or PVC water bar

450 to 600

Water bar

Main bars

Gap left for shrinkage infilled with concrete after shrinkage has taken place

Distribution bars lapped in shrinkage gap

Shrinkage joint

Figure 2.6.1 Joints and water bars.

structures, while providing an adequate barrier to the passage of water, are not always vapourproof.

DRAINED CAVITIES

This method provides an excellent barrier to moisture penetration of basements by allowing any moisture that has passed through the structural wall to drain down within a cavity formed between the inner face of the structural wall and an inner non-loadbearing wall. This internal wall is built of a floor covering of special triangular precast concrete tiles, which allows the moisture from the cavity to flow away under the tiles to a sump, where it is discharged into a drainage system either by gravity or by pumping. This method of waterproofing is usually studied in detail during advanced courses in construction technology.

MEMBRANES

A membrane is a relatively thin material placed on either the external or internal face of a basement wall or floor to provide the resistance to the passage of moisture to the inside of the basement. If the membrane is applied externally protection is also given to the structural elements and the hydrostatic pressure will keep it firmly in place, but a reasonable working space must be allowed around the perimeter of the basement. This working space will entail extra excavation and subsequent backfilling after the membrane has been applied. If adequate protection is not given to the membrane it can easily be damaged during the backfilling operation. An internally applied membrane gives no protection to the structural elements, and there is the danger that the membrane may be forced away from the surfaces by water pressure unless it is adequately loaded. These loading coats will reduce the usable volume within the basement (see Figs 2.6.2 and 2.6.3).

Suitable materials that can be used for forming membranes are fibre-reinforced bituminous felt, polythene sheet, polyisobutylene plastic, epoxy resin compounds, bituminous compounds and mastic asphalt. Also, more recent developments for impervious membrane applications include polymer-modified bitumen with polyester reinforcement. This is available as styrene–butadiene–styrene (SBS) or atactic polypropylene (APP).

ASPHALT TANKING

Asphalt is a natural or manufactured mixture of bitumen with a substantial proportion of inert mineral matter. When heated, asphalt becomes plastic and can be moulded by hand pressure into any shape. Bitumen is a complex mixture of hydrocarbons and has both waterproofing and adhesive properties. In its natural state asphalt occurs as a limestone rock impregnated with bitumen, and is mined notably in France, Switzerland and Sicily. Another source of asphalt is the asphalt lake in Trinidad in the West Indies, which was discovered by Sir Walter Raleigh in 1595 and today still yields about 100 000 tonnes annually. In the centre of this lake the asphalt is a liquid but nearer the edges it is a semi-fluid, and although large quantities are removed

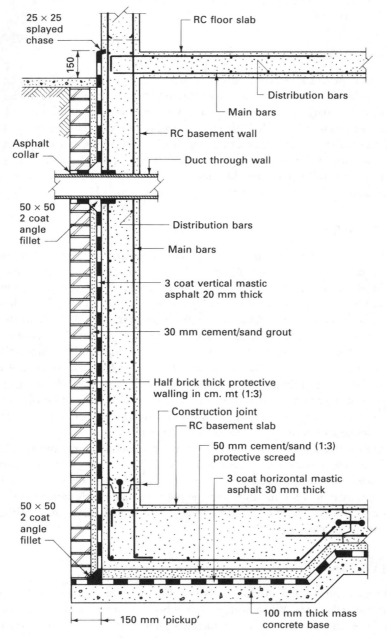

Figure 2.6.2   External tanking in mastic asphalt.

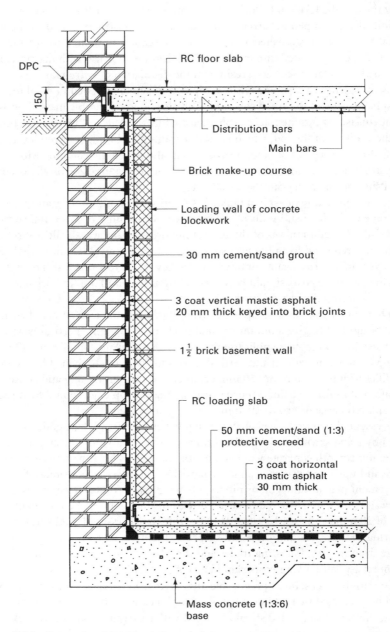

Figure 2.6.3 Internal tanking in mastic asphalt.

during the day the lake refills during the night. The lake asphalt is refined or purified in Trinidad and shipped in barrels for use in building construction all over the world. Natural rock asphalt is crushed and processed to remove unwanted mineral matter before being compounded into mastic asphalt. Bitumen for use with mastic asphalt is also made on a large scale as a residue in the distillation of petroleum.

Mastic asphalt is a type of asphalt composed of suitably graded mineral matter and asphaltic cement to form a coherent, voidless and impermeable mass. The asphalt cement consists of bitumen, lake asphalt, asphaltite or blends of these, sometimes with the addition of flux oil, which is used for softening bitumen or rendering it less viscous. Fine aggregates of natural rock asphalt and limestone combine with coarse aggregates of igneous and calcareous angular stones if required. Naturally occurring graded siliceous material can also be added.

The basic principle of asphalt tanking is to provide a continuous waterproof membrane to the base and walls of the basement. Continuity between the vertical and horizontal membranes is of the utmost importance, and as asphalt will set rapidly once removed from the heat source used to melt the blocks, it is applied in layers over small areas; again continuity is the key factor to a successful operation. Joints in successive coats should be staggered by at least 150 mm in horizontal work and at least 75 mm in vertical work.

On horizontal and surfaces up to 30° from the horizontal three coats of asphalt should be applied to give a minimum total thickness of 30 mm. Vertical work should also be a three-coat application to give a total thickness of 20 mm. The junction between horizontal and vertical work should be strengthened by a two-coat angle fillet forming a 50 mm × 50 mm chamfer. To prevent curling and consequent infiltration of moisture behind the vertical tanking the top edge should be turned into a splayed chase or groove 25 mm wide × 25 mm deep.

It is essential that vertical asphalt is suitably keyed to its background. Concrete formed by using sawn boards for the formwork will usually provide an acceptable surface, but smooth concrete will need treatment such as bush-hammering the surface and washing to remove all loose particles. Alternatively a primer of sand/cement plastic emulsion or pitch/polymer rubber emulsion can be used. Brick walls can be constructed of keyed bricks, or the joints can be raked out to a depth of 20 mm as the work proceeds to provide the necessary keyed surface.

During the construction period the asphalt tanking must be protected against damage from impact, following trades and the adverse effects of petrol and oil. Horizontal asphalt tanking coats should be covered with a fine concrete screed at least 50 mm thick as soon as practicable after laying. Vertical asphalt tanking coats should be protected by building a half brick or block wall 30 mm clear of the asphalt; the cavity so formed should be filled with a mortar grout as the work proceeds to ensure perfect interface contact. In the case of internal tanking this protective wall will also act as the loading coat.

Any openings for the passages of pipes or ducts may allow moisture to penetrate unless adequate precautions are taken. The pipe or duct should be primed and coated with three coats of asphalt so that the sleeve formed extends at least 75 mm on either side of the tanking membrane before being placed in the wall or floor. The pipe or duct is connected to the tanking by a two-coat angle fillet (see Fig. 2.6.2).

The main advantages of mastic asphalt as a waterproof membrane are as follows:

- It is a thermoplastic material and can therefore be heated and reheated if necessary to make it pliable for moulding with a hand float to any desired shape or contour.
- It is durable: bituminous materials have been used in the construction of buildings for over 5000 years and have remained intact to this day, as shown by excavations in Babylonia.
- It is impervious to both water and water vapour.
- It is non-toxic, vermin and rot proof, and is odourless after laying.
- It is unaffected by sulphates in the soil, which, if placed externally, will greatly improve the durability of a concrete structure.

The application of mastic asphalt is recognized as a specialist trade in the building industry and therefore most asphalt work is placed in the hands of specialist subcontractors, most of which are members of the Mastic Asphalt Council and Employers Federation Limited. The Federation is a non-profit-making organization whose objectives are to provide technical information and promote the use of mastic asphalt as a high-quality building material.

OTHER SHEET MEMBRANES

Plastic and bitumen sheeting materials are suited to shallow basements. The base structure of concrete or masonry is prepared with a primer of bituminous solution before sheeting, and is hot bitumen bonded with 100 mm side and 150 mm end lapping in at least two layers. Figure 2.6.4 shows application to wall and floor.

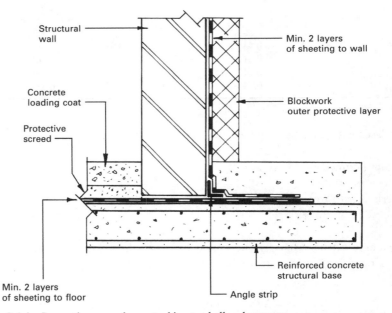

Figure 2.6.4   Impervious membrane tanking to shallow basement.

■■■ INSULATION OF BASEMENTS

Basements in heated buildings require insulation against energy loss through the walls and floor. The contribution made by the surrounding subsoil is significant, but insufficient alone to satisfy current energy conservation measures. Thermal insulation should be at least equivalent to that specified for above-ground walls and floors.

ESTIMATING U VALUES

Floors

$$U = \frac{1}{(1/U_0) + R}$$

where $U_0 = U$ value for an uninsulated floor (see table below)
  R = Thermal resistance of the applied insulation (m² K/W),
   i.e. insulation thickness/insulation conductivity

U values of uninsulated basement floors (W/m² K):

Perimeter to area ratio	Basement depth (m)				
	0.5	1.0	1.5	2.0	2.5
0.1	0.20	0.19	0.18	0.17	0.16
0.3	0.44	0.41	0.38	0.35	0.33
0.5	0.61	0.55	0.50	0.46	0.43
0.7	0.74	0.65	0.59	0.53	0.49
0.9	0.84	0.73	0.65	0.58	0.53

E.g. Basement with a floor 5 m × 4 m and a depth of 2.5 m

  Floor perimeter = 18 m
  Floor area = 20 m²
  Perimeter to area ratio = 18/20 = 0.9
  $U_0 = 0.53$ W/m² K

If 75 mm of extruded polystyrene, conductivity (λ value) 0.035 W/m K, is incorporated in the floor construction, the U value will be:

$$\frac{1}{(1/0.53) + (0.075/0.035)} = \frac{1}{4.029} = 0.248 \text{ W/m}^2 \text{ K}$$

Walls

U values of basement walls (W/m² K):

Wall thermal resistance (m² K/W)	Basement depth (m)				
	0.5	1.0	1.5	2.0	2.5
0.2	1.55	1.16	0.95	0.81	0.71
0.5	0.98	0.78	0.66	0.58	0.52
1.0	0.61	0.51	0.45	0.40	0.37
1.5	0.46	0.38	0.34	0.32	0.29
2.0	0.35	0.30	0.27	0.25	0.24
2.5	0.28	0.25	0.23	0.21	0.20

As before, basement depth is 2.5 m. Assuming a wall thermal resistance including the soil, of 2.0 m² K/W, from the table the *U* value of the wall is 0.24 W/m² K.

The preceding estimation of *U* values is for guidance only. For more detail see:

■ BS EN ISO 6946: *Building components and building elements. Thermal resistance and thermal transmittance. Calculation method*;
■ BS EN ISO 13370: *Thermal performance of buildings. Heat transfer via the ground. Calculation methods*.

Note: For calculation of *U* values and thermal resistance, see Part 8 – Insulation.

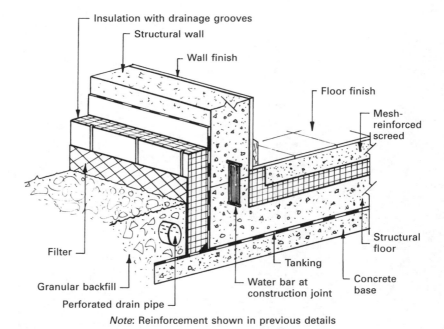

Note: Reinforcement shown in previous details

Figure 2.6.5   Basement insulation.

APPLICATION OF INSULATION

Existing construction

Assuming major excavation work and rebuilding is not viable, existing basements can be considerably improved by applying rigid insulation to the walls and floor. Surface finishes can be provided by drylining the walls with vapour check plasterboard and covering the floor with particle board and a vapour control underlay to minimise the risk of condensation.

New construction

A typical application of insulation is shown in Fig. 2.6.5. The insulation selected is usually a closed-cell water-resistant extruded polystyrene of low thermal conductivity and high compressive strength.

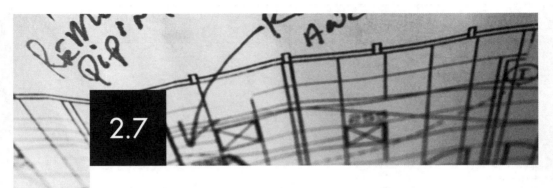

TREES: EFFECT ON FOUNDATIONS

Close proximity of trees to buildings can result in direct physical damage to foundations, drains, paths, drives, etc., by growth and expansion of root systems. With conventional strip foundations and those provided for pre-1950s dwellings, where the concrete was rarely placed deeper than 0.5 m, it is advisable to maintain a distance at least equal to the mature tree's height between building and tree. Some relaxation may be applied where tree species can be identified as having less water demand. Guidance is given in Table 2.7.1 with approximate heights of mature trees.

In the UK, shrinkable clays occur in specific areas found roughly below an imaginary line linking the Humber estuary with a point intersecting the mouth of the River Severn, as shown in Fig. 2.7.1.

Clay subsoils are particularly prone to changes in moisture content, with a high degree of shrinkage apparent during periods of drought. When added to the effect of moisture abstraction by vegetation (trees), voids and swallow holes are to be expected. Clay heave is the opposite phenomenon, attributed partly to saturation from inclement weather. It is more likely to be caused by removal of trees within the immediate 'footprint' of new buildings or the immediate surrounds. Figure 2.7.2 shows cracking patterns in buildings due to these indirect sources of damage.

The accepted depth for foundations in clay subsoils is at least 900 mm in the absence of trees. However, where trees are apparent or have been within the preceding three years, the minimum depth of foundation must be calculated. Where insurers issue structural guarantees, this depth can be determined from tables provided in procedure manuals issued by the National House Building Council or Zurich Municipal. Figure 2.7.3 provides some guidance, as an indication only, of the likely depths of foundation excavation necessary where tree distance (d) to height (h) ratios are less than unity. Beyond 1.5 m depth, conventional strip foundations will not be economical or practically viable, but trench-filled concrete will be appropriate. For foundations over 2 m, short-bored piles with ground beams are preferred.

Table 2.7.1 Tree characteristics

Species	Approximate mature height (m)	Water demand: H = high, M = medium, L = low
Elm	30	H
Oak	30	H
Poplar	30	H
Ash	30	M
Cedar	30	M
Chestnut	30	M
Hemlock	30	M
Lime	30	M
Larch	30	L
Scots pine	30	L
Cypress	25	H
Willow	25	H
Maple	25	M
Sycamore	25	M
Walnut	25	M
Beech	25	L
Birch	25	L
Holly	20	L
Cherry	15	M
Hawthorn	15	M
Rowan	15	M

Note: Where groups of trees occur and the subsoil is known to be moisture responsive (i.e. shrinkable clay), the suggested building distance is 1.5 × mature height.

Figure 2.7.1 Area of UK with zones of shrinkable clay subsoils.

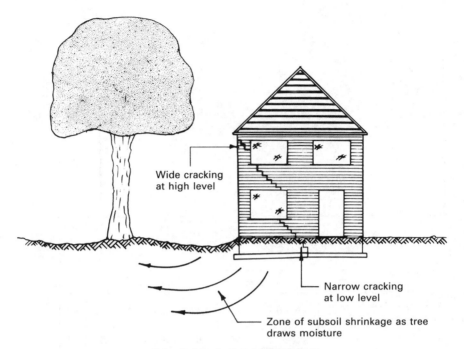

Subsoil dehydration and shrinkage

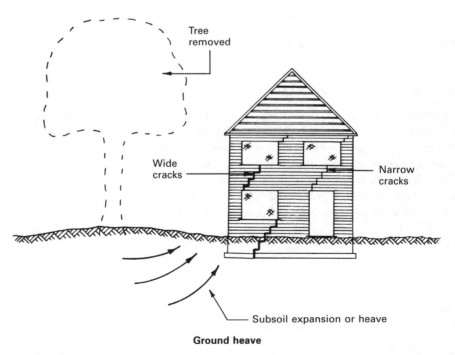

Ground heave

Figure 2.7.2 Indirect damage to buildings.

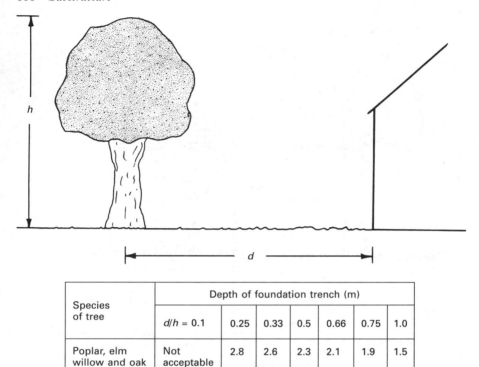

Species of tree	Depth of foundation trench (m)						
	d/h = 0.1	0.25	0.33	0.5	0.66	0.75	1.0
Poplar, elm willow and oak	Not acceptable	2.8	2.6	2.3	2.1	1.9	1.5
Others	Not acceptable	2.4	2.1	1.8	1.5	1.2	1.0

Figure 2.7.3   Tree proximity and foundation depth.

▨■■ HEAVE PRECAUTIONS

In addition to foundation depths exceeding soil movement zones, the areas closer to the surface must be provided with additional treatment. Here, prior to concreting, the trench is prepared with a lining of low-density expanded polystyrene (or other acceptable compressible material) along with an outer polythene sheet sleeving. The polystyrene is designed to absorb seasonal clay movement, and the polythene acts as a slip membrane preventing soil adhesion to the foundation. Figure 2.7.4 shows application to trench fill and piled foundations.

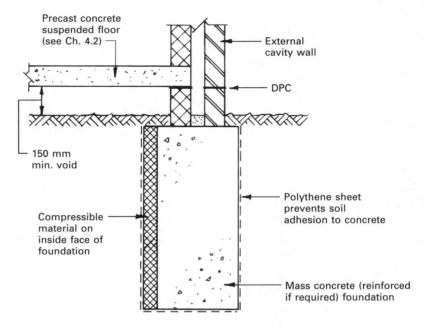

Precast concrete
suspended floor
(see Ch. 4.2)

External
cavity wall

DPC

150 mm
min. void

Polythene sheet
prevents soil
adhesion to concrete

Compressible
material on
inside face of
foundation

Mass concrete (reinforced
if required) foundation

Note: Compressible material may be required to both sides of foundation

Deep strip or trench fill

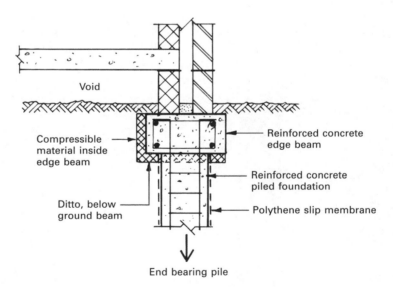

Void

Compressible
material inside
edge beam

Reinforced concrete
edge beam

Reinforced concrete
piled foundation

Ditto, below
ground beam

Polythene slip membrane

End bearing pile

Alternative piled foundation

Figure 2.7.4 Precautionary treatment to foundations in shrinkable clay.

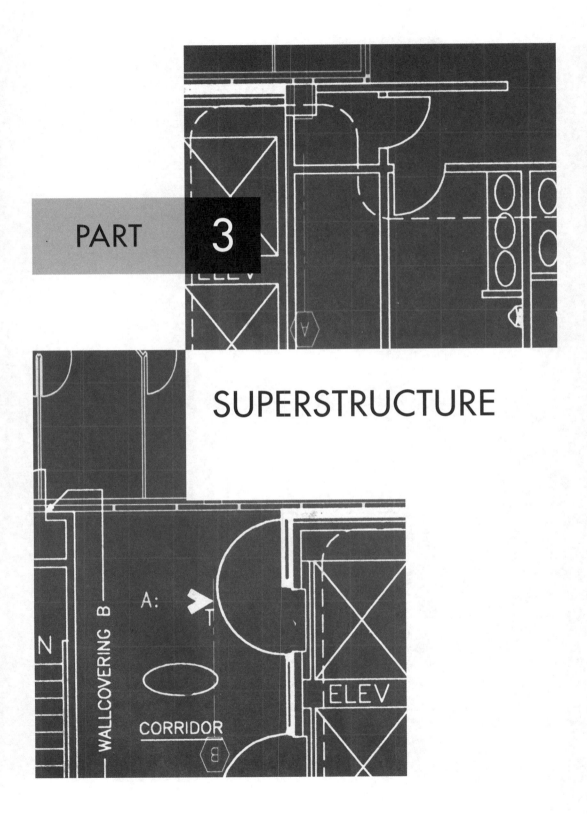

PART 3

SUPERSTRUCTURE

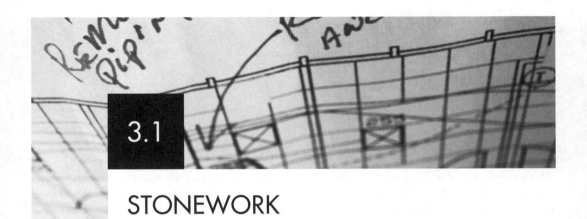

3.1

STONEWORK

The natural formation of stones or rocks is a very long process, which commenced when the earth was composed only of gases. These gases eventually began to liquefy, forming the land and the sea, the land being only a thin crust or mantle to the still molten core.

Igneous rocks originated from the molten state, plutonic rocks are those in the lower part of the earth's mantle, and hypabyssal rocks solidified rapidly near the upper surface of the crust.

Stonework terminology

- **Arris**   Meeting edge of two worked surfaces.
- **Ashlar**   A square hewn stone; stonework consisting of blocks of stone finely squared and dressed to given dimensions and laid to courses of not less than 300 mm in height.
- **Bed joint**   Horizontal joint between two consecutive courses.
- **Bonders**   Through stones or stones penetrating two-thirds of the thickness of a wall.
- **Cramp**   Non-ferrous metal or slate tie across a joint.
- **Dowel**   Non-ferrous or slate peg mortised into adjacent joints.
- **Joggle**   Recessed key filled with a suitable material, used between adjacent vertical joints.
- **Lacing**   Course of different material to add strength.
- **Natural bed**   Plane of stratification in sedimentary stones.
- **Quarry sap**   Moisture contained in newly quarried stones.
- **Quoin**   Corner stone.
- **Stool**   Flat seating on a weathered sill for jamb or mullion.
- **String course**   Distinctive course or band used mainly for decoration.
- **Weathering**   Sloping surface to part of the structure to help shed the rain.

■■■ BUILDING STONES

Stones used in building can be divided into three classes:

- igneous;
- sedimentary;
- metamorphic.

IGNEOUS STONES

These stones originate from volcanic action, being formed by the crystallisation of molten rock matter derived from deep in the earth's crust. It is the proportions of these crystals that give the stones their colour and characteristics. Granites are typical of this class of stone, being hard, durable, and capable of a fine, polished finish. Granites are composed mainly of quartz, feldspar and mica.

SEDIMENTARY STONES

These stones are composed largely of material derived from the breakdown and erosion of existing rocks deposited in layers under the waters that, at that time, covered much of the earth's surface. Being deposited in this manner, their section is stratified, and shows to a lesser or greater degree the layers as deposited. Some of these layers are visible only if viewed under a microscope. Under the microscope it will be seen that all the particles lie in one direction, indicating the flow of the current in the water. Sandstones and limestones are typical examples of sedimentary stones.

Sandstones are stratified sedimentary rocks produced by eroded and disintegrated rocks such as granite, being carried away and deposited by water in layers. The brown and yellow tints in sandstones are due to the presence of oxides of iron.

Limestones may be organically formed by the deposit of tiny shells and calcareous (containing limestone) skeletons in the seas and rivers, or they may be formed chemically by deposits of lime in ringed layers. Limestones vary considerably, from a heavy crystalline form to a friable material such as chalk.

METAMORPHIC STONES

These are stones that have altered and may have been originally igneous or sedimentary rocks, but have since been changed by geological processes such as pressure, movement, heat and chemical reaction due to the infiltration of fluids. Typical examples of this type of stone are marbles and slates.

Marbles are metamorphic limestones, their original structure having been changed by pressure. Marbles, being capable of taking a high polish, are used mainly for decorative work.

Slate is a metamorphic clay, having been subjected to great pressure and heat; being derived from a sedimentary layer it can be easily split into thin members.

Stones are obtained from quarries by blasting and wedging the blocks away from the solid mass. They are partly worked in the quarry and then sent to store yards, where they can be sawn, cut, moulded, dressed and polished to the customer's requirements.

Today, natural stones are sometimes used for facing prestige buildings, constructing boundary or similar walls and, in those areas where natural stones occur, to preserve the character of the district. The distribution of natural stones in the British Isles is shown in Fig. 3.1.1.

■ ■ ■ CAST STONES

Substitutes for natural stones are available in the form of cast stones, either as reconstructed or as artificial stones. BS 1217 defines these as a building material manufactured from cement and natural aggregate, for use in a manner similar to and for the same purpose as natural stone.

RECONSTRUCTED STONE

These types of cast stone are homogeneous throughout and therefore have the same texture and colour as the natural stones they are intended to substitute for. They are free from flaws and stratification and can be worked in the same manner as natural stone; alternatively, they can be cast into shaped moulds giving the required section.

ARTIFICIAL STONE

These stones consist partly of a facing material and partly of a structural concrete. The facing is a mixture of fine aggregate of natural stone and cement or sand and pigmented cement to resemble the natural stone colouring. This facing should be cast as an integral part of the stone and have a minimum thickness of 20 mm. They are cheaper than reconstructed stones but have the disadvantage that, if damaged, the concrete core may be exposed.

■ ■ ■ ASHLAR WALLING

This form of stone walling is composed of carefully worked stones, regularly coursed, bonded and set with thin or rusticated joints, and is used for the majority of high-class facing work in stone. The quoins are sometimes given a surface treatment to emphasise the opening or corner of the building. The majority of ashlar work is carried out in limestone varying in thickness from 100 to 300 mm and set in mason's putty, which is a mixture of stone dust, lime putty and Portland cement, a typical mix ratio being 7:5:2.

RULES FOR ASHLAR WORK

- Back faces of ashlar stones should be painted with a bituminous or similar waterproofing paint.

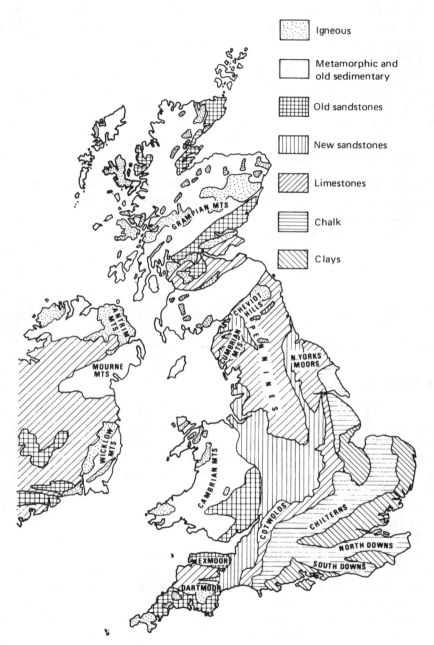

Figure 3.1.1   Distribution of natural stones.

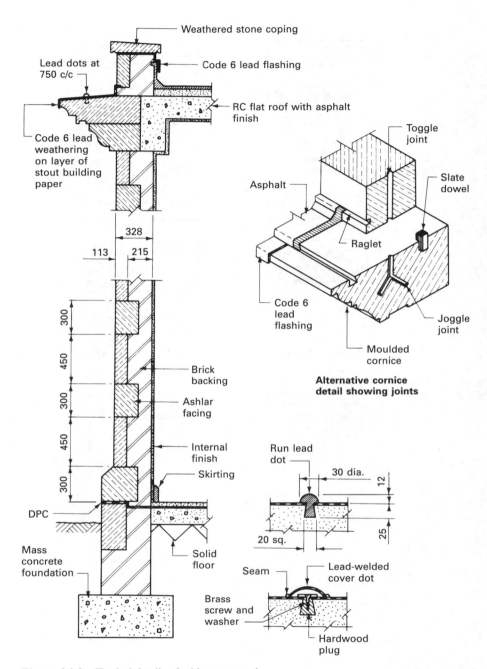

Weathered stone coping

Lead dots at 750 c/c

Code 6 lead flashing

Code 6 lead weathering on layer of stout building paper

RC flat roof with asphalt finish

328

113 215

300

450

300

450

300

Brick backing

Ashlar facing

Internal finish

Skirting

DPC

Mass concrete foundation

Solid floor

Toggle joint

Asphalt

Slate dowel

Raglet

Code 6 lead flashing

Joggle joint

Moulded cornice

Alternative cornice detail showing joints

Run lead dot

30 dia.

12

20 sq.

25

Seam

Lead-welded cover dot

Brass screw and washer

Hardwood plug

Figure 3.1.2   Typical details of ashlar stonework.

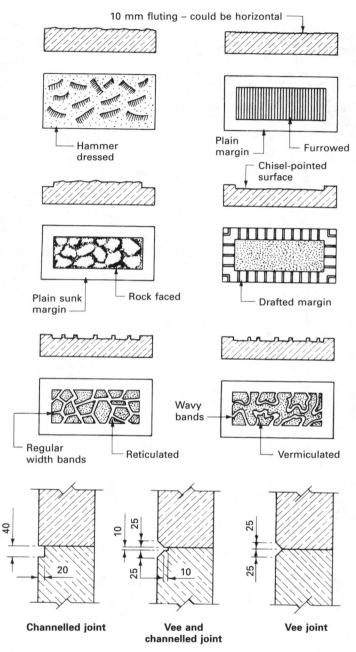

Figure 3.1.3   Typical surface and joint treatments.

- External stonework must not be taken through the thickness of the wall, because this could create a passage for moisture.
- Ledges of cornices and external projections should be covered with lead, copper or asphalt to prevent damage by rain or birds.
- Moulded cornices should be raked back at 45° to counteract the cantilever action.
- Faces of stones should be given a protective coat of slurry during construction, the slurry being washed off immediately prior to completion.

Typical details of ashlar work are shown in Figs 3.1.2–3.1.5.

▨ ▪ ▪ RUBBLE WALLING

These are walls consisting of stones that are left in a rough or uneven state, thus presenting a natural appearance to the face of the wall. These stones are usually laid with a wide joint, and are frequently used in various forms in many rural areas. They can be laid dry or bedded in earth in boundary walls, or bedded in lime mortar when used for the walls of farm outbuildings; if used in conjunction with ashlar stonework, a cement or gauged mortar is used. It is usual to build the quoins to corners, window and door openings in dressed or ashlar stones. As with ashlar work it is advisable to treat the face of any backing material with a suitable waterproofing coat to prevent the passage of moisture or the appearance of cement stains on the stone face.

Solid stone walls will behave in the same manner as solid brick walls with regard to the penetration of moisture and rain: it will therefore be necessary to take the same precautions in the form of damp-proof courses to comply with Part C of the Building Regulations.

Typical examples of stone and rubble walling are shown in Figs 3.1.4, 3.1.6 and 3.1.7.

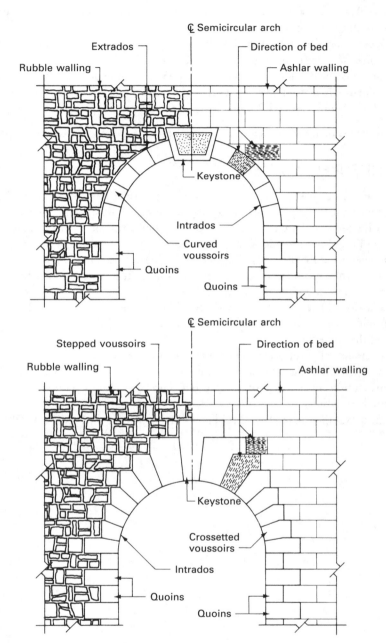

Figure 3.1.4   Typical treatments to arch openings.

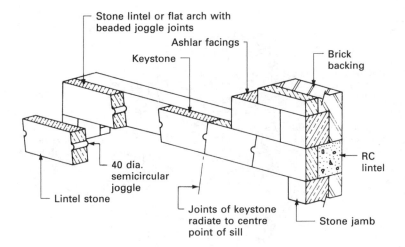

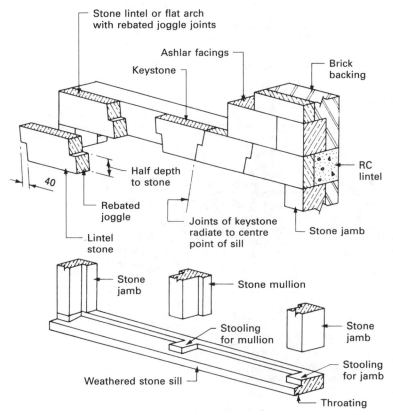

Figure 3.1.5 Typical treatments to square openings.

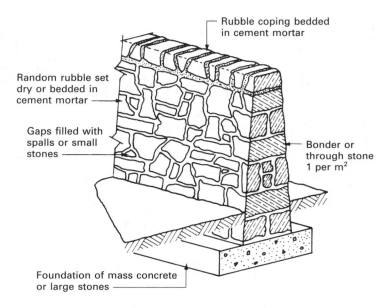

Rubble coping bedded in cement mortar

Random rubble set dry or bedded in cement mortar

Gaps filled with spalls or small stones

Bonder or through stone 1 per m²

Foundation of mass concrete or large stones

Uncoursed random rubble wall

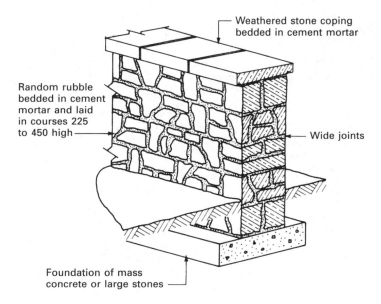

Weathered stone coping bedded in cement mortar

Random rubble bedded in cement mortar and laid in courses 225 to 450 high

Wide joints

Foundation of mass concrete or large stones

Coursed random rubble wall

Figure 3.1.6 Typical rubble walls.

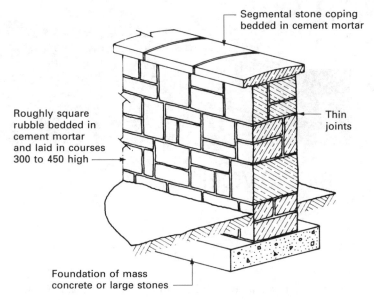

Segmental stone coping
bedded in cement mortar

Roughly square
rubble bedded in
cement mortar
and laid in courses
300 to 450 high

Thin
joints

Foundation of mass
concrete or large stones

Coursed square rubble wall

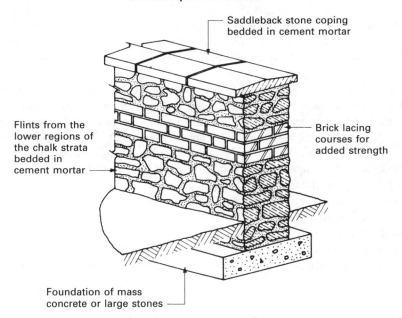

Saddleback stone coping
bedded in cement mortar

Flints from the
lower regions of
the chalk strata
bedded in
cement mortar

Brick lacing
courses for
added strength

Foundation of mass
concrete or large stones

Flint stone wall

Figure 3.1.7   Typical stone walls.

3.2

BRICKWORK

The history of the art of brickmaking and the craft of bricklaying can be traced back to before 6000 BC. It started in western Asia, spread eastwards, and was introduced into this country by the Romans. The use of brickwork flourished during the third and fourth centuries, after which the craft suffered a rapid decline until its reintroduction from Flanders towards the end of the fourteenth century. Since then it has been firmly established, and remains as one of the major building materials.

BS 3921 specifically relates to bricks manufactured from clay for use in walling. Other standards exist for units produced from calcium silicate (BS 187) and concrete (BS 6073). A brick is defined in BS 3921 as a walling unit with coordinating or format size of 225 mm length, 112.5 mm width and 75 mm height. Bricks are known by their format size – that is, the actual or work size plus a 10 mm mortar joint allowance to three faces. Therefore the standard brick of 225 mm × 112.5 mm × 75 mm has actual dimensions of 215 mm × 102.5 mm × 65 mm. The terms used for bricks and brickwork are shown in Figs 3.2.1, 3.2.2 and 3.2.3.

Brickwork is used primarily in the construction of walls by the bedding and jointing of bricks into established bonding arrangements. The term also covers the building-in of hollow and other lightweight blocks. The majority of the bricks used today are made from clay or shale conforming to the requirements of BS 3921.

■■■ MANUFACTURE OF CLAY BRICKS

The basic raw material is clay, shale or brickearth, all of which are in plentiful supply in this country. The raw material is dug, and then prepared either by weathering or by grinding before being mixed with water to the right plastic condition. It is then formed into the required brick shape before being dried and fired in a kiln.

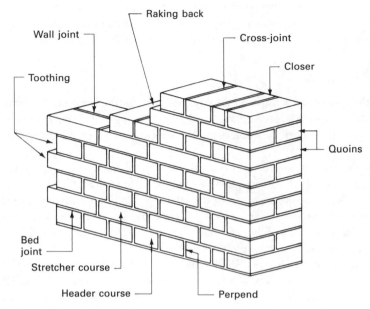

Figure 3.2.1   Brickwork terminology.

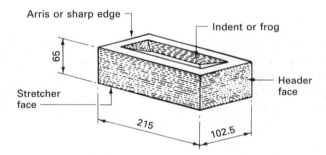

Figure 3.2.2   Standard brick.

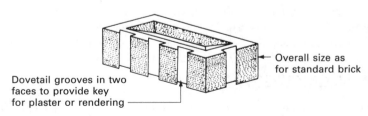

Figure 3.2.3   Keyed brick.

Different clays have different characteristics, such as moisture content and chemical composition: therefore distinct variations of the broad manufacturing processes have been developed, and these are easily recognised by the finished product.

PRESSED BRICKS

This is the type of brick most commonly used, accounting for nearly two-thirds of the 5000 million produced on average in this country each year. There are two processes of pressed brick manufacture: the semi-dry method and the stiff plastic method.

By far the greatest number of bricks are made by the **semi-dry pressed process** and are called **flettons**; these form over 40% of the total brick production in Britain. The name originates from the village of Fletton outside Peterborough, where the bricks were first made. This process is used for the manufacture of bricks from the Oxford clays, which have a low natural plasticity. The clay is ground, screened and pressed directly into the moulds.

The **stiff plastic process** is used mainly in Scotland, the north of England and south Wales. The clays in these areas require more grinding than the Oxford clays, and the clay dust needs tempering (mixing with water) before being pressed into the mould.

All pressed bricks contain **frogs**, which are sometimes pressed on both bed faces. In general, pressed bricks are more accurate in shape than other clay bricks, with sharp arrises and plain faces.

WIRE-CUT BRICKS

Approximately 28% of bricks produced in Britain are made by this process. The clay, which is usually fairly soft and of a fine texture, is extruded as a continuous ribbon and is cut into brick units by tightly stretched wires spaced at the height or depth for the required brick. Allowance is made during the extrusion and cutting for the shrinkage that will occur during firing. Wire-cut bricks do not have frogs, and on many the wire-cutting marks can be clearly seen.

SOFT MUD PROCESS BRICKS

This process is confined mainly to the south-eastern counties of England, where suitable soft clays are found. The manufacture can be carried out by machine or by hand, either with the natural clay or with a mixture of clay and lime or chalk. In both methods the brick is usually frogged, and is less accurate in shape than other forms of bricks. Sand is usually used in the moulds to enable the bricks to be easily removed, and this causes an uneven patterning or creasing on the face.

■■■ BRICK CLASSIFICATION

Bricks are a diverse product: therefore they are difficult to standardise for classification. They may be known by the terms used in BS 3921 – that is, compressive strength

(N/mm^2) and percentage by mass water absorption. Alternatively, the manufacturer's name or catalogue descriptions may be used, possibly in combination with the BS characteristics. The minimum compressive strength acceptable to the Building Regulations is 5 N/mm^2 over gross sectional area (2.8 N/mm^2 for blocks).

Some common terminology:

VARIETIES

- **Common**   Suitable for general building work but having no special claim to give an attractive appearance.
- **Facing**   Specially made or selected to have an attractive appearance when used without rendering or plaster.
- **Engineering**   Having a dense and strong semi-vitreous body conforming to defined limits for absorption and strength.

QUALITIES

- **Internal**   Suitable for internal use only; may need protection on site during bad weather or during the winter.
- **Ordinary**   Less durable than special quality but normally durable in the external face of a building. Some types are unsuitable for exposed situations.
- **Special**   For use in conditions of extreme exposure where the structure may become saturated and frozen, such as retaining walls and pavings.

TYPES

- **Solid**   Those in which small holes passing through or nearly through the brick do not exceed 25% of its volume, or in which frogs do not exceed 20% of its volume. A small hole is defined as a hole less than 20 mm wide or less than 500 mm^2 in area.
- **Perforated**   Those in which holes passing through the brick exceed 25% of its volume and the holes are small as defined above.
- **Hollow**   Those in which the holes passing through the brick exceed 25% of its volume and the holes are larger than those defined as small holes.
- **Cellular**   Those in which the holes are closed at one end and exceed 20% of the volume of the brick.

OTHER CLASSIFICATIONS

Bricks may also be classified by one or more of the following:

- place of origin, for example London;
- raw material, for example clay;
- manufacture, for example wire-cut;
- use, for example foundation;
- colour, for example blue;
- surface texture, for example sand-faced.

■■■ CALCIUM SILICATE BRICKS

These bricks are also called sandlime and sometimes flintlime bricks and are covered by BS 187, which gives several classes of brick: the higher the numbered class, the stronger is the brick. The format size of a calcium silicate brick is the same as that given for a standard clay brick.

These bricks are made from carefully selected clean sand and/or crushed flint mixed with controlled quantities of lime and water. At this stage colouring pigments can be added if required; the relatively dry mix is then fed into presses to be formed into the required shape. The moulded bricks are then hardened in sealed and steam-pressurised autoclaves. This process, which takes from 7–10 hours, causes a reaction between the sand and the lime, resulting in a strong homogeneous brick that is ready for immediate delivery and laying. The bricks are very accurate in size and shape but do not have the individual characteristics of clay bricks.

■■■ CONCRETE BRICKS

These are made from a mixture of inert aggregate and cement in a similar fashion to calcium silicate bricks, and are cured either by natural weathering or in an autoclave. Details of the types and properties available as standard concrete bricks are given in BS 6073.

■■■ MORTARS FOR BRICKWORK

The mortar used in brickwork transfers the tensile, compressive and shear stresses uniformly between adjacent bricks. To do this it must satisfy certain requirements:

- It must have adequate strength, but not greater than that required for the design strength.
- It must have good workability.
- It must retain plasticity long enough for the bricks to be laid.
- It must be durable over a long period.
- It must bond well to the bricks.
- It must be able to be produced at an economic cost.

If the mortar is weaker than the bricks, shrinkage cracks will tend to follow the joints of the brickwork, and these are reasonably easy to make good. If the mortar is stronger than the bricks, shrinkage cracks will tend to be vertical through the joints and the bricks, thus weakening the fabric of the structure.

MORTAR MIXES

Mortar is a mixture of sand and lime or a mixture of sand and cement with or without lime. Proportioning of the materials can be carried by volume, but this method is inaccurate, and it is much better to proportion by weight. The effect of the lime is to make the mix more workable, but as the lime content increases the mortar's resistance to damage by frost action decreases.

Plasticisers, by having the effect of entraining small bubbles of air in the mix and breaking down surface tension, will also increase the workability of a mortar.

Mortars should never be re-tempered, and should be used within 2 hours of mixing or be discarded.

Typical mixes (by volume)

- **Cement mortar (cement:sand):**
  - 1:3 – suitable for brickwork in exposed conditions such as parapets and for brickwork in foundations;
- **Lime mortar (lime:sand):**
  - 1:3 – for internal use only;
- **Gauged mortars (cement:lime:sand):**
  - 1:1:6 – suitable for most conditions of severe exposure;
  - 1:2:9 – suitable for most conditions except those of severe exposure;
  - 1:3:12 – internal use only.

■ ■ ■ DAMPNESS PENETRATION

It is possible for dampness to penetrate into a building through the brick walls by one or more of three ways:

1. by the rain penetrating the head of the wall and soaking down into the building below the roof level;
2. by the rain beating against the external wall and soaking through the fabric into the building;
3. by ground moisture entering the wall at or near the base and creeping up the wall by capillary action and entering the building above the ground floor level.

Numbers 1 and 3 can be overcome by the insertion of a suitable damp-proof course in the thickness of the wall. Number 2 can be overcome by one of two methods:

- applying to the exposed face of the wall a barrier such as cement rendering or some suitable cladding such as vertical tile hanging;
- constructing a cavity wall, whereby only the external skin becomes damp, the cavity providing a suitable barrier to the passage of moisture through the wall.

DAMP-PROOF COURSES

The purpose of a damp-proof course in a building is to provide a barrier to the passage of moisture from an external source into the fabric of the building, or from one part of the structure to another. Damp-proof courses may be either horizontal or vertical, and can generally be divided into three types:

- those below ground level to prevent the entry of moisture from the soil;
- those placed just above ground level to prevent moisture creeping up the wall by capillary action; this is sometimes called **rising damp**;
- those placed at openings, parapets and similar locations to exclude the entry of the rainwater that falls directly onto the fabric of the structure.

Materials for damp-proof courses

BS 743: *Specification for materials for damp-proof courses* gives seven suitable materials for the construction of damp-proof courses, all of which should have the following properties:

- be completely impervious;
- be durable, having a longer life than the other components in the building and therefore not needing replacing during its lifetime;
- be in comparatively thin sheets so as to prevent disfigurement of the building;
- be strong enough to support the loads placed upon it without exuding from the wall;
- be flexible enough to give with any settlement of the building without fracturing.

The following also apply:

- BS 6398: *Specification for bitumen damp-proof courses for masonry*;
- BS 6515: *Specification for polyethylene damp-proof courses for masonry*;
- BS 8215: *Code of practice for design and installation of damp-proof courses in masonry construction.*

Lead
This should be at least to BS EN 12588: *Lead and lead alloys. Rolled lead sheet for building purposes*; it is a flexible material supplied in thin sheets, and therefore large irregular shapes with few joints can be formed, but it has the disadvantage of being expensive.

Copper
This should have a minimum thickness of 0.25 mm; like lead it is supplied in thin sheets and is expensive.

Bitumen
This is supplied in the form of a felt, usually to brick widths, and is therefore laid quickly with the minimum number of joints. Hessian and other fibrous-based bitumens may be found in older construction, but they are now largely superseded by dpc's of plastic materials.

Mastic asphalt
Applied in two layers giving a total thickness of 25 mm; it is applied *in-situ* and is therefore jointless, but is expensive in small quantities.

Polythene
Black low-density polythene sheet of single thickness not less than 0.5 mm thick should be used; it is easily laid but can be torn and punctured easily.

Slates
These should not be less than 230 mm long nor less than 4 mm thick and laid in two courses set breaking the joint in cement mortar 1:3. Slates have limited

flexibility but are impervious and very durable. Cost depends upon the area in which the building is being erected.

Bricks

These should comply with the requirements of BS 3921 'engineering' classification. They are laid in two courses in cement mortar, and may contrast with the general appearance of other brickwork in the same wall.

■ ■ ■ BRICKWORK BONDING

When building with bricks it is necessary to lay the bricks to some recognised pattern or bond in order to ensure stability of the structure and to produce a pleasing appearance. All the various bonds are designed so that no vertical joint in any one course is directly above or below a vertical joint in the adjoining course. To simplify this requirement special bricks are produced to BS 4729: *Specification for dimensions of bricks of special shapes and sizes*. Alternatively, the bricklayer can cut from whole bricks on site. Application of some of these specials is shown in Fig. 3.2.4, with some additions in Fig. 3.2.5. The various bonds are also planned to give the greatest practical amount of lap to all the bricks, and this should not be less than a quarter of a brick length. Properly bonded brickwork distributes the load over as large an area of brickwork as possible, so that the angle of spread of the load through the bonded brickwork is 60°.

COMMON BONDS

- **Stretcher bond**   Consists of all stretchers in every course and is used for half-brick walls and the half-brick skins of hollow or cavity walls (see Fig. 3.2.6).
- **English bond**   A very strong bond consisting of alternate courses of headers and stretchers (see Fig. 3.2.7).
- **Flemish bond**   Each course consists of alternate headers and stretchers; its appearance is considered to be better than English bond, but it is not quite so strong. This bond requires fewer facing bricks than English bond, needing only 79 bricks per square metre as opposed to 89 facing bricks per square metre for English bond. This bond is sometimes referred to as **double Flemish bond** (see Fig. 3.2.8).
- **Single Flemish bond**   A combination of English and Flemish bonds, having Flemish bond on the front face with a backing of English bond. It is considered to be slightly stronger than Flemish bond. The thinnest wall that can be built using this bond is a one-and-a-half brick wall.
- **English garden wall bond**   Consists of three courses of stretchers to one course of headers.
- **Flemish garden wall bond**   Consists of one header to every three stretchers in every course; this bond is fairly economical in facing bricks and has a pleasing appearance.

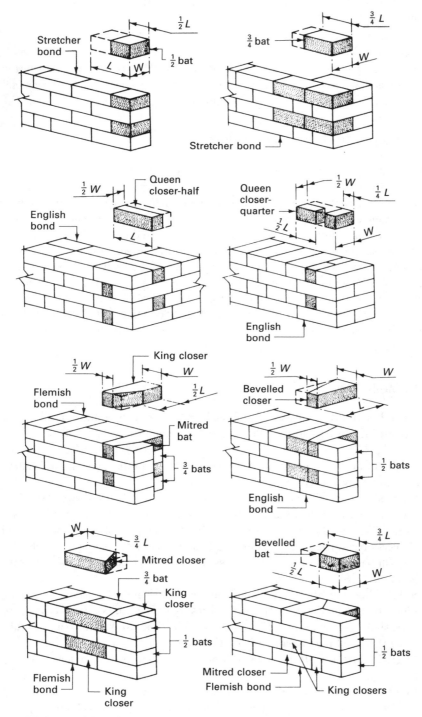

Figure 3.2.4   Special bricks.

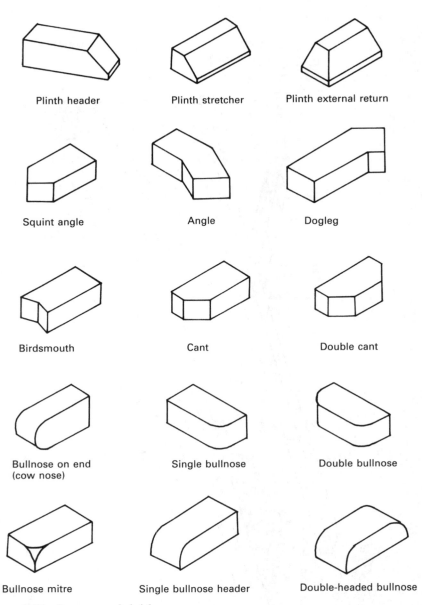

Plinth header   Plinth stretcher   Plinth external return

Squint angle   Angle   Dogleg

Birdsmouth   Cant   Double cant

Bullnose on end
(cow nose)   Single bullnose   Double bullnose

Bullnose mitre   Single bullnose header   Double-headed bullnose

Figure 3.2.5 Purpose–made bricks.

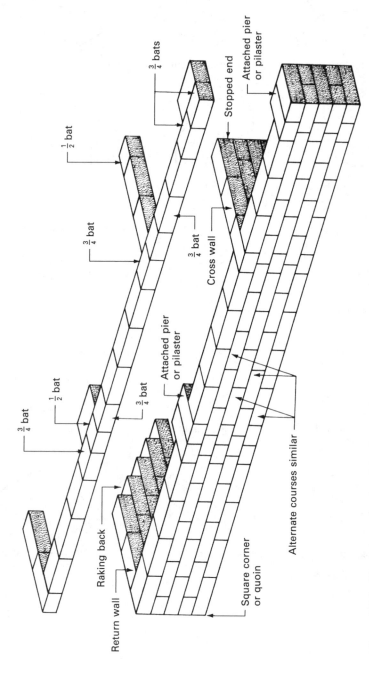

Figure 3.2.6 Typical stretcher bond details.

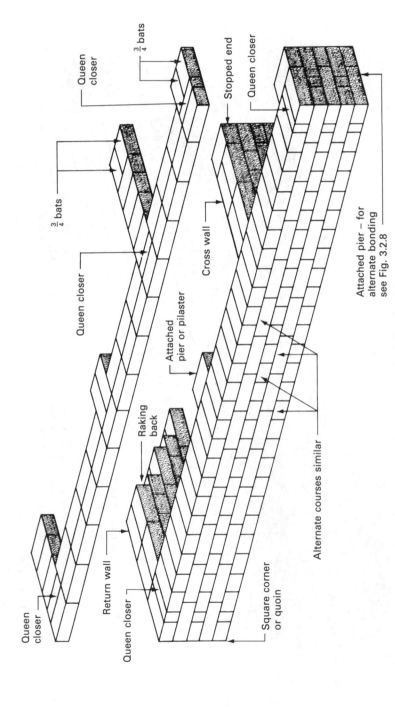

Figure 3.2.7 Typical English bond details.

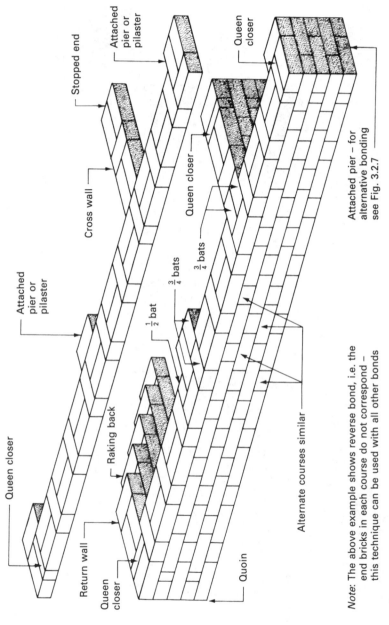

Figure 3.2.8   Typical Flemish bond details.

Note: The above example shows reverse bond, i.e. the end bricks in each course do not correspond – this technique can be used with all other bonds

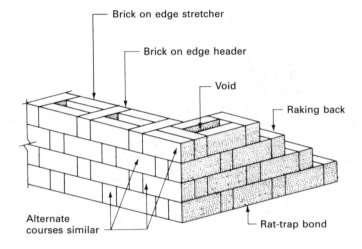

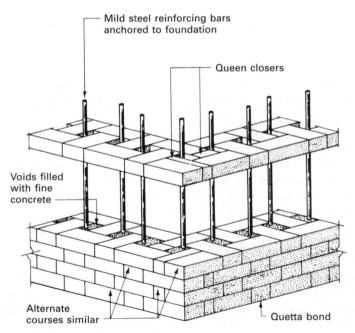

Figure 3.2.9 Special bonds.

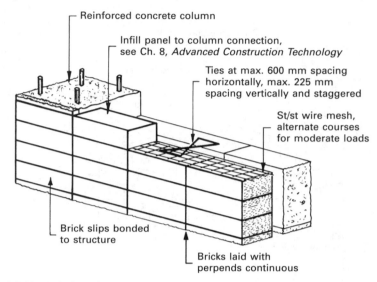

Reinforced concrete column

Infill panel to column connection,
see Ch. 8, *Advanced Construction Technology*

Ties at max. 600 mm spacing
horizontally, max. 225 mm
spacing vertically and staggered

St/st wire mesh,
alternate courses
for moderate loads

Brick slips bonded
to structure

Bricks laid with
perpends continuous

Figure 3.2.10   Reinforced stack bond.

SPECIAL BONDS

- **Rat-trap bond**   This is a brick on edge bond and gives a saving on materials
  and loadings, suitable as a backing wall to a cladding such as tile hanging
  (see Fig. 3.2.9).
- **Quetta bond**   Used on one-and-a-half brick walls for added strength; suitable
  for retaining walls (see Fig. 3.2.9).
- **Stack bond**   A brickwork feature used for partitions and infill panels.
  Bricks may be laid on bed or end with continuous vertical joints. Figure 3.2.10
  shows steel-mesh-reinforced horizontal bed joints to compensate for the lack
  of conventional bond.

METRIC MODULAR BRICKWORK

The standard imperial format brick does not fit reasonably well into the
system of dimensional coordination, with its preferred dimension of 300 mm.
Metric modular bricks have been designed and produced with four different
formats (see Fig. 3.2.11), but BS 6649: *Specification for clay and calcium
silicate modular bricks* provides only for a single format of 200 mm × 100 mm
× 75 mm.

The bond arrangements are similar to the well-known bonds but are based on
third bonding: that is, the overlap is one-third of a brick and not one-quarter as
with the standard format brick. Examples of metric modular brick bonding are
shown in Fig. 3.2.11.

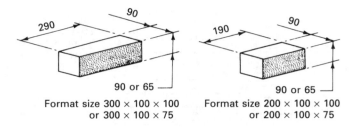

Metric modular bricks

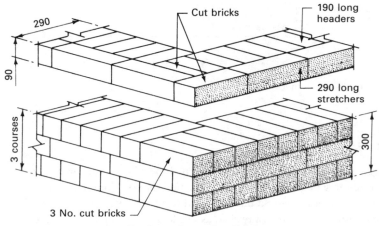

Header and stretcher bond

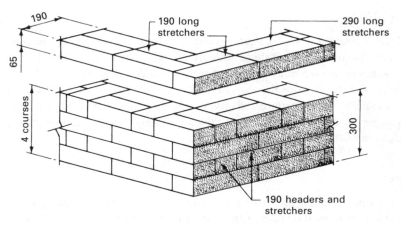

Header stretcher bond

Figure 3.2.11 Metric modular brickwork.

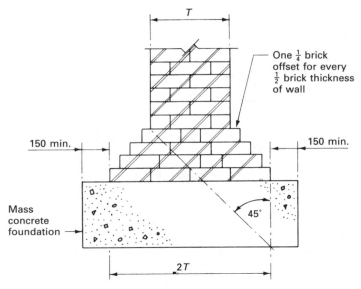

Figure 3.2.12 Typical footings.

FOOTINGS

These are wide courses of bricks placed at the base of a wall to spread the load over a greater area of the foundations. This method is seldom used today; instead the concrete foundation would be reinforced to act as a beam or reinforced strip. The courses in footings are always laid as headers as far as possible; stretchers, if needed, are laid in the centre of the wall (see Fig. 3.2.12).

BOUNDARY WALLS

These are subjected to severe weather conditions and therefore should be correctly designed and constructed. If these walls are also acting as a retaining wall the conditions will be even more extreme, but the main design principle of the exclusion of water remains the same. The presence of water in brickwork can lead to frost damage, mortar failure and efflorescence. The incorporation of adequate damp-proof courses and overhanging throated copings is of the utmost importance in this form of structure (see Fig. 3.2.13).

Efflorescence

This is a white stain appearing on the face of brickwork caused by deposits of soluble salts formed on or near the surface of the brickwork as a result of evaporation of the water in which they have been dissolved. It is usually harmless, and disappears within a short period of time; dry brushing or with clean water

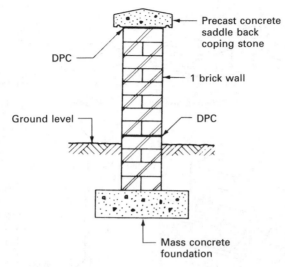

Figure 3.2.13 Typical boundary wall.

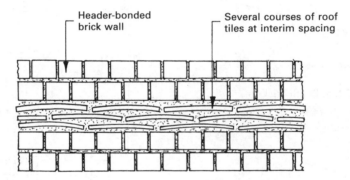

Figure 3.2.14 Lacing course (see also Fig. 3.1.7).

may be used to remove the salt deposit, but the use of acids should be left to the expert.

■■■ FEATURE BRICKWORK

Repetitive brickwork can be visually very monotonous. Some relief can be provided with coloured mortars and method of pointing. Greater opportunity for artistic expression is achieved by varying the colours and textures of the bricks to produce patterns and images on a wall. A further dimension is obtained by using cut and purpose-made bricks as projecting features. Some possible applications are shown in Figs 3.2.14–3.2.18.

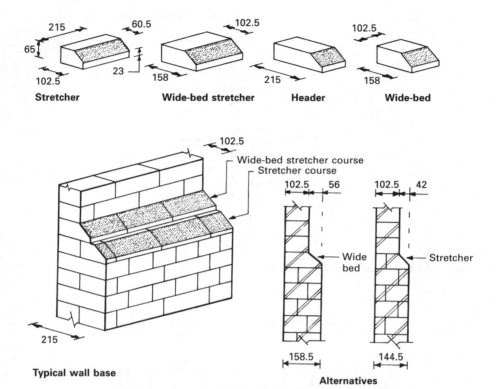

Figure 3.2.15   Plinth brickwork.

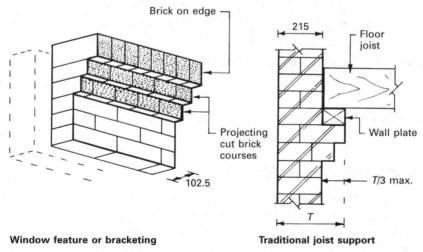

Figure 3.2.16   Corbelling.

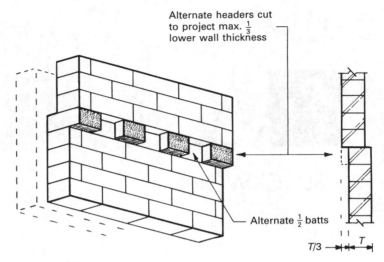

Alternate headers cut to project max. $\frac{1}{3}$ lower wall thickness

Alternate $\frac{1}{2}$ batts

$T/3$ T

Figure 3.2.17 Dentil coursing.

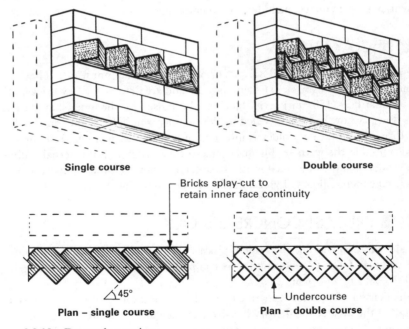

Single course

Double course

Bricks splay-cut to retain inner face continuity

$45°$

Plan – single course

Undercourse

Plan – double course

Figure 3.2.18 Dogtooth coursing.

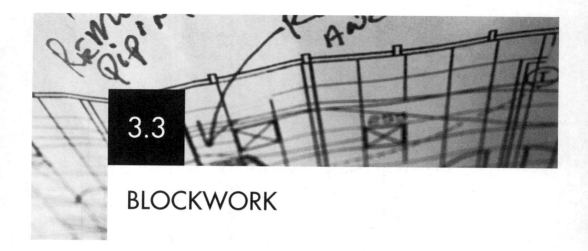

3.3

BLOCKWORK

A block can be defined as a walling unit exceeding the BS dimensions specified for bricks, and its height should not exceed either its length or six times its thickness. This avoids confusion with slabs or panels. Blocks are produced in clay and various cement/aggregate mixes, including wood chippings.

■■■ CLAY BLOCKS

These have format size of $300 \times 225 \times 62.5$, 75, 100 or 150 mm width. They are manufactured hollow by an extrusion process and are fired as for clay bricks.
The standard six (150 mm) cavity block has been used for the inner skin of cavity walls, but has now been superseded by developments in lightweight concrete blocks. The three (75 mm) cavity block is intended primarily for partitions. 150 mm blocks are also used in the warmer climates of the world as a single-leaf external walling with rendered and emulsioned finish. Chapter 4.6 shows another application to structural concrete floors. Typical details are shown in Fig. 3.3.1.

■■■ PRECAST CONCRETE BLOCKS

The specification and manufacture of precast concrete blocks or masonry units is covered in BS 6073. Classification is by compressive strength categories: 2.8, 3.5, 5, 7, 10, 15, 20 and 35 N/mm^2.

The density of a precast concrete block gives an indication of its compressive strength – the greater the density the stronger the block. Density will also give an indication as to the thermal conductivity and acoustic properties of a block. The lower the density the lower is the thermal conductivity factor, whereas the higher the density the greater is the reduction of airborne sound through the block.

The actual properties of different types of precast concrete block can be obtained from manufacturers' literature together with their appearance classification such as plain, facing or special facing.

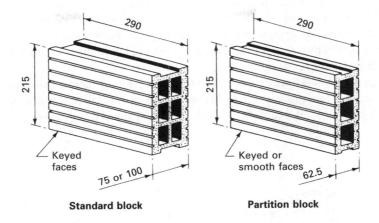

Standard block   **Partition block**

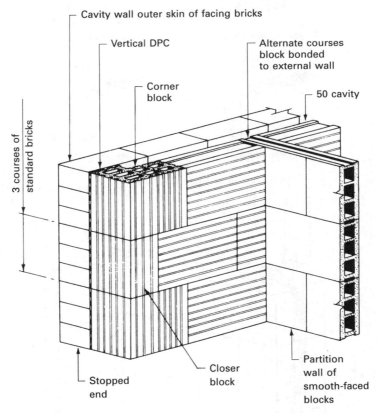

Figure 3.3.1   Hollow clay blocks and blockwork.

Aerated concrete for blocks is produced by introducing air or gas into the mix so that, when set, a uniform cellular block is formed. The usual method employed is to introduce a controlled amount of fine aluminium powder into the mix; this reacts with the free lime in the cement to give off hydrogen, which is quickly replaced by air and so provides the aeration.

Precast concrete blocks are manufactured to a wide range of standard sizes, the most common face format sizes being 400 mm × 200 mm and 450 mm × 225 mm, with thicknesses of 75, 90, 100, 140, 190 and 215 mm. Typical details are shown in Fig. 3.3.2.

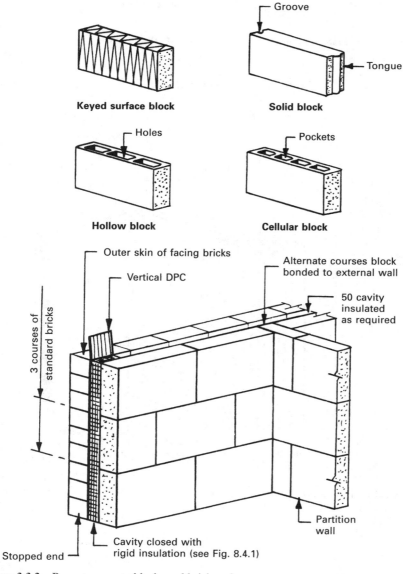

Figure 3.3.2   Precast concrete blocks and brickwork.

Concrete blocks are laid in what is essentially stretcher bond and joined to other walls by block bonding or by leaving metal ties or strips projecting from suitable bed courses. As with brickwork the mortar used in blockwork should be weaker than the material of the walling unit; generally a 1:2:9 gauged mortar mix will be suitable for work above ground level.

Concrete blocks shrink on drying out: therefore they should not be laid until the initial drying shrinkage has taken place (usually this is about 14 days under normal drying conditions), and should be protected on site to prevent them from becoming wet, expanding and causing subsequent shrinkage, possibly resulting in cracking of the blocks and any applied finishes such as plaster. Where length of wall exceed 6 m or thereabouts, it is necessary to incorporate vertical movement joints. These comprise a stainless steel or galvanised steel former incorporating horizontal bed joint ties at 450 mm spacing. Figure 3.3.3 shows the installation with

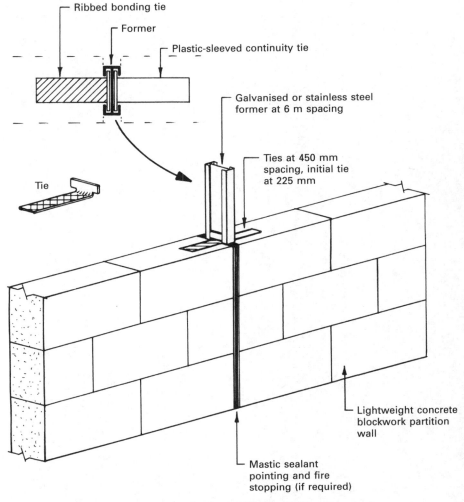

Figure 3.3.3   Blockwork movement joint (BS 5628: Part 3: *Use of masonry*).

bonded profiled or perforated ties to one side and plastic-sleeved ties to the other to maintain continuity and facilitate movement.

The main advantages of blockwork over brickwork are:

- labour saving – easy to cut, larger units;
- easier fixings – most take direct fixing of screws and nails;
- higher thermal insulation properties;
- lower density;
- provides a suitable key for plaster and cement rendering.

The main disadvantages are:

- lower strength;
- less resistance to rain penetration;
- loadbearing properties less (one- or two-storey application);
- lower sound insulation properties.

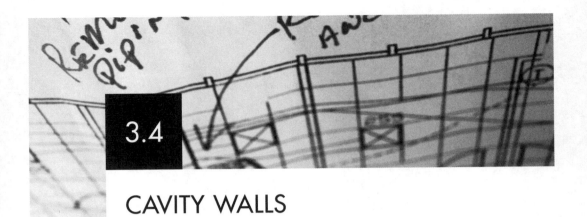

3.4

CAVITY WALLS

A wall constructed in two leaves or skins with a space or cavity between them is called a **cavity wall**, and it is the most common form of external wall used in domestic building today. The main purpose of constructing a cavity wall is to prevent the penetration of rain to the internal surface of the wall. It is essential that the cavity is not bridged in any way, as this would provide a passage for the moisture.

Air bricks have been used to ventilate the cavity, and these would be built in at the head and base of the cavity wall in order that a flow of air might pass through the cavity thus drying out any moisture that had penetrated the outer leaf. Unless the wall is exposed to very wet conditions the practice of inserting air bricks to ventilate the cavity is not now recommended, because it lowers the thermal and sound insulation values of the wall.

The main consideration in the construction of a cavity wall above ground-level damp-proof course is the choice of a brick or block that will give the required durability, strength and appearance and also conform to Building Regulation requirements. The main function of the wall below ground-level damp-proof course is to transmit the load safely to the foundations; in this context the two half-brick or block leaves forming the wall act as retaining walls. There is a tendency for the two leaves to move towards each other because of the pressure of the soil and the space provided by the cavity. To overcome this problem the cavity below ground level is filled with a weak mix of concrete, thus creating a solid wall in the ground (see Fig. 3.4.1). Alternatively, the substructural wall may be built solid. It is also advisable to leave out every fourth vertical joint in the external leaf at the base of the cavity and above the cavity fill, to allow any moisture trapped in the cavity a means of escape.

Parapets, whether solid or of cavity construction, are exposed to the elements on three sides and need careful design and construction. They must be provided with adequate barriers to moisture in the form of damp-proof courses, as dampness

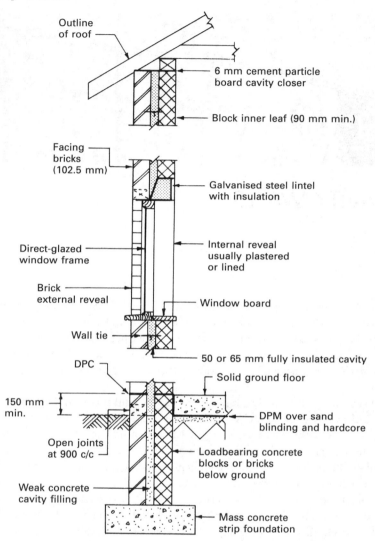

Figure 3.4.1   Typical cavity wall details.

could penetrate the structure by soaking down the wall and bypassing the roof and entering the building below the uppermost ceiling level. A solid parapet wall should not be less than 150 mm thick and not less than the thickness of the wall on which it is carried, and its height should not exceed four times its thickness. The recommended maximum heights of cavity wall parapets are shown in Fig. 3.4.2.

■■■■ BUILDING REGULATIONS

Regulation A1 requires that a building shall be so constructed that the combined dead, imposed and wind loads are sustained and transmitted to the ground safely

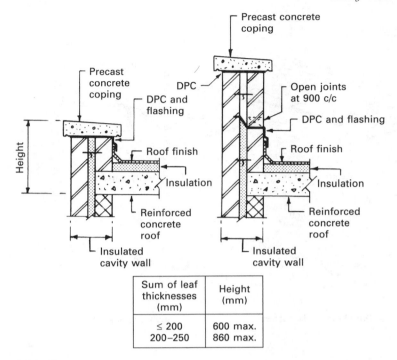

Sum of leaf thicknesses (mm)	Height (mm)
≤ 200	600 max.
200–250	860 max.

Figure 3.4.2   Parapets.

and without causing any movement that will impair the stability of any part of another building. Guidance to meet these requirements for cavity walls is given in Approved Document A.

Section 2C of this document considers full storey-height cavity walls for residential buildings of up to three storeys and requires that:

1. The compressive strengths of bricks and blocks should be not less than 5 N/mm^2 and 2.8 N/mm^2 respectively for buildings up to two storeys. Greater strength classifications are required for three-storey buildings (see Approved Document A, Section 2C, Diagram 7).
2. Cavities (gap between masonry faces) should be at least 50 mm, but may be up to 300 mm. Both leaves in cavity walls should have ties embedded at least 50 mm into adjacent masonry.
3. Wall ties should comply with BS EN 845-1: *Specification for ancillary components for masonry. Ties, tension straps, hangers and brackets.* Plastic and galvanised steel ties have been used, but now the preferred material is austenitic stainless steel. Maximum spacings are shown in Fig. 3.4.3. Cavities < 75 mm may be tied with butterfly pattern ties, but in excess of this they should be provided with twisted pattern ties.
4. Cavity walls normally have leaves at least 90 mm thick. An exception is with leaves of 65–90 mm thickness and cavity up to 75 mm with ties at 450 mm max. horizontal spacing.

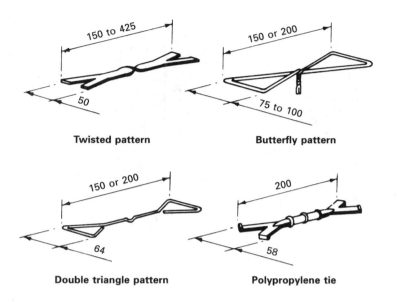

Twisted pattern   Butterfly pattern

Double triangle pattern   Polypropylene tie

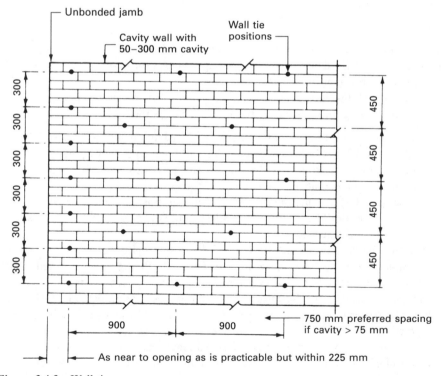

Figure 3.4.3   Wall ties.

5. The combined thickness of the two leaves of a cavity wall should be not less than 190 mm for a maximum wall height of 3.5 m and length not exceeding 12 m, and also for heights between 3.5 m and 9 m in wall lengths not exceeding 9 m. Wall lengths and heights up to 12 m require a minimum thickness of 290 mm. Variations:

Wall height	Wall length	Wall thickness
≤ 12 m	≤ 9 m	290 mm for one storey
		190 mm for remainder
≤ 12 m	≤ 12 m	290 mm for two storeys
		190 mm for remainder

6. Mortar should be as given for mortar designation (iii) in BS 5628-1: *Structural use of unreinforced masonry*, or a gauged mortar mix of 1:1:6 by volume, or its equivalent.
7. Cavity walls of any length need to be provided with roof lateral support. Walls over 3 m length will also require floor lateral support at every floor forming a junction with the supported wall. If roof lateral support is not provided by type of covering (tiles or slates), a pitch of 15° or more plus a minimum wall plate bearing of 75 mm, durable metal straps with a minimum cross-section of 30 mm × 5 mm will be needed at not more than 2 m centres (see Fig. 6.1.9). If the floor does not have at least a 90 mm bearing on the supported wall, lateral support should be provided by similar straps at not more than 2 m spacing, or the joists should be fixed using restraint-type joist hangers (see Fig. 4.7.1).

PREVENTION OF DAMP IN CAVITY WALLS

Approved Document C recommends a cavity to be carried down at least 150 mm below the lowest damp-proof course, and that any bridging of the cavity, other than a wall tie or closing course protected by the roof, is to have a suitable damp-proof course to prevent the passage of moisture across the cavity. Where the cavity is closed at the jambs of openings a vertical damp-proof course should be inserted unless some other suitable method is used to prevent the passage of moisture from the outer leaf to the inner leaf of the wall.

Approved Document C recommends a damp-proof course to be inserted in all external walls at least 150 mm above the highest adjoining ground or paving to prevent the passage of moisture rising up the wall and into the building, unless the design is such that the wall is protected or sheltered.

■■■ ADVANTAGES AND DISADVANTAGES

ADVANTAGES OF CAVITY WALL CONSTRUCTION

These can be listed as follows:

- able to withstand a driving rain in all situations from penetrating to the inner wall surface;
- gives good thermal insulation, keeping the building warm in winter and cool in the summer;

- no need for external rendering;
- enables the use of cheaper and alternative materials for the inner construction;
- a nominal 255 mm cavity wall has a higher sound insulation value than a standard one-brick-thick wall;
- can accommodate supplementary insulation material.

DISADVANTAGES OF CAVITY WALL CONSTRUCTION

These can be listed as follows:

- requires a high standard of design and workmanship to produce a soundly constructed wall – this will require good supervision during construction;
- the need to include a vertical damp-proof course to all openings;
- slightly dearer in cost than a standard one-brick-thick wall.

■■■ JOINTING AND POINTING

These terms are used for the finish given to both the vertical and the horizontal joints in brickwork irrespective of whether the wall is of brick, block, solid or cavity construction.

Jointing is the finish given to the joints when carried out as the work proceeds.

Pointing is the finish given to the joints by raking out to a depth of approximately 20 mm and filling in on the face with a hard-setting cement mortar, which could have a colour additive. This process can be applied to both new and old buildings. Typical examples of jointing and pointing are shown in Fig. 3.4.4.

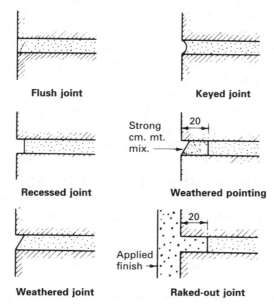

Figure 3.4.4   Brick joints.

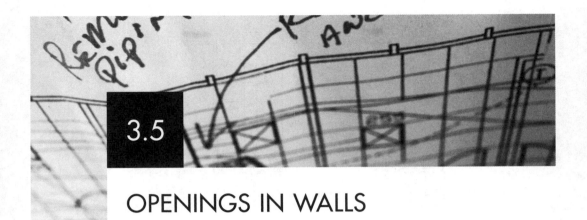

3.5

OPENINGS IN WALLS

An opening in an external wall consists of a head, jambs or reveals, and a sill or threshold.

■■■ HEAD

The function of a head is to carry the load of brickwork over the opening and transmit this load to the jambs at the sides. To fulfil this task it must have the capacity to support the load without unacceptable deflection. A variety of materials and methods is available in the form of a lintel or beam, such as:

- **Timber**   Suitable for light loads and small spans; the timber should be treated with a preservative to prevent attack by beetles or fungi.
- **Steel**
    - For small openings – a mild steel flat or angle section can be used to carry the outside leaf of a cavity wall, the inner leaf being supported by a concrete or steel lintel.
    - For medium spans – a channel or joist section is usually suitable.
    - For large spans – a universal beam section to design calculations will be needed. Steel lintels and beams that are exposed to the elements should be either galvanised or painted with several coats of bituminous paint to give them protection against corrosion.
- **Concrete**   These can be designed as *in-situ* or precast reinforced beams or lintels and can be used for all spans. Prestressed concrete lintels are available for the small and medium spans (see Chapter 10, *Advanced Construction Technology*).
- **Stone**   These can be natural, artificial or reconstructed stone but are generally used as a facing to a steel or concrete lintel (see Fig. 3.1.5).
- **Brick**   Unless reinforced with mild steel bars or mesh, brick lintels are suitable only for small spans up to 1 m but, like stone, bricks are also employed as a facing to a steel or concrete lintel.

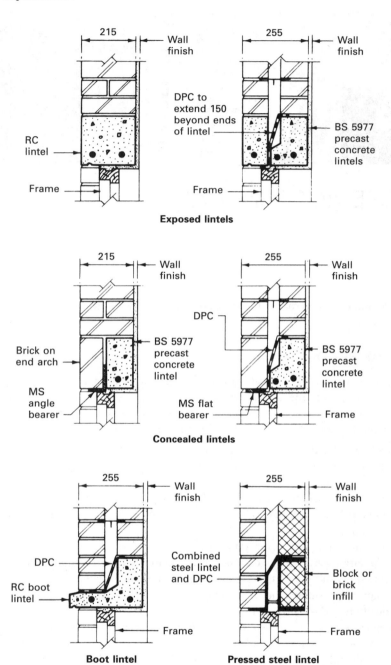

Figure 3.5.1   Typical head treatments to openings (pre-1985 Building Regulations).

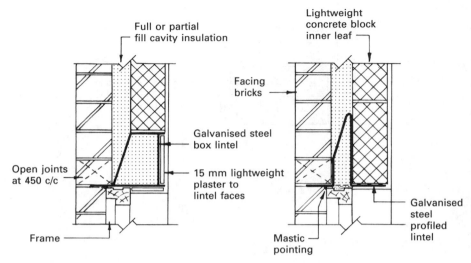

Figure 3.5.2   Typical head treatments to openings.

Lintels require a bearing at each end of the opening; the amount will vary with the span, but generally it will be 100 mm for the small spans and up to 225 mm for the medium and large spans. In cavity walling a damp-proof course will be required where the cavity is bridged by the lintel, and this should extend at least 150 mm beyond each end of the lintel. Open joints are sometimes used to act as weep holes; these are placed at 900 mm centres in the outer leaf immediately above the damp-proof course. Typical examples of head treatments to openings are shown in Figs 3.5.1 and 3.5.2.

▧▧■ ARCHES

These are arrangements of wedge-shaped bricks designed to support each other and carry the load over the opening round a curved profile to abutments on either side. Full details of arch construction are given in the next chapter.

▧▧■ JAMBS

In solid walls these are bonded to give the required profile and strength; examples of bonded jambs are shown in Fig. 3.2.4. In cavity walls the cavity can be closed at the opening by using a suitable frame, or by turning one of the leaves towards the other, forming a butt joint in which is incorporated a vertical damp-proof course as required by the Building Regulations. Typical examples of jamb treatments to openings are shown in Figs 3.5.3 and 3.5.4.

▧▧■ SILL

The function of a sill is to shed rainwater, which has run down the face of the window or door and collected at the base, away from the opening and the face of the wall. Many methods and materials are available; appearance and durability are the

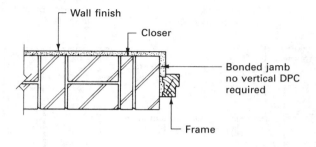

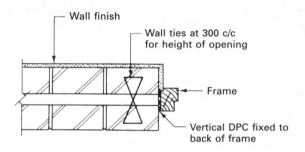

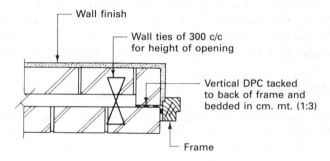

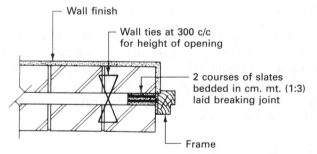

Figure 3.5.3 Typical jamb treatments to openings (pre–1985 Building Regulations).

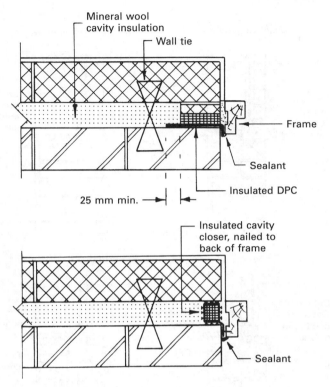

Figure 3.5.4 Typical jamb treatments to openings.

main requirements, as a sill is not a member that is needed to carry heavy loads. Sills, unlike lintels, do not require a bearing at each end. Typical examples of sill treatments to openings are shown in Figs 3.5.5 and 3.5.6.

■■■ CONSERVATION OF FUEL ENERGY

The details shown in Figs 3.5.1, 3.5.3 and 3.5.5 are no longer acceptable by UK building standards for new housing, because of inadequate thermal insulation and the opportunity for thermal bridging (see Chapter 8.3). However, much of our existing housing stock is constructed in this manner, and therefore these illustrations are retained as historic reference.

■■■ CHECKED OPENINGS

REVEAL

That part of an opening returning at right angles from the front face of a wall. The traditional construction for door and window reveals includes a check or recess to accommodate the frame. It also provides a barrier against draught and rain penetration. Examples of this dated, but nevertheless existing, construction are shown in Fig. 3.5.7.

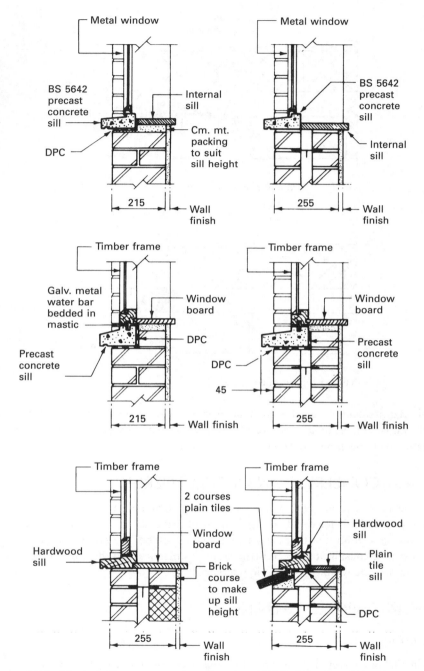

Figure 3.5.5 Typical sill treatments to openings (pre–1985 Building Regulations).

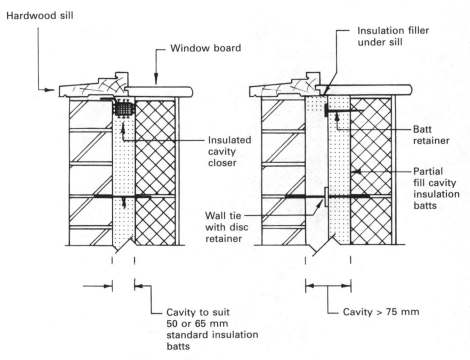

Hardwood sill

Window board

Insulation filler under sill

Insulated cavity closer

Batt retainer

Partial fill cavity insulation batts

Wall tie with disc retainer

Cavity to suit 50 or 65 mm standard insulation batts

Cavity > 75 mm

Figure 3.5.6   Typical sill treatments to openings.

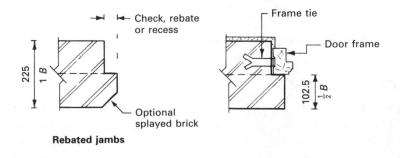

Check, rebate or recess

225   1 B

Optional splayed brick

Rebated jambs

Frame tie

Door frame

102.5 $\frac{1}{2}$ B

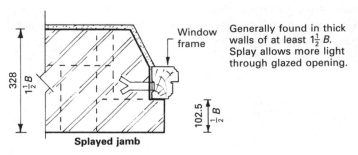

Window frame

Generally found in thick walls of at least $1\frac{1}{2}$ B. Splay allows more light through glazed opening.

328   $1\frac{1}{2}$ B

102.5 $\frac{1}{2}$ B

Splayed jamb

Figure 3.5.7   Traditional checked openings (sectional plans).

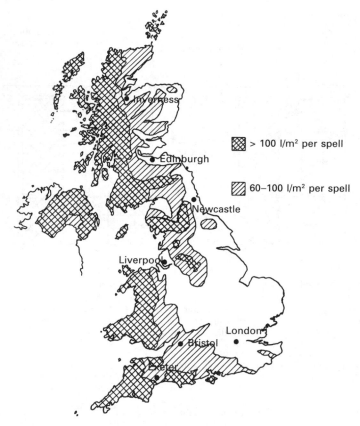

Figure 3.5.8   Principal areas exposed to driving rain.

BUILDING REGULATIONS

Approved Document C: *Site preparation and resistance to contaminants and moisture*
identifies the need to incorporate checked rebates at openings exposed to driving
rain. In the UK, driving rain exposure zones are mainly to the west and south-west
coasts, as shown in Fig. 3.5.8. Isolated situations also occur, and location of these
can be obtained from the local authority building control department. In keeping
with the traditional construction shown, the frame is set behind the outer leaf of
masonry. Contemporary practice also requires an insulated cavity, vertical dpc and
cavity closer, as shown in Fig. 3.5.9.

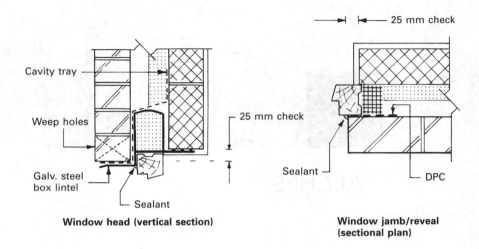

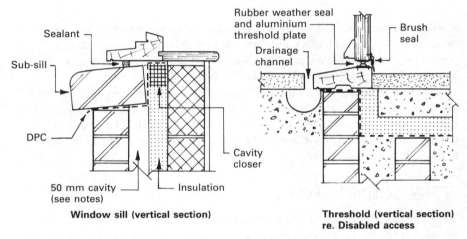

Window head (vertical section)

Window jamb/reveal (sectional plan)

Window sill (vertical section)

Threshold (vertical section) re. Disabled access

Notes:
1. In areas most exposed to driving rain, a 50 mm air gap in the cavity between inner and outer leaf is necessary to prevent dampness bridging the insulation.
2. The detail shown has a 100 mm cavity part insulated and a 100 mm insulating block inner leaf. This will provide a *U* value of about 0.35. A 100 mm fully insulated cavity in a brick and insulating block wall has a *U* value of about 0.27.
3. To achieve the lesser *U* value and still retain a 50 mm air gap, the thickness of the inner leaf of insulating blockwork may be increased and/or additional insulation applied to the inside of the inner leaf behind a protective plasterboard lining.
4. See Part 8 – Insulation, for definition and calculation of *U* values.

Figure 3.5.9 Contemporary checked openings.

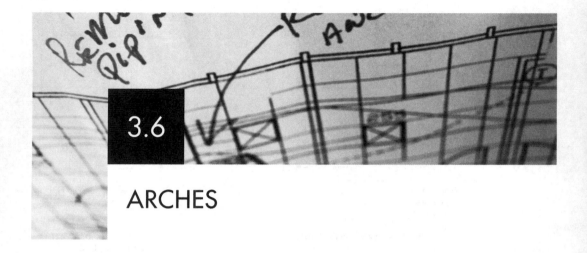

ARCHES

These are arrangements of wedge-shaped bricks called **voussoirs**, which are designed to support each other and carry the load over the opening, round a curved profile, to abutments on either side. An exception to this form is the flat or 'soldier' arch constructed of bricks laid on end or on edge.

When constructing an arch it must be given temporary support until the brick joints have set and the arch has gained sufficient strength to support itself and carry the load over the opening. These temporary supports are called **centres** and are usually made of timber; their design is governed by the span, load and thickness of the arch to be constructed.

■■■ SOLDIER ARCHES

This type of arch consists of a row of bricks showing on the face either the end or the edge of the bricks. Soldier arches have no real strength, and if the span is over 1000 mm they will require some form of permanent support such as a metal flat or angle (see Fig. 3.5.1). If permanent support is not given, the load will be transferred to the head of the frame in the opening instead of the jambs on either side. Small spans can have an arch of bonded brickwork by inserting into the horizontal joints immediately above the opening some form of reinforcement such as expanded metal or bricktor, which is a woven strip of high-tensile steel wires designed for the reinforcement of brick and stone walls. It is also possible to construct a soldier arch by inserting metal cramps in the vertical joints and casting these into an *in-situ* backing lintel of reinforced concrete.

■■■ ROUGH ARCHES

These arches are constructed of ordinary uncut bricks; being rectangular in shape, they give rise to wedge-shaped joints. To prevent the thick end of the joint from becoming too excessive, rough arches are usually constructed in header courses. The rough arch is used mainly as a backing or relieving arch to a gauged brick or stone arch, but they are sometime used in facework for the cheaper form of building or where appearance is of little importance.

■■■ GAUGED ARCHES

These are the true arches and are constructed of bricks cut to the required wedge shape, called **voussoirs**. The purpose of voussoirs is to produce a uniform thin joint that converges onto the centre point or points of the arch. There are two methods of cutting the bricks to the required wedge shape: **axed** and **rubbed**. If the brick is of a hard nature it is first marked with a tin saw, to produce a sharp arris, and then it is axed to the required profile. For rubbed brick arches a soft brick called a **rubber** is used; the bricks are first cut to the approximate shape with a saw, and are then finished off with an abrasive stone or file to produce the sharp arris. In both cases a template of plywood or hardboard to the required shape will be necessary for marking out the voussoirs. Typical examples of stone arches are shown in Fig. 3.1.4; the terminology and setting out of simple brick arches is shown in Fig. 3.6.1.

CENTRES

These are temporary structures, usually of light timber construction, which are strong enough to fulfil their function of supporting arches of brick or stone while they are being built and until they are sufficiently set to support themselves and the load over the opening. Centres can be an expensive item to a builder: therefore their design should be simple and adaptable so that as many uses as possible can be obtained from any one centre. A centre is always less in width than the soffit of an arch to allow for plumbing – that is, alignment and verticality of the face with a level or rule.

The type of centre to be used will depend upon:

■ the weight to be supported;
■ the span;
■ the width of the soffit.

Generally soffits not wider than 150 mm will require one rib at least 50 mm wide, and are usually called **turning pieces**. Soffits from 150 to 350 mm require two ribs, which are framed together using horizontal tie members called **laggings**. Soffits over 350 mm require three or more sets of ribs. The laggings are used to tie the framed ribs together and to provide a base upon which the arch can be built. Close laggings are those that are touching each other, forming a complete seating for a gauged arch; open laggings, spaced at twice the width of the laggings, centre to centre, are used for rough arches.

If the arch is composed of different materials – for example a stone arch with a relieving arch of brickwork – a separate centre for each material should be used. Typical examples of centres for brick arches are shown in Figs 3.6.2, 3.6.3 and 3.6.4.

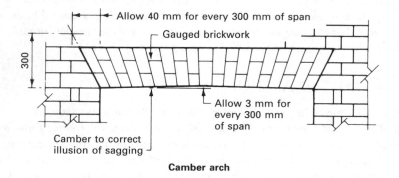

Allow 40 mm for every 300 mm of span

Gauged brickwork

300

Allow 3 mm for
every 300 mm
of span

Camber to correct
illusion of sagging

Camber arch

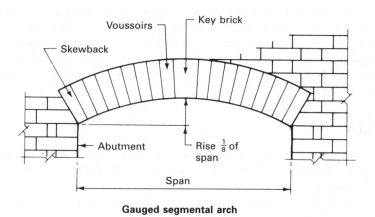

Voussoirs — ⌐ Key brick

— Skewback

— Abutment   └ Rise $\frac{1}{8}$ of
span

Span

Gauged segmental arch

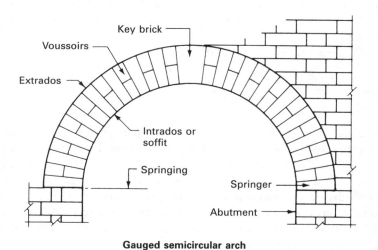

Key brick —

Voussoirs —

Extrados —

— Intrados or
soffit

⌐ Springing

Springer —

Abutment —

Gauged semicircular arch

Figure 3.6.1   Typical examples of brick arches.

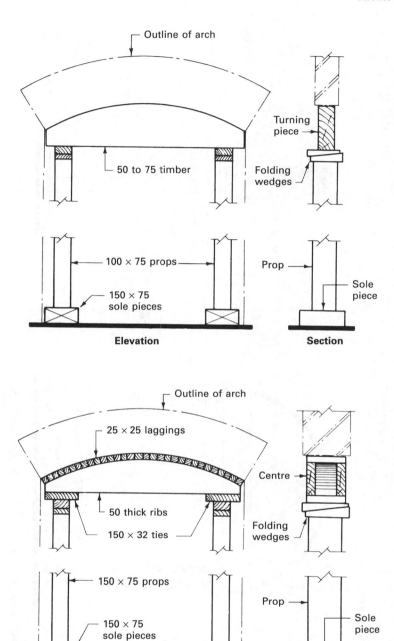

Figure 3.6.2 Centres for small–span arches.

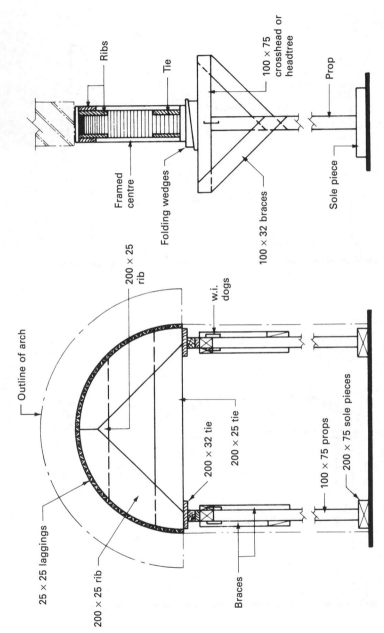

Ribs

Tie

100×75 crosshead or headtree

Prop

Framed centre

Folding wedges

100×32 braces

Sole piece

200×25 rib

w.i. dogs

Outline of arch

200×32 tie

200×25 tie

100×75 props

200×75 sole pieces

25×25 laggings

200×25 rib

Braces

Figure 3.6.3　Typical framed centre for spans up to 1500 mm.

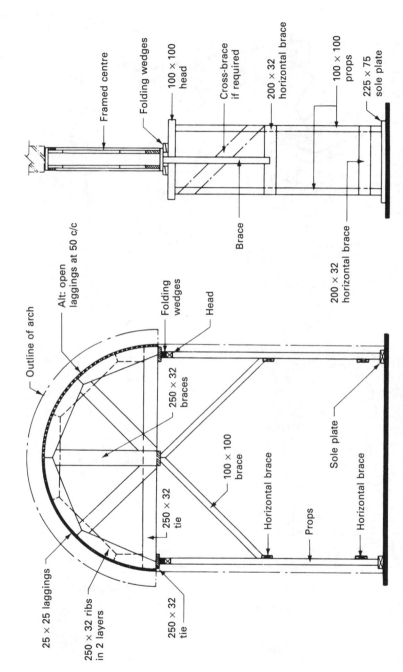

Figure 3.6.4 Typical framed centre for spans up to 4000 mm.

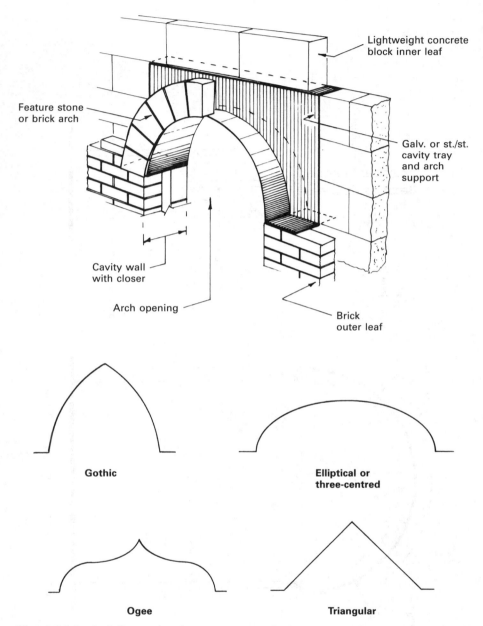

Lightweight concrete block inner leaf

Feature stone or brick arch

Galv. or st./st. cavity tray and arch support

Cavity wall with closer

Arch opening

Brick outer leaf

Gothic

Elliptical or three-centred

Ogee

Triangular

Figure 3.6.5   Arch form and cavity tray.

CAVITY TRAYS

Arch construction can be simplified with structural support from a proprietary galvanised or stainless steel combined cavity tray and damp-proof course. Figure 3.6.5 shows the application and omission of conventional centring, saving considerable construction time as temporary supports are not required. The tray unifies the structural support with the damp-proof course and remains an integral part of the structure. Arch profiles are limited to tray manufacturers' designs and spans, but most will offer semicircular and segmental as standard, with gothic, elliptical, ogee and triangular styles available to order. Clear spans up to about 2 m are possible.

3.7

TIMBER-FRAMED HOUSING

Timber framing of softwood members and sheet facings has a long association with non-loadbearing partition walls (see timber stud partitions – Chapter 7.7). Loadbearing timber walls for small buildings such as dwelling houses may be constructed by creating a framework of relatively small timbers (up to ex. 150 mm × 50 mm (The 'ex' means 'out of'.)) spaced vertically at 400 or 600 mm. These are known as **studs** and have short struts of timber or noggins placed between them at 1 m maximum spacing to prevent distortion. Some panel prefabricators provide thin galvanised steel diagonal bracing as an alternative to struts. Head and sole plates complete the framing.

With the benefit of quality control procedures and efficient industrial manufacturing processes, the high cost of imported timber in the UK is offset by effective factory prefabrication of complete wall, floor and roof units. Furthermore, rationalised site assembly with mobile crane and semi-skilled operatives provides for considerable saving in construction time.

As a construction material, timber is comparatively light; it is therefore easy to handle and is less of a structural dead weight factor than traditional masonry. It has a high strength to weight ratio and is also very stiff when related to its strength. Prefabrication of structural units eliminates the need for skilled carpentry as all joints are simply nailed.

Apart from the cost of timber, the other disadvantages are poor resistance to fire in sizes under 150 mm × 100 mm, as there is insufficient substance to char-protect the inner structure, and its hygroscopic nature, which can support decay in damp situations. The effect of fire is limited by cladding with non-combustible material such as plasterboard internally and facing brick externally. Provided the timber is installed dry and assembled with correct protection (dpc, vapour control layers, etc.), dampness penetration will not be an issue.

Manufacture and construction techniques derive from the applications of **platform** or **balloon** framing. Figure 3.7.1 shows the principle of storey-height platform-framed walls made up of ex. 100 mm × 50 mm timber studding at 400 or 600 mm centres. These are supported at intermediate floor levels or platforms by

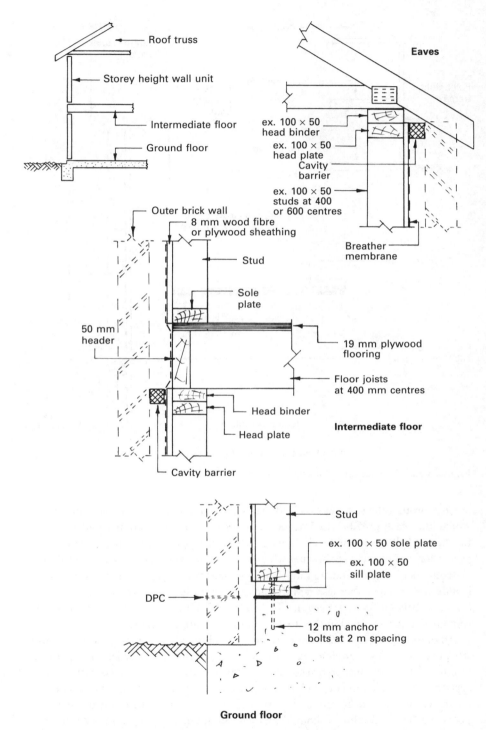

Figure 3.7.1 Timber frame construction – platform frame.

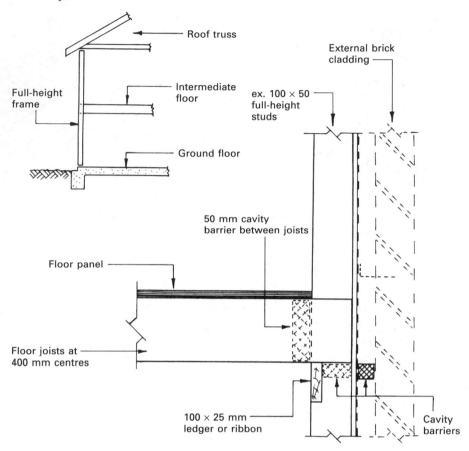

Figure 3.7.2   Timber frame construction – balloon frame.

nailing through the sole plates. Head plates are similarly fixed to a head binder, which links each prefabricated panel. Figure 3.7.2 shows the ground floor to eaves full-height balloon-framed wall. This is independent of the intermediate floor except for complementary bracing where the floor joists fix to studs and a **ribbon** or **ledger** let into the stud framing. Storey-height platform framing is simpler to handle and transport but less rapidly constructed on site than the one lift to roof level that balloon framing permits. The effects of timber movement are more easily contained and restrained in the smaller units of a platform frame.

An outer cladding of brickwork provides a traditional appearance, with galvanised steel angle wall ties spanning a cavity to secure the masonry outer leaf to the timber framing. The possibility of fire spread through the cavity is restrained with cavity barriers at strategic intervals. These normally occur at intersections of elements of construction, e.g. wall to roof and party wall to external wall. Fire resistance of cavity barriers in dwellings should be at least 30 minutes, i.e. 30 minutes' integrity and 15 minutes' insulation (see Building Regulations: Approved Document B3: Section 9, BS 476: *Fire tests on building materials and structures*, and Fig. 3.7.3).

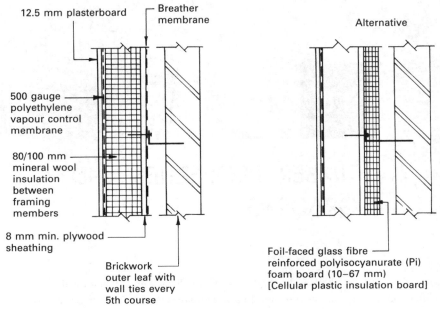

12.5 mm plasterboard

Breather membrane

Alternative

500 gauge polyethylene vapour control membrane

80/100 mm mineral wool insulation between framing members

8 mm min. plywood sheathing

Brickwork outer leaf with wall ties every 5th course

Foil-faced glass fibre reinforced polyisocyanurate (Pi) foam board (10–67 mm) [Cellular plastic insulation board]

External wall construction

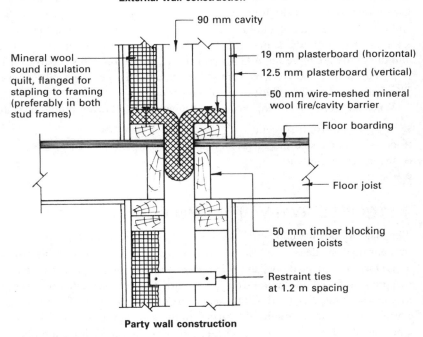

90 mm cavity

Mineral wool sound insulation quilt, flanged for stapling to framing (preferably in both stud frames)

19 mm plasterboard (horizontal)

12.5 mm plasterboard (vertical)

50 mm wire-meshed mineral wool fire/cavity barrier

Floor boarding

Floor joist

50 mm timber blocking between joists

Restraint ties at 1.2 m spacing

Party wall construction

Figure 3.7.3   Timber frame construction – further details.

3.8

TIMBER: PROPERTIES AND GRADING

Timber derives from thousands of different tree species, but the botanical classification divides timber into **hardwood** and **softwood**. Hardwoods are from broad-leaved trees, most of which are deciduous, holly being an exception. Not all hardwoods are hard; balsa is an example of a *soft* hardwood. Softwoods are from coniferous trees, and not all of these are relatively soft: for example, yew is a very strong, dense and durable material. For structural purposes, strength characteristics and durability are more important means for classification. This is particularly important with softwood, which is used predominantly for structural timbers in the UK, because of the prohibitive cost of hardwood.

When specifying softwood structural timbers, the following should be stated:

- any preservative treatment;
- cross-sectional size and surface finish;
- the moisture content;
- the strength class.

▓▓▓ PRESERVATIVE TREATMENT

The majority of softwood contains a high proportion of sapwood. This is less resilient than the denser heartwood found closer to the core of a timber trunk. Therefore, as a precaution against dampness and possible fungal decay (wet or dry rot), plus potential infestation from wood-boring insects, suitable preservatives may be required. The need will be determined by the position of timbers within a building and the geographical location. In parts of southern England there is a perceived risk of softwood timber infestation by the **house longhorn beetle**. The Building Regulations, Approved Document to support Regulation 7, define the particular boroughs where softwoods must be treated if applied in any aspect of roof construction. See also BS 8417: *Preservation of timber. Recommendations*, for clarification of acceptable chemical treatment.

■■■ CROSS-SECTIONAL SIZE AND SURFACE FINISH

Softwood cross-sectional size has previously been specified under three classifications: sawn, planed (machined), and regularised. Sawn is the result of converting a timber log into commercially accepted sizes, e.g. 50 mm × 100 mm, 75 mm × 225 mm. Application of these sections is largely for unseen structural use, such as floor and roof members. Planed involves machining about 1.5 mm off each surface to provide a smooth finish: for example, 50 mm × 100 mm sawn becomes 47 mm × 97 mm, otherwise known as ex. 50 × 100. Regularising has been used to achieve a uniform width in timber to provide a more even final product. The timber may be sawn or planed to, say, 47 mm for uniformity in timber studding.

However, these terms are now redefined, with regularising no longer featured, in the requirements of BS EN 336: *Structural timber. Sizes, permitted deviations.* This standard specifies **target sizes** in order to simplify the previous confusion over the finished dimensions. Target size is in only two tolerance classes: T1, applicable to sawn surfaces; and T2, applicable to planed timber. For example, if a section of timber is required 50 mm wide sawn and 197 mm in depth planed, it will be specified as 50 mm (T1) × 197 mm (T2).

■■■ MOISTURE CONTENT

When timber is felled it is saturated. In this state it is unworkable and, if allowed to remain so, will be prone to shrinkage, distortion and the effects of fungal growths. Natural drying or seasoning is possible in some climates, but most commercial organisations use kilns to reduce the moisture content of rough sawn softwood to between 12% and 20%. This is expressed as the weight of water in the timber as a percentage of the weight of the timber dry:

$$\text{moisture content } \% = \frac{(\text{wet weight}) - (\text{dry weight})}{\text{dry weight}} \times \frac{100}{1}$$

After seasoning it is essential that the product remains in a well-ventilated, atmospherically stable environment. Timber is hygroscopic, i.e. made up of cells that will absorb moisture readily. Therefore significant changes in temperature and humidity may cause it to react by expanding, contracting, deforming or twisting. Moisture content of stored timber can be checked regularly with a surveyor's moisture meter. These have two pointed probes, and respond to conductance due to the amount of electrical moisture in the material.

■■■ STRENGTH CLASSES

Grading of timber may be visual or by the more efficient use of computerised grading machines. Individual pieces are assessed against permissible defects limits and marked accordingly. Some examples of grade markings are shown in Fig. 3.8.1.

Grading occurs in this country or the country of origin. BS EN 518 and 519 (European standards for visual and machine grading, respectively) effectively

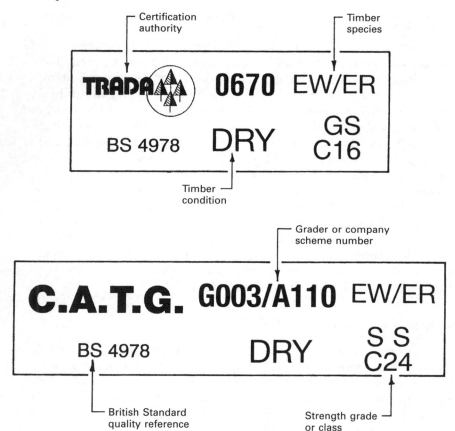

Figure 3.8.1   Examples of grading stamps on softwood timber.

rationalise much of the timber graded in Europe and many surrounding countries, e.g. Scandinavia, but confusion may occur with timber graded elsewhere in the absence of an international standard. Visual grading in the UK may be in accordance with BS 4978. This provides two quality standards: GS (general structural) and SS (special structural). The letter M preceding GS or SS indicates grading by machine. GS and SS grades will meet the requirements of the aforementioned European standards.

Timber imported from North America is independently graded to rules applicable in Canada and the USA. The Canadian National Lumber Grades Authority (NLGA) and the American National Grading Rules for Dimension Lumber (NGRDL) grade in accordance with product application: see Table 3.8.1.

Comparisons can be drawn, and Table 3.8.2 provides an indication of the similarity of product grades and strength classes to BS EN 338: *Structural timber. Strength classes.*

As can be seen from Table 3.8.2, using GS or SS graded timber from different sources may not result in the same strength characteristics, as these can vary between species. The full listings of BS EN 338 strength classes are shown in

Table 3.8.1 North American timber grades and markings

Timber mark	Grade	Application	Range of nominal widths
Sel str No. 1 No. 2 No. 3	Select structural Number 1 Number 2 Number 3	Structural joists and planks	150 mm and over
Sel str No. 1 No. 2 No. 3	Select structural Number 1 Number 2 Number 3	Structural light framing	50 mm to 100 mm
Const Std Util Stud	Construction Standard Utility Stud	Light framing	50 mm to 100 mm

Table 3.8.2 Timber grading and strength class comparisons

Species	Standard							
	BS EN 338	C14	C16	C18	C22	C24	C27/TR26	C30
	Bldg Regs		SC3			SC4	SC5	
Whitewood or redwood	BS 4978		GS			SS		
British spruce	BS 4978	GS		SS				
British pine	BS 4978	GS			SS			
Canadian S-P-F or hem-fir	NLGA or NGRLD (J&P/SLF)		No. 1 No. 2			Sel		
	NLGA or NGRLD (LF)	Const						
	NLGA or NGRLD (Stud)	Stud						

Key: S-P-F = Spruce, pine, fir.
   hem   = Hemlock.
   Others, see Table 3.8.1.

Notes
1. North American machine stress rated (MSR) grades are also accepted in the UK and correspond across the BS EN 338 range given above.
2. For a full and more detailed comparison, refer to BS5268-2: *Structural use of timber. Code of practice for permissible stress design, materials and workmanship.*

Table 3.8.3   Grade stresses and moduli of elasticity for various strength classes: for service classes 1 and 2

Strength class	Bending parallel to grain (N/mm²)	Tension parallel to grain (N/mm²)	Compression parallel to grain (N/mm²)	Compression perpendicular to grain* (N/mm²)		Shear parallel to grain (N/mm²)	Modulus of elasticity		Characteristic density, ρ_k† (kg/m³)	Average density, ρ_{mean}† (kg/m³)
							Mean (N/mm²)	Minimum (N/mm²)		
C14	4.1	2.5	5.2	2.1	1.6	0.60	6 800	4 600	290	350
C16	5.3	3.2	6.8	2.2	1.7	0.67	8 800	5 800	310	370
C18	5.8	3.5	7.1	2.2	1.7	0.67	9 100	6 000	320	380
C22	6.8	4.1	7.5	2.3	1.7	0.71	9 700	6 500	340	410
C24	7.5	4.5	7.9	2.4	1.9	0.71	10 800	7 200	350	420
TR26‡	10.0	6.0	8.2	2.5	2.0	1.10	11 000	7 400	370	450
C27	10.0	6.0	8.2	2.5	2.0	1.10	12 300	8 200	370	450
C30	11.0	6.6	8.6	2.7	2.2	1.20	12 300	8 200	380	460
C35	12.0	7.2	8.7	2.9	2.4	1.30	13 400	9 000	400	480
C40	13.0	7.8	8.7	3.0	2.6	1.40	14 500	10 000	420	500
D30	9.0	5.4	8.1	2.8	2.2	1.40	9 500	6 000	530	640
D35	11.0	6.6	8.6	3.4	2.6	1.70	10 000	6 500	560	670
D40	12.5	7.5	12.6	3.9	3.0	2.00	10 800	7 500	590	700
D50	16.0	9.6	15.2	4.5	3.5	2.20	15 000	12 600	650	780
D60	18.0	10.8	18.0	5.2	4.0	2.40	18 500	15 600	700	840
D70	23.0	13.8	23.0	6.0	4.6	2.60	21 000	18 000	900	1080

Note: Strength classes C14 to C40 and TR 26 are for softwoods, D30 to D70 are for hardwoods.

*   When the specification specifically prohibits wane at bearing areas, the higher values of compression perpendicular to grain stress may be used, otherwise the lower values apply.

†   The values of characteristic density given above are for use when designing joints. For the calculation of dead load, the average density should be used.

‡   The strength class TR26 is essentially for the manufacture of trussed rafters but may be used for other applications with the grade stresses and moduli given above. For joints, the tabulated permissible loads for strength class C27 should be used. When used with the provisos given in BS 5268: Part 3 the grade stresses are similar to the former M75 redwood/whitewood so timber and trussed rafter designs to this M75 grade/species combination are interchangeable with timber and trussed rafter designs using the TR26 strength class.

Table 3.8.3, extracted from BS 5268-2, with kind permission of the British Standards Institution. It provides a comprehensive comparison of timber strength properties and characteristics, which may be used as a basis for structural design and timber selection.

Note: Service classes 1 and 2 in Table 3.8.3 refer to timber used in moderate climates where ambient temperature is about 20 °C and relative humidity does not exceed 65% and 85% respectively for more than a few weeks per year. Respective moisture contents will not exceed 12% and 20%. Service class 3 exists where climatic conditions impose higher moisture contents than service class 3.

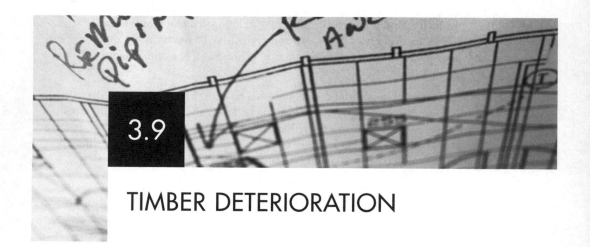

3.9

TIMBER DETERIORATION

Timber is a very resilient and robust material in normal atmospheric conditions. It can even retain its structure in a permanently wet environment, as evidenced by stanchions and wooden pier foundations that have remained intact for centuries. As well as being very durable, timber will outlast many other building materials succumbing to the effects of rain, frost and chemicals. However, the weakness of both hard and softwoods is its source as a food for plant growths in the form of fungi and to certain species of insect. With each, enzymes digest the cellulose fibres and lignin adhesive comprising the structure of wood.

▦▪■ FUNGAL ATTACK

Fungi, unlike other plant growths, have no leaves, and do not require chlorophyll or sunlight, but will readily consume organic material. For a fungal growth on timber to succeed, the timber will require a moisture content in excess of 20% (see Chapter 3.8). Generally, softwoods for internal structural use will be commercially seasoned to a moisture content of about 12–16%. Internal softwood for joinery and hardwoods will be seasoned to a lower figure of around 10%.

Certain species of fungus will attack trees during growth, but these are an issue for the timber producer. The principal concern for builders, building owners and users is the fungi known to develop on seasoned timber in service.

CATEGORIES OF ROT ATTRIBUTED TO FUNGI

Brown rot

So named because the wood becomes dark in colour. It is further characterised by the timber drying and cracking to produce small surface squares, manifesting as cubes within the depth of the wood. This shrinking is caused by destruction of the cellulose tissues.

White rot

This leaves the wood pale in colour, soft, and fibrous to touch. The timber structure decays as both cellulose tissues and lignin adhesive are destroyed.

DRY AND WET ROT

Generally, timber deterioration due to fungal attack is more conveniently categorised as dry or wet rot, depending on the extent of exposure to dampness. The initial indication of rot infestation is usually a stale or musty smell emanating from a damp source below floors, in cellars or within a roof space. Closer investigation often reveals whitish/grey fungal plant growths and possibly coloured fruiting bodies depending on how far the fungus has advanced.

Dry rot

Dry rot is known biologically by the Latin name *Serpula lacrymans* or *Merulius lacrymans*. It is a 'brown rot', leaving the timber dry and friable. Fungal growth is caused when red/rusty-coloured spores from an established fungus drift through the air to settle and germinate on damp timber. The spores develop into white strands (hyphae) appearing as cotton–wool-like patches (mycelium) growing flesh-textured fruiting bodies (sporophores), in turn producing more spores.

Early detection can eradicate the problem. The source of dampness must be removed or rectified. Undetected leaking plumbing is often the cause, or condensation due to inadequate underfloor or roof space ventilation. Eliminating the initial dampness may not always eradicate dry rot, as it can still thrive by developing small vein-like tubing of 2–3 mm diameter (rhizomorphs) to extract dampness from areas adjacent to the timber food source. This can include moisture in brickwork, render, plaster or concrete. Moisture from the air can be sufficient for dry rot to live, particularly in areas of high humidity.

Optimum growth conditions are a combination of dampness and warmth. Temperatures between 13 °C and 24 °C are ideal. At freezing temperatures the fungus becomes dormant, but it will die in temperatures above 40 °C.

Treatment of dry rot
- Eliminate the source of dampness.
- Dry out the building area affected.
- Cut out all affected timber at least 500 mm beyond the decay.
- Remove all affected plaster and other finishes within the vicinity of attack.
- Sterilise the affected area by applying heat from a blowtorch (carefully!) to all adjacent masonry, concrete and sound plaster.
- Where the attack is severe, drill close-spaced inclined holes of about 12 mm diameter into the adjacent masonry and liberally feed the holes with patent fungicide to saturate the structure.
- Make good jointing and pointing to masonry and plasterwork with a zinc oxychloride additive or similar fungicide in the mix. Specialist paints containing fungicide are also available.
- Replace all affected timber with well-seasoned, preservative-treated wood.

Following on from remedial work, regular monitoring is essential for several months to ensure successful eradication of dry rot. Even though the source of dampness may be removed, dry rot has an irritating habit of reappearing, somehow managing to thrive on only a nominal amount of moisture.

Wet rot

The most common types of wet rot fungus in the UK are known biologically by the Latin names of *Coniophora puteana* or *Coniophora cerebella* and *Phellinus contiguus* or *Poria contigua*. The growth and development of both fungi are similar to those of dry rot, but fruiting bodies are rare.

Coniophora, otherwise known as cellar fungus, prefers very damp conditions. It is a 'brown rot', splitting the internal structure of the host timber longitudinally and laterally into small cube shapes, but leaving the surface largely intact. The timber crumbles when prodded, and it is dull dark brown or black in colour.

Phellinus has become more associated with decay in external joinery. It is a 'white rot', attacking poor-quality sapwood where used in door and window frames, fascia boards, cladding, etc. Inadequate treatment of the timber, poor jointing techniques, inappropriate adhesives and lack of aftercare will all contribute to rainwater penetration. Wood in these situations is often painted, and the first sign of fungal decay is usually surface irregularities. Closer inspection will reveal splitting within the body of wood, breaking into soft strands.

Optimum growth conditions are as for dry rot, but wet rots are more easily controlled as they require greater exposure to dampness to thrive. Hence, once the moisture source is removed, the decayed timber can be replaced. Adjacent timber, brickwork, plaster, etc. are unlikely to be affected, but as a precaution the whole area should be brush-treated with a fungicide.

■ ■ ■ INSECT ATTACK

SPECIES

There are several species of wood-boring insect that are capable of seriously damaging structural timber and joinery in buildings. During their relatively short time in adult form they are classified as beetles. Most of their life is in the larval stage, gnawing and burrowing through timber – hence the general descriptive term of wood borer or woodworm. Most species of larvae have a preference for the less dense sapwood growth areas as their source of food.

The most common species in the UK are:

- common furniture beetle (*Anobium punctatum*);
- death watch beetle (*Xestobium rufovillosum*);
- powder post beetle (*Lyctus* family);
- house longhorn beetle (*Hylotrupes bajulus*).

Figure 3.9.1 compares the species proportionally. The largest is the adult house longhorn beetle at about 25 mm overall.

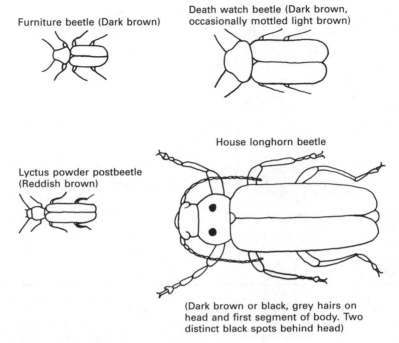

Furniture beetle (Dark brown)

Death watch beetle (Dark brown, occasionally mottled light brown)

House longhorn beetle

Lyctus powder postbeetle
(Reddish brown)

(Dark brown or black, grey hairs on head and first segment of body. Two distinct black spots behind head)

Figure 3.9.1   Common wood-boring beetles.

HABITAT

Most species prefer a slightly damp, draught-free environment. Unventilated roof spaces and recesses behind eaves cupboards are ideal areas in which to thrive. Other areas suitable to the lifestyle of the woodworm include understair cupboards and voids within enclosed baths and other sanitary fittings. Timber of intermediate floors in housing is also vulnerable to attack, but raised timber ground floors are less vulnerable if peripheral air vents remain clear for air circulation.

LIFE CYCLE

Wood-boring insects are most active during the warmer spring and summer months. The life cycle shown in Fig. 3.9.2 is similar for all species, progressing from egg, through larva (grub) and pupa (chrysalis), to adult (beetle). Adult female beetles seek rough crevices of sawn timber or former borehole exits to deposit their eggs. As the larvae hatch they bore into the wood, using it as food and shelter. The tunnelling effect of hundreds of larvae from each batch of eggs can be extremely damaging. The larval stage is the predominant part of an insect's life, extending for several years before maturing to a chrysalis. The chrysalis develops into an adult beetle just beneath the timber surface, from where it emerges to reproduce, lay eggs and generate more damage. Table 3.9.1 provides some comparison of the behavioural characteristics of species.

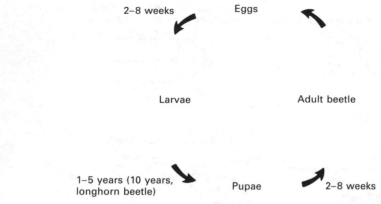

Figure 3.9.2   Wood borers – life cycle.

Table 3.9.1   Common wood borers

Name	Egg quantity	Egg maturity	Laval period	Pupal stage	Emergence as a beetle	Location	Exit holes
Furniture beetle	20–50	4–5 weeks	2–3 years	4–8 weeks	May–Sept	Sapwood of soft or hard wood	1–2 mm dia.
Death watch beetle	40–80	2–8 weeks	4–6 years	2–4 weeks	Mar–June	Hardwoods, preferably decayed oak	3 mm dia.
Powder post beetle	70–200	2–3 weeks	1–2 years	3–4 weeks	May–Sept	Sapwood of new hardwood	1–2 mm dia.
Longhorn beetle	< 200	2–3 weeks	3–10 years	About 3 weeks	July–Sept	Sapwood of softwood	5 mm × 10 mm oval

Note: Periods given will vary relative to ambient temperatures.

RECOGNITION AND TREATMENT

Woodworm is usually first recognised by the distinctive flight holes and powdery deposits (**frass**) on the surface of timber. By this stage the internal structure of the timber may have suffered considerable damage. The extent of damage may be established by chiselling away the timber surface to examine the borings or galleries produced by the larvae.

If the damage is slight and there are only a few exit holes, remedial spraying or brushing with liberal applications of a proprietary insecticide will control the problem. Structural timbers seriously damaged must be removed and burnt. All new timber must be pre-treated, preferably with a vacuum/pressure-impregnated insecticide sourced from a commercial timber supplier.

In new construction, prevention of woodworm infestation by using pre-treated timber is the only effective method of control. The Building Regulations, Approved Document A: *Structure* (Section 2B) provides specific reference and guidance for the use of treated timber for roof construction in parts of Surrey and adjacent areas. This requirement originated several decades ago following a very serious attack by the house longhorn beetle on roof timbers in Camberley.

Reference sources for acceptable methods of timber preservation include:

- The British Wood Preserving and Damp-Proofing Association's Manual;
- BS 5589: *Code of practice for preservation of timber*;
- BS 5268-5: *Structural use of timber. Code of practice for the preservative treatment of structural timber*;
- BS 5707: *Specification for preparations of wood preservatives in organic solvents*.

3.10

STEEL-FRAMED HOUSING

Steel-framed housing is similar to timber frame in that standard sections are used to make up a structural frame. The frame is lined with plasterboard and infilled with insulating material. An outer leaf of brick cladding retains a traditional image and protection from the elements.

Advantages
- dimensionally stable (timber warps);
- will not rot;
- will not burn;
- does not absorb moisture;
- unpalatable to insects and animals;
- predictable consistent expansion and movement characteristics;
- requires only semi-skilled labour to assemble;
- components factory-manufactured to quality-controlled standards;
- relatively large span to weight/size possible.

Disadvantages
- unpredictable behaviour in fire;
- risk of corrosion.

■■■ BACKGROUND

Steel-framed dwellings were first built in the UK c.1920. They did not catch on, and were soon displaced as 'breeze' block and brick technology developed. During the post World War II building boom, when traditional building materials were scarce and demand for housing was high, there was a resurgence of interest. At this time a variety of structural steel frame systems for housing were produced, but they were labour intensive, and components were costly. Technically the concept was sound, and these buildings have performed well in the long term.

■■■ DEVELOPMENT

Apart from some ventures into the manufacture of prefabricated modular units, timber frame has been the only real alternative to traditional masonry construction. However, since about 1990 there have been many new initiatives based on the use of economical cold-formed steel sections (see Part 10) for fabrication of simple frames. This has provided a cost-competitive alternative to timber frame and traditional masonry construction.

■■■ COMPONENTS

WALLS

Basically a stud frame of cold-formed galvanised steel channels produced by rolling, notching, holing and folding. These can be delivered to site for cutting and bending to shape, and fixing with self-drilled, self-tapping screws. Generally, panels are made up in a factory to modular sizes (600 mm) and assembled on site to make up a structural frame.

ROOF TRUSSES

Channel sections can be used, but some manufacturers prefer sigma profile sections for roof members.

COMPONENT SIZES

Sizes differ marginally from various system producers, but are generally as indicated in Fig. 3.10.1. Figures 3.10.2 and 3.10.3 show typical construction details.

SUPPLEMENTARY EARTH BONDING

Otherwise known as cross-bonding. The steel frame must not be used for local earthing of electrical services. The frame is regarded as extraneous metalwork and could conduct electricity if contacted by an electrical fault. It must be earthed at one point and all electrical earth connections in the circuit wired back to this point.

**Prefabricated galvanised steel
channel section (typical dimensions)**

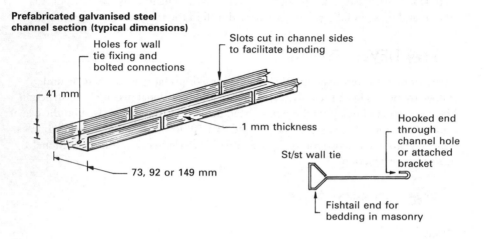

Holes for wall
tie fixing and
bolted connections

Slots cut in channel sides
to facilitate bending

41 mm

1 mm thickness

73, 92 or 149 mm

St/st wall tie

Hooked end
through
channel hole
or attached
bracket

Fishtail end for
bedding in masonry

Stud frame

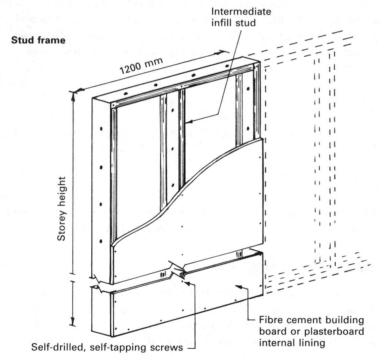

Intermediate
infill stud

1200 mm

Storey height

Self-drilled, self-tapping screws

Fibre cement building
board or plasterboard
internal lining

Figure 3.10.1   Steel frame sections.

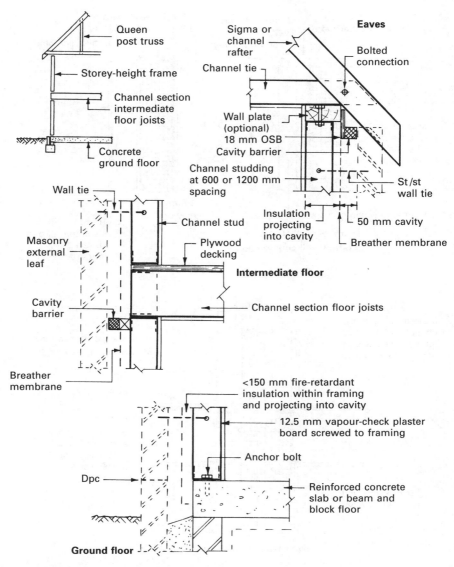

Figure 3.10.2 Typical steel frame construction.

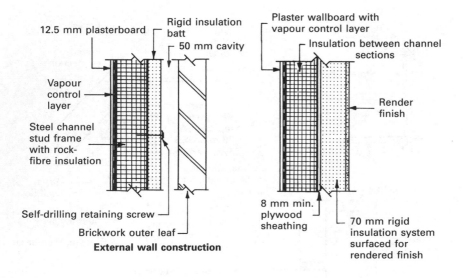

12.5 mm plasterboard

Rigid insulation batt

50 mm cavity

Vapour control layer

Steel channel stud frame with rock-fibre insulation

Self-drilling retaining screw

Brickwork outer leaf

External wall construction

Plaster wallboard with vapour control layer

Insulation between channel sections

Render finish

8 mm min. plywood sheathing

70 mm rigid insulation system surfaced for rendered finish

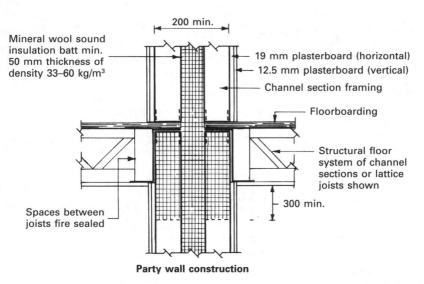

200 min.

Mineral wool sound insulation batt min. 50 mm thickness of density 33–60 kg/m³

19 mm plasterboard (horizontal)

12.5 mm plasterboard (vertical)

Channel section framing

Floorboarding

Structural floor system of channel sections or lattice joists shown

300 min.

Spaces between joists fire sealed

Party wall construction

Figure 3.10.3   Steel frame – further details.

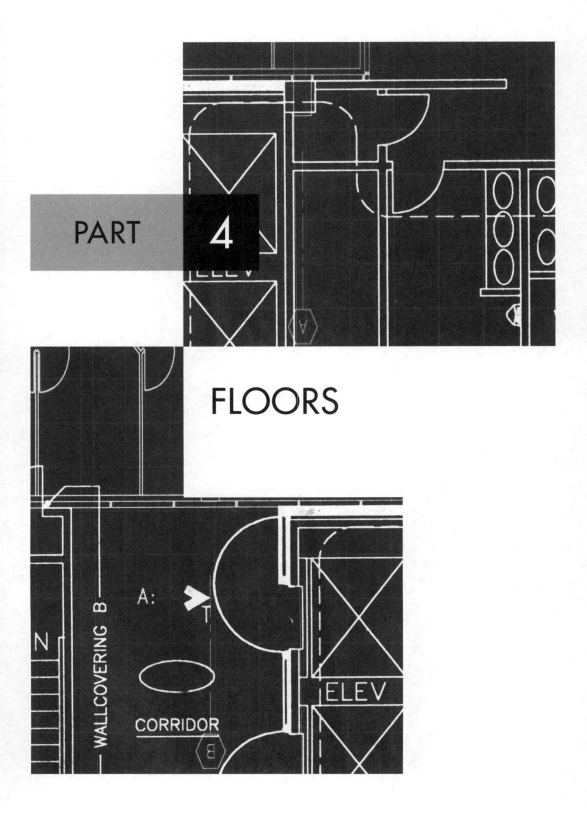

PART 4

FLOORS

4.1

SOLID CONCRETE GROUND FLOOR CONSTRUCTION

The construction of a solid ground floor can be considered under three headings:

- hardcore;
- blinding;
- concrete bed or slab.

■■■ HARDCORE

The purpose of hardcore is to fill in any small pockets that have formed during oversite excavations, to provide a firm base on which to place a concrete bed and to help spread any point loads over a greater area. It also acts against capillary action of moisture within the soil. Hardcore is usually laid in 100–150 mm layers to the required depth, and it is important that each layer is well compacted, using a roller if necessary, to prevent any unacceptable settlement beneath the solid floor.

Approved Document C recommends that no hardcore laid under a solid ground floor should contain water-soluble sulphates or other harmful matter in such quantities as to be liable to cause damage to any part of the floor. This recommendation prevents the use of any material that may swell upon becoming moist, such as colliery shale, and furthermore it is necessary to ascertain that brick rubble from demolition works and clinker furnace waste intended for use as hardcore does not have any harmful water-soluble sulphate content.

■■■ BLINDING

This is used to even off the surface of hardcore if a damp-proof membrane is to be placed under the concrete bed or if a reinforced concrete bed is specified. First, it will prevent the damp-proof membrane from being punctured by the hardcore and, second, it will provide a true surface from which the reinforcement can be positioned. Blinding generally consists of a layer of sand 25–50 mm thick or a 50–75 mm layer of weak concrete (1:12 mix usually suitable) if a true surface for reinforced concrete is required.

■■■ CONCRETE BED

Thicknesses generally specified are:

- unreinforced or plain *in-situ* concrete, 100–150 mm thick;
- reinforced concrete, 150 mm minimum.

Suitable concrete mixes are produced to BS EN 206-1: *Concrete. Specification, performance, production and conformity*:

- plain *in-situ* concrete – 50 kg cement:0.11 m^3 fine aggregate:0.16 m^3 coarse aggregate or mix specification ST2;
- reinforced concrete – 50 kg cement:0.08 m^3 fine aggregate:0.13 m^3 coarse aggregate or mix specification ST4.

The reinforcement used in concrete beds for domestic work is usually in the form of a welded steel fabric to BS 4483. Sometimes a light square mesh fabric is placed 25 mm from the upper surface of the concrete bed to prevent surface crazing and limit the size of any cracking.

In domestic work the areas of concrete are defined by the room sizes, and it is not usually necessary to include expansion or contraction joints in the construction of the bed.

PROTECTION OF FLOORS NEXT TO THE GROUND

Building Regulation C2 requires that such part of a building as is next to the ground shall have a floor so constructed as to prevent the passage of moisture from the ground to the upper surface of the floor. The requirements of this regulation can be properly satisfied only by the provision of a suitable barrier in the form of a damp-proof membrane within the floor. The membrane should be turned up at the edges to meet and blend with the damp-proof course in the walls to prevent any penetration of moisture by capillary action at edges of the bed.

Suitable materials for damp-proof membranes are:

- polyethylene (LDPE) 1200 gauge (0.3 mm) sheet with sealed joints, which is acceptable and will also give protection against moisture vapour as well as moisture;
- hot-poured bitumen, which should be at least 3 mm thick;
- cold-applied bitumen/rubber emulsions, which should be applied in not less than three coats;
- asphalt/pitchmastic, which could be dual-purpose finish and damp-proof membrane.

Note: Prevention of the ingress of radon and/or methane gases from the ground will require a wire- or fibre-reinforced LDPE membrane of up to 1 mm thickness. Radon is a naturally occurring radioactive gas originating from uranium and radium deposits in certain rock subsoils found primarily in parts of the West Country, northern England and areas of Scotland. Methane is an explosive gas, which can build up in the ground as a result of deposited decaying organic materials. Where either of these gases is prevalent, it is now standard practice to construct suspended precast concrete floors with natural draught underfloor ventilation – see Chapter 4.5.

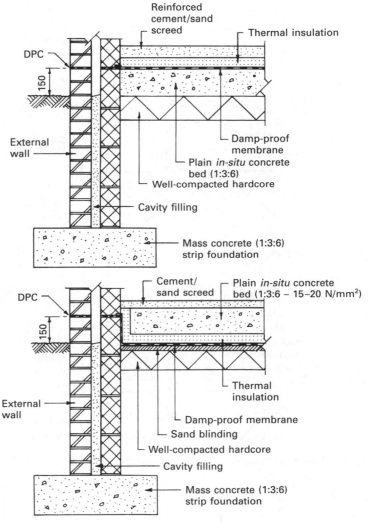

Figure 4.1.1   Typical solid floor details at external walls.

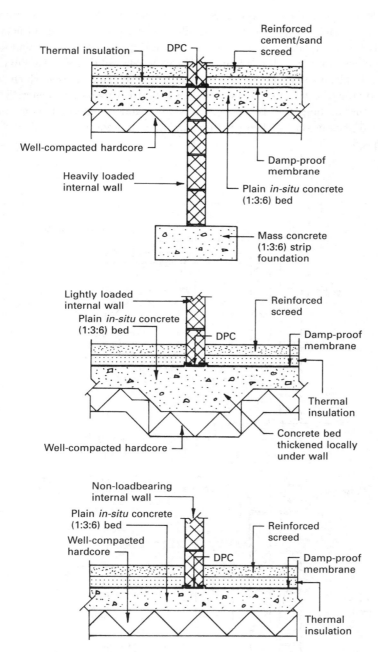

Figure 4.1.2 Typical solid floor details at internal walls.

The position of a damp-proof membrane, whether above or below the concrete bed, is a matter of individual choice. A membrane placed above the bed is the easiest method from a practical aspect and is therefore generally used. A membrane placed below the bed has two advantages: first, it will keep the concrete bed dry and in so doing will make the bed a better thermal insulator and, second, during construction it will act as a separating layer preventing leakage of the cement matrix into the hardcore layer, which could result in a weak concrete mix. Typical details of solid floor construction are shown in Figs 4.1.1 and 4.1.2.

4.2

SUSPENDED CONCRETE GROUND FLOOR CONSTRUCTION

■ ■ ■ BEAM AND BLOCK FLOOR

This type of domestic floor system is derived from the principles of the precast hollow and composite floor systems used for commercial buildings and apartments, as detailed in Chapter 4.5. It has developed into a cost- and time-effective means of constructing domestic ground and upper floors, by incorporating precast concrete beams with lightweight concrete blocks as an infilling. The benefits of quality-controlled factory manufacture of components and simple site assembly with the aid of a mobile crane to hoist the beams add to the following advantages:

■ potential to span over unsound infilling, common to sloping sites.
■ application over movable subsoils such as shrinkable clay (see Chapter 2.7).
■ suitability where ventilation under the ground floor is required to dilute intrusive gases.

The Building Regulations, Approved Document C, requires a minimum clear void depth of 75 mm below these floors, but it is usual practice to leave at least 150 mm. Ventilation of the void is advisable to dilute and prevent concentration of gases from the ground (radon and/or methane) and possible leakage from piped services. Figure 4.2.1 shows typical construction of a domestic suspended beam and block floor, where the stripped topsoil leaves the underfloor surface lower than adjacent ground. This is acceptable only if the soil is free draining. Also, ground differentials should be minimal, otherwise the external wall becomes a retaining wall and will require specific design calculations. All organic material should be removed from the void, and the surface should be treated with weedkiller. Void depth may need to be as much as 225 mm in the presence of heavy clay subsoil and nearby trees (see Chapter 2.7).

Upper floors follow the same principles of assembly, with purpose-made trimmer shoes providing support to concrete beams around stair openings. Span potential is

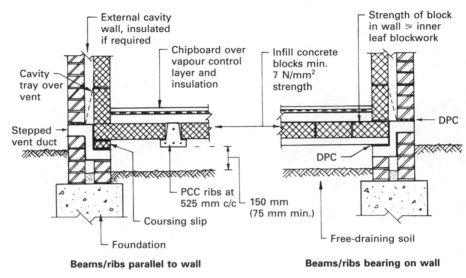

Beams/ribs parallel to wall   Beams/ribs bearing on wall

Figure 4.2.1   Beam and block domestic ground floor.

only about 5 m: therefore intermediate support from a loadbearing partition or a steel beam is acceptable. Direct flange bearing is possible, but if a deep section is required the floor structure may be accommodated on ledger angles. Typical upper floor details are shown in Fig. 4.2.2.

■ ■ ■ BEAM AND EXPANDED POLYSTYRENE (EPS) BLOCK

Precast concrete beams with EPS block infill units have developed from the beam and block principles applied to domestic floor construction. As a construction technique it has the advantages of speed and simplicity with exceptional thermal performance. Thermal insulation U values for the floor as a whole are about 0.20 W/m^{2}K, depending on the thickness and amount of EPS relative to rib spacing. The system is in effect a structurally adequate floor, with integral insulation.

The construction principles are the same as described for suspended beam (rib) and block, with some variation on rib spacing to suit EPS block width. Figures 4.2.3 and 4.2.4 show different block forms and applications. Figure 4.2.3 shows EPS units functioning as both insulation and permanent shuttering to an in-situ reinforced concrete diaphragm suspended ground floor. Figure 4.2.4 shows typical dry construction, using a moisture-resistant chipboard surface finish.

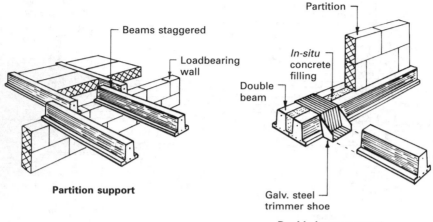

Partition

Beams staggered

Loadbearing wall

In-situ concrete filling

Double beam

Partition support

Galv. steel trimmer shoe

Double beam support

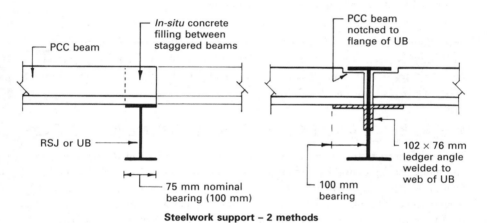

PCC beam

In-situ concrete filling between staggered beams

RSJ or UB

75 mm nominal bearing (100 mm)

PCC beam notched to flange of UB

102 × 76 mm ledger angle welded to web of UB

100 mm bearing

Steelwork support – 2 methods

Figure 4.2.2   Beam and block – upper floor intermediate support.

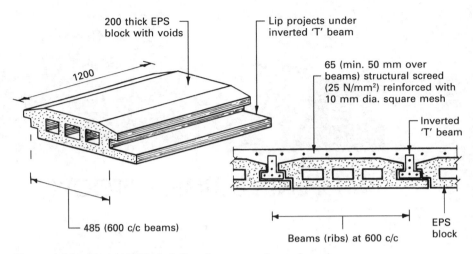

Figure 4.2.3   Typical EPS block floor for structural screed topping.

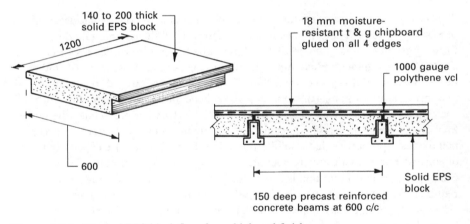

Figure 4.2.4   Typical EPS block floor for a chipboard finish.

4.3

SUSPENDED TIMBER FLOORS

■■■ SUSPENDED TIMBER GROUND FLOORS

This type of floor consists of timber boards or other suitable sheet material fixed to joists spanning over sleeper walls, and was until 1939 the common method of forming ground floors in domestic buildings. The Second World War restricted the availability of suitable timber, and solid ground floors replaced suspended timber floors. Suspended precast concrete floors have gained popularity in recent times, not only as a precaution against radon and methane, but as a less expensive form of construction. These are considered in more detail in Chapter 4.2. Today the timber floor is still used on occasions because it has some flexibility and will easily accept nail fixings – properties that a solid ground floor lacks. It is a more expensive form of construction than a concrete floor and can only be justified to match existing construction where a building is extended. It could be used on sloping sites that require a great deal of filling to make up the ground to the specified floor level, although a precast concrete flooring system could also be used, and this is likely to be much cheaper.

Suspended timber ground floors are susceptible to dry rot and draughts, and are said to be colder than other forms of flooring. If the floor is correctly designed and constructed these faults can be eliminated.

The problem of dry rot, which is a fungus that attacks damp timber, can be overcome by adequate ventilation under the floor and the correct positioning of damp-proof courses to keep the under floor area and timber dry. Through-ventilation is essential to keep the moisture content of the timber below that which would allow fungal growth to take place: that is, 20% of its oven-dry weight. The usual method is to allow a free flow of air under the floor covering by providing air bricks in the external walls. These are sited near the corners and at approximately 2 m centres around the perimeter of the building. They must have an equivalent of 1500 mm^2/m run of wall; alternatively, 500 mm^2/m^2 of floor area will also be acceptable (take greater value). If a suspended timber floor is used with an adjacent solid ground floor, pipes of 100 mm diameter are used under the solid floor to convey air to and from the external walls to the suspended floor.

▨ ■ ■ BUILDING REGULATIONS

Building Regulation C2 applies as with solid floors and recommended provisions are given in Approved Document C. Figure 4.3.1 shows the minimum dimensions recommended in Approved Document C but in practice a greater space between the concrete bed and the timber is usual. The honeycomb sleeper walls are usually built two or three courses high to allow good through-ventilation. Sleeper walls spaced at 2000 mm centres will give an economic joist size. The width of joists is usually taken as 50 mm: this will give sufficient width for the nails securing the covering, and the depth can be obtained by reference to design tables recommended in Approved Document A or by design calculations. The usual joist depth for domestic work is 125 mm.

LAYOUT

The most economic layout is to span the joists across the shortest distance of the room: this means that joists could be either parallel or at right angles to a fireplace. The fireplace must be constructed of non-combustible materials and comply with Building Regulation J3. Typical examples are shown in Figs 4.3.2 and 4.3.3.

INSULATION IN GROUND FLOORS

Rigid polystyrene or high-density mineral wool can be incorporated within a solid floor structure to comply with the Building Regulations Approved Document L: *Conservation of fuel and power*. Location may be above the damp-proof membrane and below the concrete slab, or between the screed and concrete slab. If the latter, the screed should be at least 75 mm thickness and wire reinforced against cracking. Suspended timber floors can incorporate mineral fibre insulation draped on nylon netting between the joists, or have rigid insulation boards spanning the gap between joists. More detail of thermal insulation in floors is provided in Chapter 8.2.

■ ■ ■ SUSPENDED TIMBER UPPER FLOORS

Timber, being a combustible material, is restricted by Part B of the Building Regulations to small domestic buildings as a structural flooring material. Its popularity in this context is due to its low cost in relationship to other structural flooring methods and materials. Structural softwood is readily available at a reasonable cost, is easily worked and has a good strength to weight ratio, and is therefore suitable for domestic loadings.

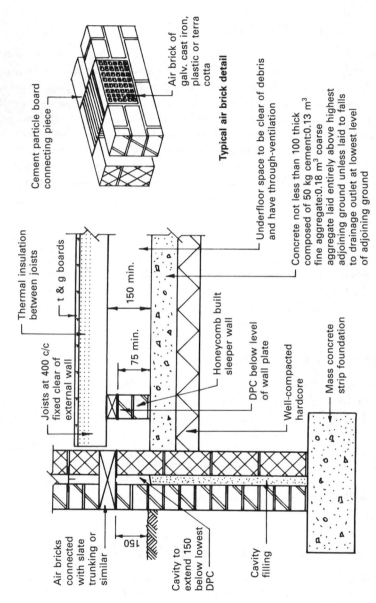

Typical air brick detail

Air brick of galv. cast iron, plastic or terra cotta

Cement particle board connecting piece

Underfloor space to be clear of debris and have through-ventilation

Concrete not less than 100 thick composed of 50 kg cement:0.13 m³ fine aggregate:0.18 m³ coarse aggregate laid entirely above highest adjoining ground unless laid to falls to drainage outlet at lowest level of adjoining ground

Thermal insulation between joists

t & g boards

150 min.

75 min.

Joists at 400 c/c fixed clear of external wall

Honeycomb built sleeper wall

DPC below level of wall plate

Well-compacted hardcore

Mass concrete strip foundation

Air bricks connected with slate trunking or similar

150

Cavity to extend 150 below lowest DPC

Cavity filling

Figure 4.3.1 Building Regulations and suspended timber floors.

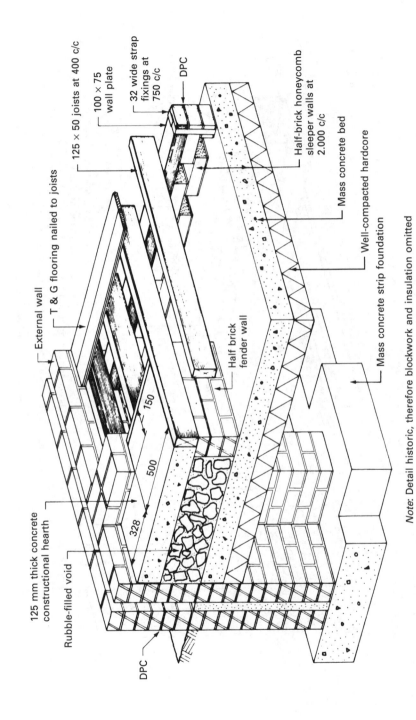

External wall

T & G flooring nailed to joists

125 × 50 joists at 400 c/c

100 × 75 wall plate

32 wide strap fixings at 750 c/c

DPC

Half-brick honeycomb sleeper walls at 2.000 c/c

Mass concrete bed

Well-compacted hardcore

Mass concrete strip foundation

Half brick fender wall

150

500

328

125 mm thick concrete constructional hearth

Rubble-filled void

DPC

Note: Detail historic, therefore blockwork and insulation omitted

Figure 4.3.2 Typical details of suspended floor – joists parallel to fireplace.

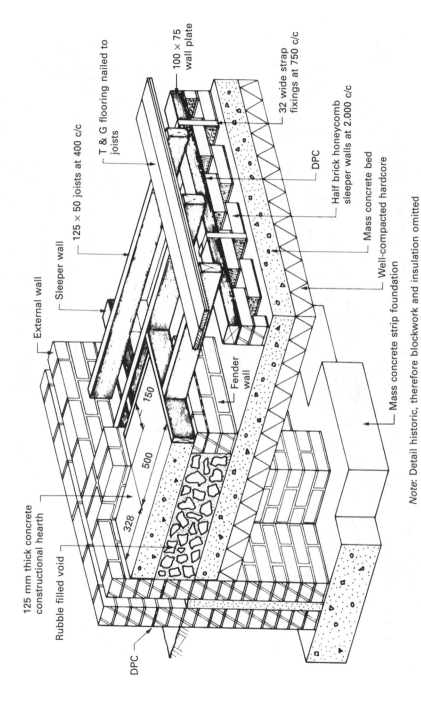

External wall

Sleeper wall

T & G flooring nailed to joists

125 × 50 joists at 400 c/c

100 × 75 wall plate

32 wide strap fixings at 750 c/c

DPC

Half brick honeycomb sleeper walls at 2.000 c/c

Mass concrete bed

Well-compacted hardcore

Mass concrete strip foundation

Fender wall

150

500

328

125 mm thick concrete constructional hearth

Rubble filled void

DPC

Note: Detail historic, therefore blockwork and insulation omitted

Figure 4.3.3   Typical details of suspended timber floor – joists at right angles to fireplace.

Terminology
- **Common joist** A joist spanning from support to support.
- **Trimming joist** Span as for common joist, but it is usually 25 mm thicker and supports a trimmer joist.
- **Trimmer joist** A joist at right angles to the main span supporting the trimmed joists; is usually 25 mm thicker than a common joist.
- **Trimmed joist** A joist cut short to form an opening, and supported by a trimmer joist; it spans in the same direction as common joists and is of the same section size.

JOIST SIZING

There are three ways of selecting a suitable joist size for supporting a domestic type floor:

1. Rule of thumb for joists of 50 mm width, spaced at 400 mm centres:

$$\frac{\text{span in mm}}{24} + 50 \text{ mm} = \text{depth in mm}$$

For example, for a 3.6 m span:

$$\frac{3600}{24} + 50 = 170 \text{ mm}$$

Therefore commercial size of joist chosen is 175 mm × 50 mm @ 400 mm c/c.

Some comparisons can be made with the methods shown in 2 and 3. However, the 'rule' is limited as a simple means of guidance for joist spacing at 400 mm only.

2. Calculation:

$$\text{BM} = \frac{fbd^2}{6} = \frac{WL}{8}$$

where BM = bending moment
f = maximum fibre stress in N/mm^2 (see Chapter 3.8)
b = breadth in mm
d = depth in mm
W = total load on joist in newtons (N)
L = clear length or span of joist in mm

For example, see Fig. 4.3.4.

Loading on each joist:
Imposed load = 4.5 m × 0.45 m × 1.5 kN = 3.04 kN
Dead loads:
Floorboards (4.5 m × 0.45 m × 10 kg)
+ Plasterboard (4.5 m × 0.45 m × 11 kg)
= 42.53 kg × 9.81 × 10^{-3} = 0.42 kN

Total loading on each joist = 3.04 kN + 0.42 kN = 3.46 kN (3460 N)

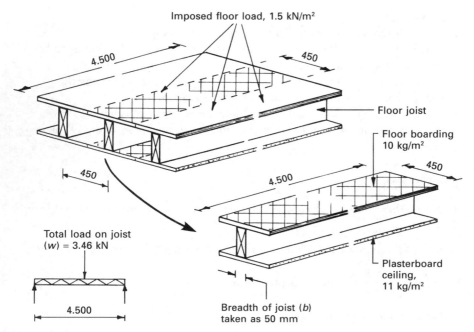

Figure 4.3.4   Joist calculations.

Note: The self-weight of the joist is usually omitted, as this is small compared with the load carried. Moreover, the commercial joist size selected is normally greater than the calculated size, and this will cover the self-weight.

$$\frac{fbd^2}{6} = \frac{WL}{8}$$

From Table 3.8.3, GS (general structural) grade timber of strength class C16 has a fibre stress (f) of 5.3 N/mm^2.

Transposing the bending moment formula to make d the subject and taking b, the breadth, as 50 mm:

$$d^2 = \frac{6WL}{8fb} = \frac{6 \times 3460 \times 4500}{8 \times 5.3 \times 50} = 44\ 066$$

$$d = \sqrt{44\ 066} = 210 \text{ mm}$$

225 mm is the nearest commercial timber size above 210 mm. Therefore 225 mm × 50 mm @ 450 mm spacing will be selected. 225 mm × 50 mm in Table 4.3.1 is comparable with a maximum clear span of 4.47 m for joists at 450 mm spacing.

3. Building Regulations. Approved Document A: *Timber intermediate floors for dwellings* and *Span tables for solid timber members*, both published by TRADA.

Table 4.3.1 provides some guidance where selection of timber is general structural (GS).

Table 4.3.1 Guide to span and loading potential for GS grade timber floor joists.

	Dead weight of floor and ceiling, excluding self-weight of joist (see notes)								
	< 25 kg/m² < 0.25 kN/m²			25–50 0.25–0.50			50–125 0.50–1.25		
	Spacing of joists mm (c/c)								
Sawn size of joist (mm)	400	450	600	400	450	600	400	450	600
	Maximum clear span (m)								
38 × 100	1.83	1.69	1.30	1.72	1.56	1.21	1.42	1.30	1.04
38 × 125	2.48	2.39	1.93	2.37	2.22	1.76	1.95	1.79	1.45
38 × 150	2.98	2.87	2.51	2.85	2.71	2.33	2.45	2.29	1.87
38 × 175	3.44	3.31	2.87	3.28	3.10	2.69	2.81	2.65	2.27
38 × 200	3.94	3.75	3.26	3.72	3.52	3.06	3.19	3.01	2.61
38 × 225	4.43	4.19	3.65	4.16	3.93	3.42	3.57	3.37	2.92
50 × 100	2.08	1.97	1.67	1.98	1.87	1.54	1.74	1.60	1.29
50 × 125	2.72	2.62	2.37	2.60	2.50	2.19	2.33	2.17	1.77
50 × 150	3.27	3.14	2.86	3.13	3.01	2.69	2.81	2.65	2.27
50 × 175	3.77	3.62	3.29	3.61	3.47	3.08	3.21	3.03	2.63
50 × 200	4.31	4.15	3.73	4.13	3.97	3.50	3.65	3.44	2.99
50 × 225	4.79	4.66	4.17	4.64	4.47	3.91	4.07	3.85	3.35
63 × 100	2.32	2.20	1.92	2.19	2.08	1.82	1.93	1.84	1.53
63 × 125	2.93	2.82	2.57	2.81	2.70	2.45	2.53	2.43	2.09
63 × 150	3.52	3.39	3.08	3.37	3.24	2.95	3.04	2.92	2.58
63 × 175	4.06	3.91	3.56	3.89	3.74	3.40	3.50	3.37	2.95
63 × 200	4.63	4.47	4.07	4.44	4.28	3.90	4.01	3.85	3.35
63 × 225	5.06	4.92	4.58	4.91	4.77	4.37	4.51	4.30	3.75
75 × 125	3.10	2.99	2.72	2.97	2.86	2.60	2.68	2.58	2.33
75 × 150	3.72	3.58	3.27	3.56	3.43	3.13	3.22	3.09	2.81
75 × 175	4.28	4.13	3.77	4.11	3.96	3.61	3.71	3.57	3.21
75 × 200	4.83	4.70	4.31	4.68	4.52	4.13	4.24	4.08	3.65
75 × 225	5.27	5.13	4.79	5.11	4.97	4.64	4.74	4.60	4.07

Notes:
1. The table allows for up to 1.5 kN/m² of imposed loading due to furniture and people. For imposed loading greater than this, calculations must be used.
2. For dead loading in excess of 1.25 kN/m² calculations must also be used.
3. Softwood floor boards should be at least 16 mm finished thickness for joist spacing up to 450 mm. 19 mm min. thickness will be required for joist spacing of 600 mm.
4. Joists should be duplicated below a bath.
5. Dead loading is usually summated in kg. To convert to a force in newtons (N), multiply by the gravitational factor 9.81 (say 10). For example: 25 kg × 10 = 250 N, or 0.25 kN.
6. See BS 648: *Schedule of weights of building materials* for material dead loads.

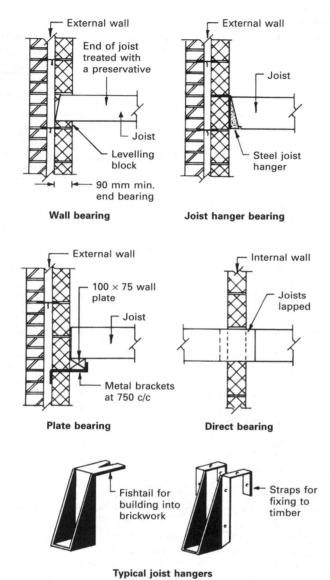

Figure 4.3.5 Typical joist support details.

JOISTS

If the floor is framed with structural softwood joists of a size not less than that required by the Approved Document, the usual width is taken as 50 mm. The joists are spaced at 400–600 mm centre to centre, depending on the width of the ceiling boards that are to be fixed on the underside. Maximum economy of joist size is obtained by spanning in the direction of the shortest distance to keep within the deflection limitations allowed. The maximum economic span for joists is between 3500 and 4500 mm; for spans over this a double floor could be used.

Support

The ends of the joists must be supported by loadbearing walls. The common methods are to build in the ends, or to use special metal fixings called **joist hangers**; other methods are possible but are seldom employed. Support on internal loadbearing walls can be by joist hangers or direct bearing when the joists are generally lapped (see Fig. 4.3.5).

Trimming

This is a term used to describe the framing of joists around an opening or projection. Various joints can be used to connect the members together, all of which can be substituted by joist hangers. Trimming around flues and upper floor fireplaces should comply with the recommendations of Approved Document J. It should be noted that, as central heating is becoming commonplace, the provision of upper-floor fireplaces is seldom included in modern designs, because they are considered to be superfluous. However, they are frequently encountered in renovation and maintenance work to older properties. Typical trimming joints and arrangements are shown in Figs 4.3.6 and 4.3.7.

Strutting

Shrinkage in timber joists will cause twisting to occur: this will result in movement of the ceiling below, and could cause the finishes to crack. To prevent this, strutting is used between the joists if the total span exceeds 2400 mm, the strutting being placed at mid-span (see Fig. 4.3.8).

DOUBLE FLOORS

These can be used on spans over 4500 mm to give a lower floor area free of internal walls. They consist of a steel beam or timber binder spanning the shortest distance, which supports common joists spanning at right angles. The beam reduces the span of the common joists to a distance that is less than the shortest span to allow an economic joist section to be used. The use of a timber binder was a popular method, but it is generally considered to be uneconomic when compared with a standard steel beam section. Typical details are shown in Fig. 4.3.9.

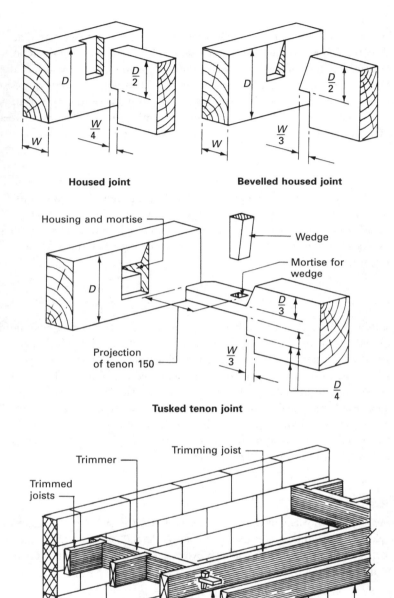

Housed joint

Bevelled housed joint

Tusked tenon joint

Typical stairwell trimming

Figure 4.3.6 Floor trimming joints and details.

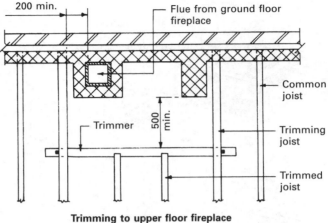

Trimming to upper floor fireplace

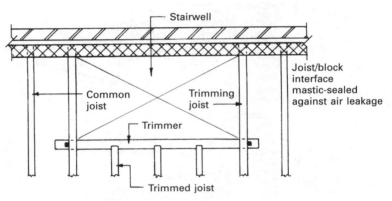

Trimming to stairwell

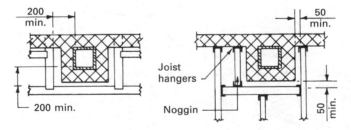

Trimming around flues

Note: Recess constructed of solid non-combustible material.

Figure 4.3.7 Typical trimming arrangements.

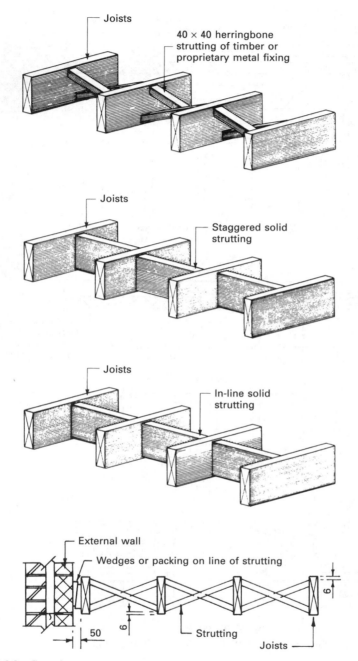

Figure 4.3.8   Strutting arrangements.

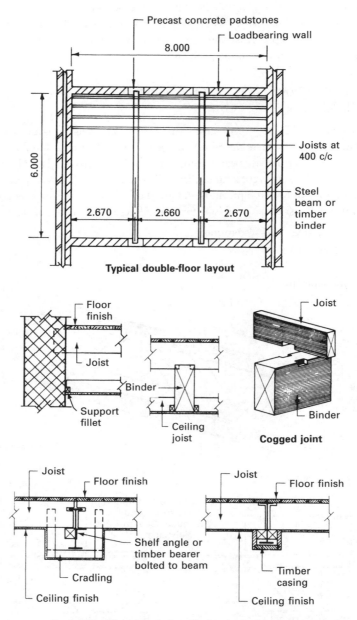

Figure 4.3.9   Double floor details.

If the span is such that a double floor is deemed necessary it would be a useful exercise to compare the cost with that of other flooring methods, such as *in-situ* reinforced concrete and precast concrete systems, which, overall, could be a cheaper and more practical solution to the problem.

COMPOSITE BEAMS/JOISTS FOR SUSPENDED FLOOR CONSTRUCTION

A double floor can be used for modest spans using a steel joist or large cross-section timber beam as intermediate floor joist support. For greater spans double flooring is uneconomic, as intermediate beams would need to be excessively large to carry the floor structure while still retaining minimal deflection (max. $0.003 \times$ span).

Spans up to about 8000 mm can be achieved cost-effectively by using composite beams at normal joist spacing of 400–600 mm. These are categorised as follows:

- open-web or steel web system joists;
- solid or boarded web joists.

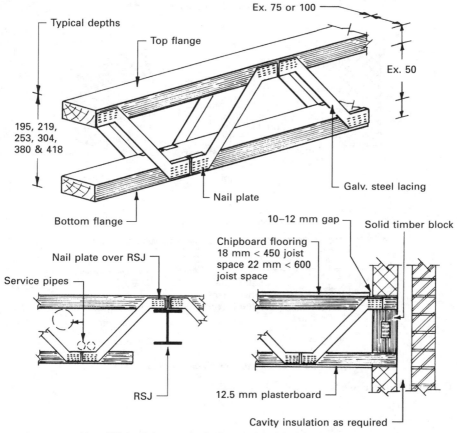

Note: Where joists are built directly into blockwork, they must be mastic sealed against air leakage.

Figure 4.3.10 Open-web joist.

Open-web joists

These are a type of lattice frame, as shown in Fig. 4.3.10. They consist of a pair of parallel stress-graded timber flanges, spaced apart with V-shaped galvanised steel web members. Open-web joists are manufactured off-site to specific spans in a factory quality-controlled environment.

Solid or boarded web joists

These are an alternative high-strength, relatively lightweight factory-produced joist system. A typical manufactured boarded web joist is shown in Fig. 4.3.11. This comprises a pair of stress-graded timber flanges separated by a central solid web of plywood or oriented strand board (OSB). At comparable joist depths,

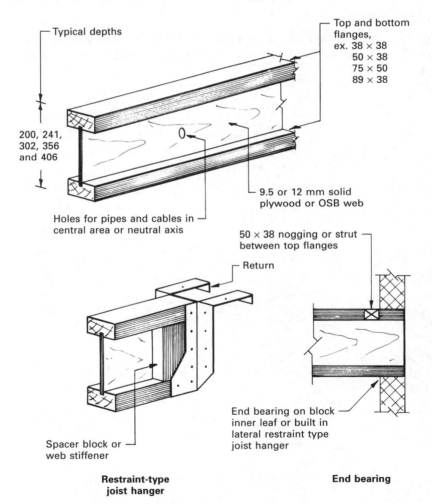

Figure 4.3.11 Solid or boarded web joist.

span potential is greater than with open web joists, with up to 10 000 mm possible. Solid web joists are less convenient for accommodating pipes and cables running at right angles to the joists. As for traditional timber-joisted floors, care and consideration must be exercised when cutting holes. The central area or neutral axis is the preferred location, away from the shear stress areas near supports. Flanges should never be notched.

Advantages of composite beams/joists
- Also suitable for other applications, e.g. roofing members.
- Factory produced to individual site/design dimensions; less site waste.
- High strength-to-weight ratio.
- Minimal shrinkage and movement, therefore limited use of strutting.
- Generally deeper than conventional joists, therefore less deflection.
- Wide section (typically ex. 75 mm flanges) provides large bearing area for floor deck and ceiling.

Open web joists have the additional benefit of providing virtually uninterrupted space for services, and they do not require end bearing; that is, joists can be top flange suspended.

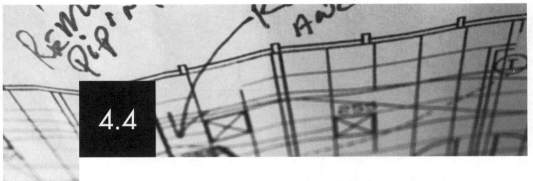

4.4

RAISED ACCESS FLOORS

These are a system of elevated platforms suspended over the structural floor of *in-situ* or precast concrete. They are not new in concept, and form the basis of deck construction popular about the outside of North American homes. Used internally, they evolved primarily in response to the information technology revolution that occurred in the latter part of the twentieth century. Although the huge volume of power, telephone and data cabling in modern offices has been largely responsible for developments in raised platforms, location for service pipes and ventilation ducts has also been influential. The dominance of services issues has promoted a re-think for designers, with the need for greater floor-to-floor heights to incorporate these floor voids (and suspended ceilings).

▇▇▇ PARTIAL ACCESS

This is where the floor finish is secured, and access is limited – possibly into simple ducts or trunking, as shown in Fig. 4.4.1. They are most suited for access to mains cabling for power and lighting, with void depths limited to about 100 mm. Accesses should be positioned to avoid furniture and be strategically located over junction boxes. This provides flexibility for changes in cable distribution where workstations and work functions may change.

▇▇▇ FULL ACCESS

These comprise standard-size (600 mm × 600 mm) interchangeable floor panels elevated on adjustable pedestals to provide multi-directional void space. Heights usually vary between 100 and 600 mm, but extremes of as little as 50 mm and as much as 2 m are possible. Adjustable-height pedestals are manufactured from steel or polypropylene, with support plates containing four lugs or projections to locate loose-fit decking panels. Threaded flat support plates may be specified where panels are to be screw-fixed in place (see Fig. 4.4.2).

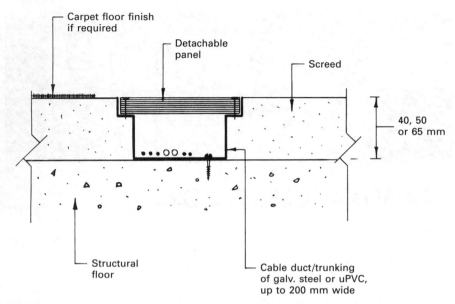

Figure 4.4.1   Floor duct.

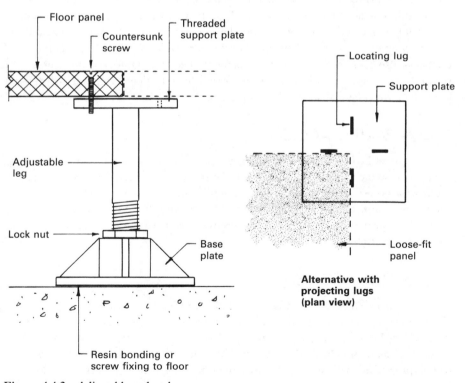

Figure 4.4.2   Adjustable pedestal.

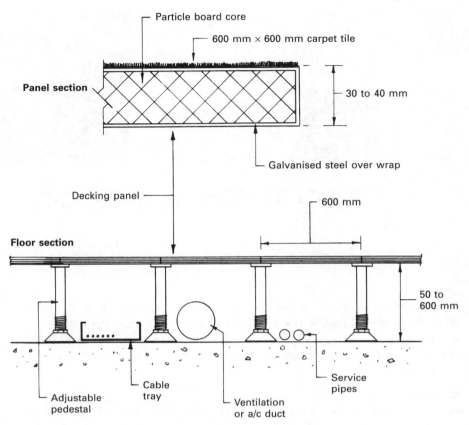

Figure 4.4.3   Raised access floor.

Floor panels or decking materials can vary considerably to suit application. Timber joists or battens may be used to support softwood boards or plywood/chipboard sheets, particularly if matching a traditional suspended floor. Office specification usually requires a fully interchangeable system with standard 600 mm × 600 mm panels. These are available in various gradings including light, medium and heavy, with corresponding maximum loadings of 10, 20 and 50 kN/m^2. Panel construction and application are shown in Fig. 4.4.3.

The combination of fully bonded chipboard within a steel casing provides structural soundness and integrity in the event of fire. The chipboard acts as a fire, acoustic and thermal insulator, and the steel resists tensile loading, also protecting the core against spread of flame. The presence of extraneous metal requires contact protection by electrical earth continuity through the complete floor system.

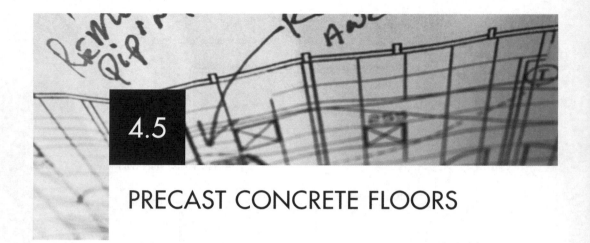

4.5

PRECAST CONCRETE FLOORS

The function of any floor is to provide a level surface that is capable of supporting all the live and dead loads imposed. Reinforced concrete, with its flexibility in design, good fire resistance and sound-insulating properties, is widely used for the construction of suspended floors for all types of building. The disadvantages of *in-situ* concrete are:

■ the need for formwork;
■ the time taken for the concrete to cure before the formwork can be released for reuse and the floor made available as a working area;
■ the very small contribution by a large proportion of the concrete to the strength of the floor.

Floors composed of reinforced precast concrete units have been developed over the years to overcome some or all of the disadvantages of *in-situ* reinforced concrete slab. To realise the full economy of any one particular precast flooring system the design of the floors should be within the span, width, loading and layout limitations of the units under consideration, coupled with the advantages of repetition.

■■■ CHOICE OF SYSTEM

Before any system of precast concrete flooring can be considered in detail the following factors must be taken into account:

■ maximum span;
■ nature of support;
■ weight of units;
■ thickness of units;
■ thermal insulation properties;

- sound insulation properties;
- fire resistance of units;
- speed of construction;
- amount of temporary support required.

The systems available can be considered as either precast hollow floors or composite floors; further subdivision is possible by taking into account the amount of temporary support required during the construction period.

PRECAST HOLLOW FLOORS

Precast hollow floor units are available in a variety of sections such as box planks or beams; tee sections, I beam sections and channel sections (see Fig. 4.5.1). The economies that can reasonably be expected over the *in-situ* floor are:

- 50% reduction in the volume of concrete;
- 25% reduction in the weight of reinforcement;
- 10% reduction in size of foundations.

The units are cast in precision moulds, around inflatable formers or foamed plastic cores. The units are laid side by side, with the edge joints being grouted together; a structural topping is not required, but the upper surface of the units is usually screeded to provide the correct surface for the applied finishes (see Fig. 4.5.1). Little or no propping is required during the construction period, but usually some means of mechanical lifting is required to offload and position the units. Hollow units are normally the cheapest form of precast concrete suspended floor for simple straight spans with beam or wall supports up to a maximum span of 20.000 m. They are not considered suitable where heavy point loads are encountered unless a structural topping is used to spread the load over a suitable area.

The hollow beams or planks give a flat soffit, which can be left in its natural state or be given a skim coat of plaster; the voids in the units can be used to house the services that are normally incorporated in the depth of the floor. The ribbed soffit of the channel and tee units can be masked by a suspended ceiling; again, the voids created can be utilised to house the services. Special units are available with fixing inserts for suspended ceilings, service outlets and edges to openings.

COMPOSITE FLOORS

These floors are a combination of precast units and *in-situ* concrete. The precast units, which are usually prestressed or reinforced with high-yield steel bars, are used to provide the strength of the floor with the smallest depth practicable and at the same time act as permanent formwork to the *in-situ* topping, which provides the compressive strength required. It is essential that an adequate bond is achieved between the two components. In most cases this is provided by the upper surface texture of the precast units; alternatively a mild steel fabric can be fixed over the units before the *in-situ* topping is laid.

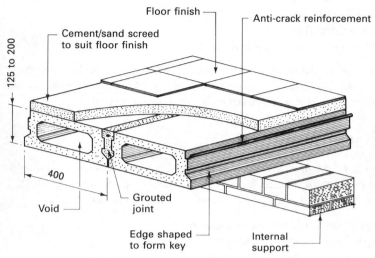

Floor finish

Anti-crack reinforcement

Cement/sand screed
to suit floor finish

125 to 200

400

Void

Grouted
joint

Edge shaped
to form key

Internal
support

Spans up to 13.000

Typical hollow floor unit details

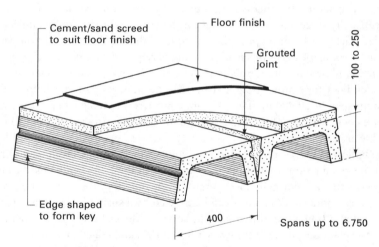

Cement/sand screed
to suit floor finish

Floor finish

Grouted
joint

100 to 250

Edge shaped
to form key

400

Spans up to 6.750

Anti-crack reinforcement required if units
are continuous over internal supports

Typical channel section floor unit details

Figure 4.5.1   Precast concrete hollow floors.

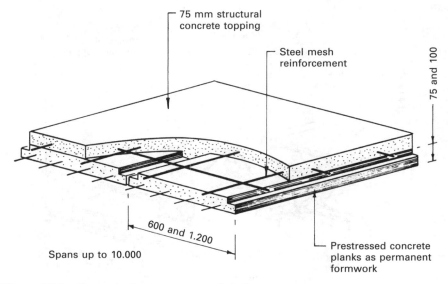

Figure 4.5.2 Composite floors – prestressed plank.

Composite floors generally take one of two forms:

- thin prestressed planks with a side key and covered with an *in-situ* topping (see Fig. 4.5.2).
- reinforced or prestressed narrow beams, which are placed at 600 mm centres and are bridged by concrete filler blocks known as **pots**. The whole combination is covered with *in-situ* structural concrete topping. Most of the beams used in this method have a shear reinforcing cage projecting from the precast beam section (see Fig. 4.5.3).

In both forms temporary support should be given to the precast units by props at 1.800–2.400 m centres until the *in-situ* topping has cured.

◼◼◼ COMPARISON OF SYSTEMS

Precast hollow floors are generally cheaper than composite; *in-situ* concrete is not required, and therefore the need for mixing plant and storage of materials is eliminated. The units are self-centring, therefore temporary support is not required; the construction period is considerably shorter; and generally the overall weight is less.

Composite floors will act in the same manner as an *in-situ* floor and can therefore be designed for more complex loadings. The formation of cantilevers is easier with this system, and support beams can be designed within the depth of the floor, giving a flat soffit. Services can be housed within the structural *in-situ* topping, or within the voids of the filler blocks. Like the precast hollow floor, composite floors are generally cheaper than a comparable *in-situ* floor, within the limitations of the system employed.

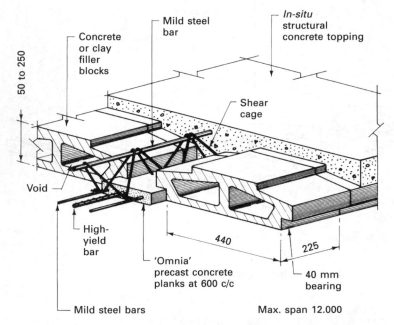

Typical composite floor using PCC planks

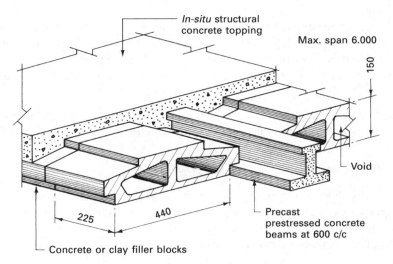

Typical composite floor using PCC beams

Figure 4.5.3   Composite floors – beam or pot.

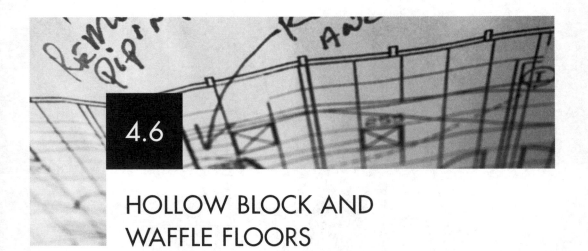

HOLLOW BLOCK AND WAFFLE FLOORS

Precast concrete suspended floors are generally considered to be for light to medium loadings spanning in one direction. Hollow block, or hollow pot floors as they are sometimes called, and waffle or honeycomb floors can be used as an alternative to the single spanning precast floor, because they can be designed to carry heavier loadings. They are in fact ribbed floors consisting of closely spaced narrow and shallow beams giving an overall reduction in depth of the conventional reinforced concrete *in-situ* beam and slab floor.

■■■ HOLLOW BLOCK FLOORS

These are formed by laying over conventional floor soffit formwork a series of hollow lightweight clay blocks or pots in parallel rows, with a space between these rows to form the ribs. The blocks act as permanent formwork, giving a flat soffit suitable for plaster application, and impart good thermal insulation and fire resistance to the floor. The ribs formed between the blocks can be reinforced to suit the loading conditions of the floor, thus providing flexibility of design (see Fig. 4.6.1). The main advantages of this system are its light weight, which is generally less than comparable floors of concrete construction, and its relatively low cost.

■■■ WAFFLE OR HONEYCOMB FLOORS

These are used mainly as an alternative to an *in-situ* flat slab or a beam and slab suspended floor, because they require less concrete, less reinforcement, and can be used to reduce the number of beams and columns required, with resultant savings on foundations. The honeycomb pattern on the underside can add to the visual aspect of the ceiling by casting attractive shadow patterns.

The floor is cast over lightweight moulds or pans made of glass fibre, polypropylene or steel, forming a two-directional ribbed floor (see Fig. 4.6.2).

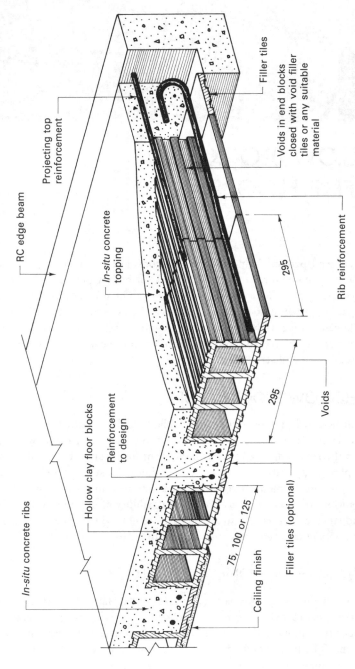

Filler tiles

Voids in end blocks closed with void filler tiles or any suitable material

Projecting top reinforcement

RC edge beam

In-situ concrete topping

295

Rib reinforcement

295

Voids

Hollow clay floor blocks

Reinforcement to design

In-situ concrete ribs

75, 100 or 125

Filler tiles (optional)

Ceiling finish

Figure 4.6.1   Hollow block floor.

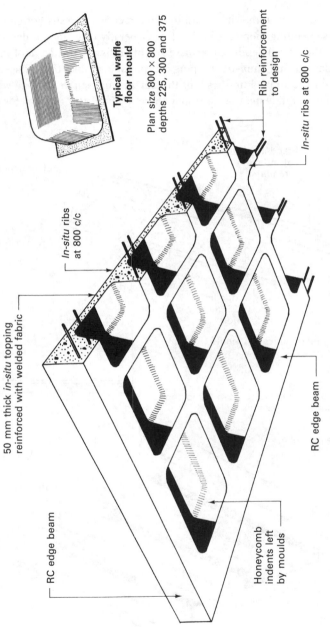

Typical waffle floor mould

Plan size 800 × 800
depths 225, 300 and 375

Rib reinforcement to design

In-situ ribs at 800 c/c

In-situ ribs at 800 c/c

50 mm thick *in-situ* topping reinforced with welded fabric

Honeycomb indents left by moulds

RC edge beam

RC edge beam

Figure 4.6.2 Waffle or honeycomb floor.

The moulds are very strong, lightweight, and are capable of supporting all the normal loads encountered in building works. Support is reduced to the minimum because the moulds are arranged in parallel rows and span between the parallel lines of temporary supports.

The reinforcement in the ribs is laid in two directions to resist both longitudinal and transverse bending moments in the slab. Generally three mould depths are available, but the overall depth can be increased by adding to the depth of the topping.

With all floors using an *in-situ* topping it is possible to float the surface in preparation for the applied finishes, but this surface may suffer damage while being used by the following trades. It may therefore be better to allow for a floor

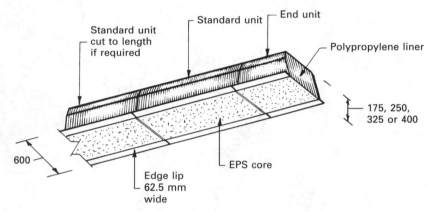

Figure 4.6.3   Inverted trough mould.

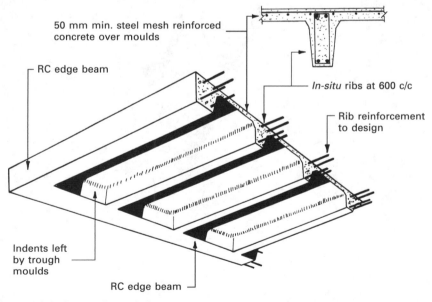

Figure 4.6.4   Inverted trough floor.

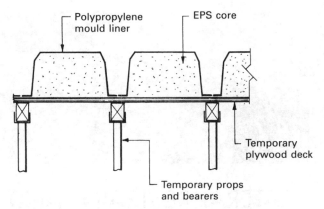

Figure 4.6.5 Temporary support to inverted trough floor.

screed to be applied to the *in-situ* topping at a later stage in the contract, prior to the fixing of the applied finish.

■■■ INVERTED TROUGH FLOOR

The inverted trough moulds shown in Fig. 4.6.3 are used to produce an elongated reinforced concrete variation of the waffle trough type of floor. The moulds ensure that *in-situ* reinforced concrete is in specific locations only, saving considerably on concrete costs and the dead weight that the concrete would otherwise contribute. The one-way spanning floor is viewed from the underside in Fig. 4.6.4. Lateral beams may also be incorporated as interim support in large-span situations. The continuous longitudinal voids provide an unobtrusive location for service pipes and cables. Brackets or threaded inserts set in the underside of the concrete can provide support for services and for suspending a ceiling.

TEMPORARY SUPPORT

Support to troughs is from a framework of props, bearers and decking, as shown in Fig. 4.6.5. Prop spacing is determined by anticipated loading of operatives and wet concrete. After the concrete has gained strength and the temporary support is removed, the expanded polystyrene core and polypropylene mould are struck with a flat-bladed implement. With careful handling, cores and moulds have several reuses before being discarded.

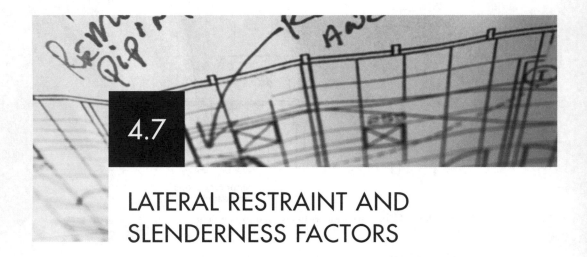

4.7

LATERAL RESTRAINT AND SLENDERNESS FACTORS

Walls have been designed such that their mass, together with the loads carried from floors and roofs, provide adequate resistance to buckling. Buckling can be controlled by using buttresses or piers, but these may encroach on space and be visually unattractive. Contemporary design favours more slender construction: therefore stability and resistance to tensile lateral forces such as wind, and possibly impact, is achieved by using support from the floor and roof (see Part 6).

Floor structures function as a stable horizontal platform supported by perimeter and intermediate walls. They complement that support by providing restraint and stability for the wall against lateral overturning forces, in conjunction with adequate structural connections. Connections to external, compartment (fire), separating (party), and internal loadbearing walls in houses where walls exceed 3 m length may be achieved by:

1. 90 mm minimum direct bearing of the floor by timber joists (see Fig. 4.3.5), or from a continuous concrete slab;
2. BS EN 845-1 approved galvanised steel restraint type joist hangers with a minimum 100 mm bearing, spaced at no more than 2 m;
3. galvanised steel purpose-made straps of minimum cross-section 30 mm × 5 mm, spaced at no more than 2 m, secured to the inner leaf of masonry and over at least 3 joists;
4. galvanised steel straps, as 3 above, applied to adjacent floors. Continuity of contact between the floor and wall is also acceptable.

[*Note*: Items 1 and 2 assume that joist spacing does not exceed 1.2 m and the house is not over two storeys. If over two storeys, supplementary strapping should be provided to joists in the longitudinal direction, as shown in Fig. 4.7.1.]

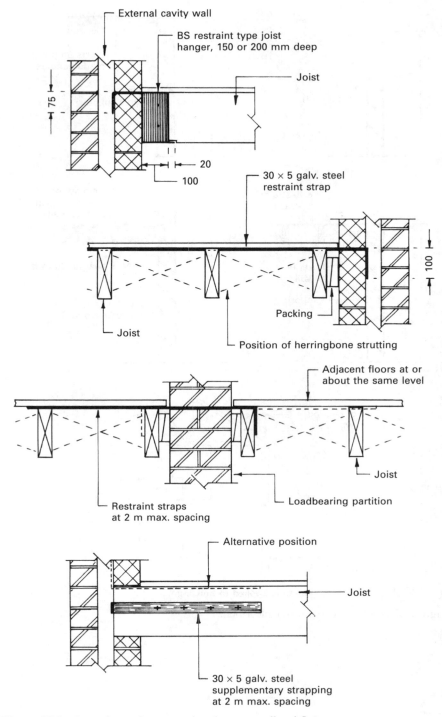

Figure 4.7.1 Lateral restraint connections between wall and floor.

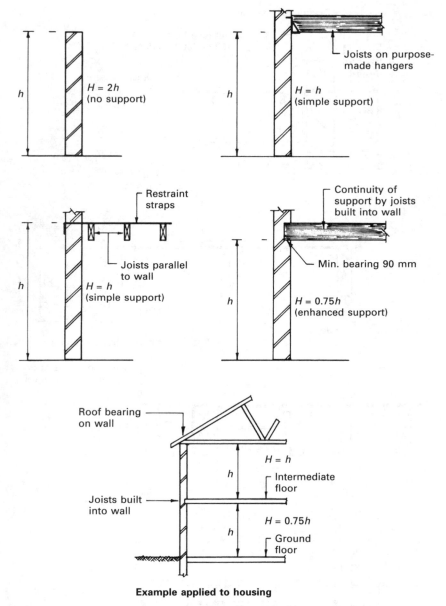

Example applied to housing

Figure 4.7.2   Effective height (*H*) calculation.

▧ ■ ■ SLENDERNESS RATIO

This is an acceptable standard for measuring dimensional stability. Actual wall height and thickness (see Chapters 3.1–3.5) can be varied for structural calculation purposes. This is known as the effective height or thickness and depends on the method (if any) of restraint. The effectiveness of connecting a wall to an adjacent structure is shown in Fig. 4.7.2.

Calculations to determine effective wall thickness are also possible for different wall constructions. Single leaf/solid walls without piers have an effective thickness equal to the actual thickness. Attached piers will cause some variation, and reference to BS 5628-1: *Structural use of unreinforced masonry* will provide appropriate coefficients. Cavity walls have an effective thickness equal to two-thirds of the sum of the actual thickness of both leaves, or the actual thickness of the thicker leaf (whichever is greater).

Effective length can also be important for structural calculations. This depends on the extent of support from abutments, as shown in Fig. 4.7.3.

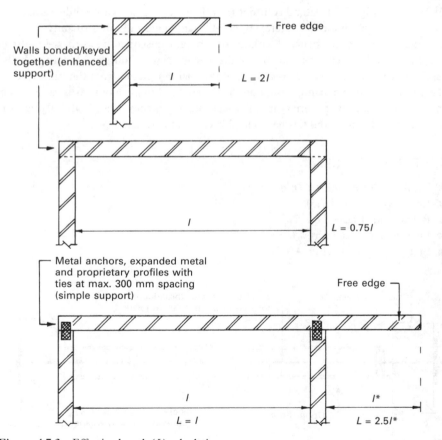

Figure 4.7.3 Effective length (*L*) calculation.

EXAMPLE CALCULATION OF SLENDERNESS RATIO

A cavity wall consists of 102.5 mm brick outer leaf and 150 mm concrete block inner leaf. The wall height from floor to ceiling is 3 m, with floor joists built into the inner leaf (enhanced support).

Effective thickness $= 0.67 (102.5 + 150) = 169$ mm
Effective height $\quad = 0.75 \times 3000 = 2250$ mm

$$\text{Slenderness ratio} \quad = \frac{\text{effective height}}{\text{effective thickness}} = \frac{2250}{169} = 13.30$$

Maximum acceptable slenderness ratios vary with applications in accordance with BS 5628-1: *Structural use of unreinforced masonry*. The main objective is to evaluate and compare numerically different construction methods: the lower the slenderness ratio, the less prone a structure is to buckling under load due to insufficient rigidity.

■■■ LATERAL RESTRAINT – RETRO-FIX

As shown in Fig. 4.7.1, lateral restraint from floors is necessary to stabilise adjacent walls (see Fig. 6.1.9 for roof lateral support). Many older buildings (mainly housing) have been constructed without strapping. Consequently, as these buildings age, there can be a tendency for their walls to show signs of outward bulging. Walls rarely bulge inwards, because the effect of thermal movement from the sun causes the outer surface to expand. Also, floor loads will tend to push outwards rather than inwards, particularly if there is an inner leaf carrying greater load. Wall bulging can also be caused where wall ties have failed because of corrosion.

REMEDIAL MEASURES

- Traditional through-tie (Fig. 4.7.4).
- Retro-strap (Fig. 4.7.5).
- Retro-stud (Fig. 4.7.6).
- Driven resin-fixed screw (Fig. 4.7.7).
- Self-tapping helix (Fig. 4.7.8).

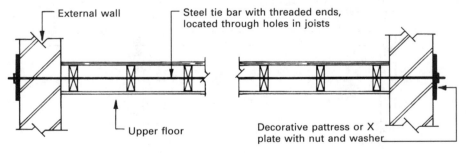

Figure 4.7.4 Traditional through-tie.

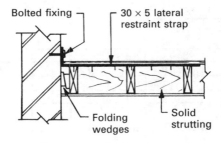

Figure 4.7.5   Retro-strap.

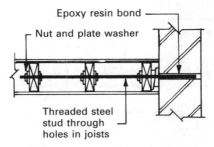

Figure 4.7.6   Retro-stud.

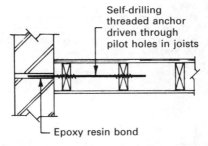

Figure 4.7.7   Driven resin-fixed screw.

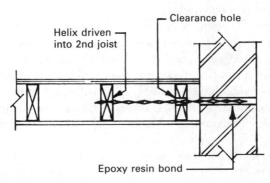

Figure 4.7.8   Self-tapping helix.

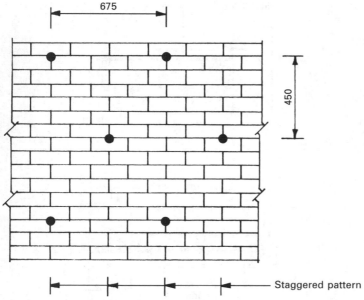

Figure 4.7.9   Standard spacing of polyurethane injection holes.

Notes:

1. At door and window openings, hole spacing is 300 mm vertically (200 mm horizontally from openings) and 400 mm horizontally above the opening.
2. At party walls, vertical holes are placed at 300 mm spacing, 300 mm from the party wall line.
3. Any unsleeved air bricks are replaced with sleeved, to prevent foam from leakage to where it is not required, e.g. under suspended floors.
4. At upper floor level, supplementary stainless steel anchors are positioned at 1000 mm horizontal intervals.

WALL TIE DECAY

Wall ties should be specified as manufactured from corrosion-resistant stainless steel. Historically, they have been produced from iron or steel, at best galvanised with a coating of zinc. Iron and steel ties can corrode and break, particularly where the zinc coating to steel is of poor quality, or it has been damaged during construction. Ties will have greater exposure to corrosion where pointing has broken down, and where old lime mortars have softened by decomposition due to atmospheric pollutants and frost damage. Inevitably there will be loss of stability between the two leaves of masonry wall, creating areas of outward bulging from the outer leaf.

A solution is to demolish the outer leaf, one small area at a time, and rebuild it complete with new ties. A simpler, less dramatic approach is to apply chemical cavity stabilisation. The material used is structural polyurethane foam of closed cell characteristic. The polyurethane is created by on-site mixing of an isocyanate with a resin to produce a liquid for injection by hand gun through 12 mm diameter holes. Holes are pre-drilled in the outer leaf at intervals as shown in Fig. 4.7.9. Within the cavity the fluid mix reacts and expands to produce a rigid foam mass. This foam combines as an adhesive agent and as an insulant.

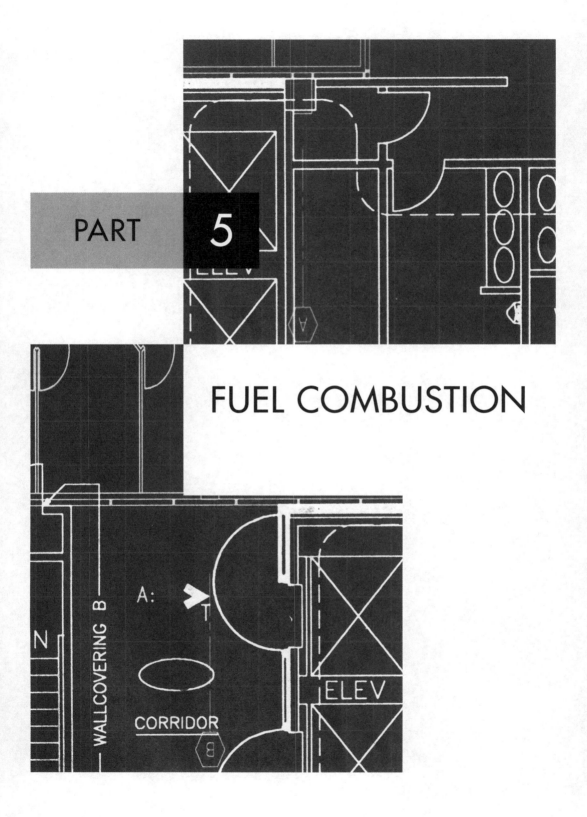

PART 5

FUEL COMBUSTION

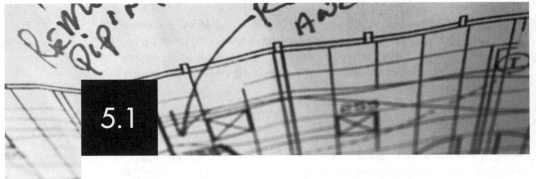

5.1

FIREPLACES, CHIMNEYS AND FLUES

The traditional method of providing heating in a domestic building was the open fire burning coal and/or wood, but because of its low efficiency compared with modern heating appliances its use as the primary source of heating has declined. It must be noted that all combustible fuels, such as coal, smokeless fuels, oils, gas and wood, require some means of conveying the products of combustion away from the appliance or fireplace to the open air. They must discharge in a manner that does not create an environmental hazard or fire risk.

Terminology
- **Fireplace**   This is the area in which combustion of the fuel takes place. It may be in the form of an open space with a fret or grate in which wood or coal is burnt, or a free-standing appliance such as a slow combustion stove, room heater, oil-burning appliance, gas fire or gas boiler.
- **Chimney**   The fabric surrounding the flue and providing it with the necessary strength and protection.
- **Flue**   This is strictly a void through which the products of combustion pass. It is formed by lining the inside of a chimney with a suitable lining material to give protection to the chimney fabric from the products of combustion, as well as forming a flue of the correct shape and size to suit the type of fuel and appliance being used.

▧ ▣ ■ HEAT-PRODUCING APPLIANCES

Building Regulations, Approved Document J, Sections 1 to 4 are concerned with heat-producing appliances, and require that they are installed with an adequate air supply for efficient working, are provided with a suitable means of discharging the products of combustion to the outside air, and are installed so as to reduce the risk of the building catching fire. Approved Document J gives practical guidance to satisfy these regulations, and makes particular reference to solid-fuel and oil-burning appliances with rated outputs up to 50 kW and 45 kW respectively. The Document also refers to gas-burning appliances with rated input up to 70 kW. Input and output ratings can be used to calculate the efficiency of appliances. For example, if a gas fire has input and output ratings of 12 kW and 9 kW respectively, the appliance efficiency is

$$\frac{9}{12} \times \frac{100}{1} = 75\%$$

When a fuel is burnt to provide heat it must have an adequate air supply to provide the necessary oxygen for combustion to take place. The air supply is drawn through the firebed in the case of solid fuels and injected in the case of oil-burning appliances. A secondary supply of air is also drawn over the flames in both cases. In rooms where an open fire is used the removal of air induces ventilation, and this helps to combat condensation. The air used in combustion is drawn through the appliance by suction provided by the flue, and this is affected by the temperature, pressure, volume, cross-sectional area and surface lining of the flue.

Openings for air entry to rooms containing appliances must satisfy the following:

- Solid-fuel open appliance: must have an area at least 50% of the appliance throat opening.
- Other solid-fuel and oil-burning appliances (unless fitted with a balanced flue): must have an area at least 550 mm^2 per kW of rated output above 5 kW. If a flue draught stabiliser is fitted to a solid-fuel appliance, an extra 300 mm^2 is required for each kW of rated output.
- Gas appliances with an open flue: must have an area at least 500 mm^2 per kW of input rating over 7 kW.
- Gas appliances with room-sealed-balanced flues do not require special provision, but gas fires and solid-fuel-effect fires should have ventilation as for solid-fuel appliances or in accordance with BS 5871-1 to 3.

[*Note*: Further details of gas appliance flue installation are considered in the next chapter.]

To improve the efficiency of a flue it is advantageous for the chimney to be situated on an internal wall, because this will reduce the heat losses to the outside air; also, the flue gas temperature will be maintained, and this will in turn reduce the risk of condensation within the flue. The condensation could cause corrosion of the chimney fabric due to the sulphur compounds that are a by-product of the processes of combustion.

CHIMNEYS

A chimney should be built as vertical as is practicable to give maximum flue gas flow – if bends are required the angle should be 30° maximum with the vertical.

If an appliance is connected to a chimney the flue pipe should be as short as possible and continuous with an access for cleaning included at the base of the main flue. The chimney should be terminated above the roof so that it complies with the Building Regulations and is unaffected by adverse pressures that occur generally on the windward side of a roof and may cause down-draughts. The use of a chimney terminal or pot could be to transfer the rectangular flue section to a circular cross-section, giving better flue flow properties, or to provide a cover to protect the flue from the entry of rain or birds. A terminal should be fixed so that it does not impede the flow of flue gases in any way. The construction should be from solid non-combustible materials capable of withstanding temperatures of 1100 °C, without impairing stability. Brickwork and blockwork must be lined with suitable materials as defined in Approved Document J, Sections 1.27 and 1.28. Unlined blockwork chimneys of refractory material or a combination of high-alumina cement and kiln-burnt or pumice aggregates are acceptable.

BUILDING REGULATIONS

Combustion appliances and fuel storage systems are covered by Part J of the Regulations, with Approved Documents in five sections. The main requirements for solid fuel are shown in Figs. 5.1.1–5.1.4.

Firebacks

The functions of a fireback are to contain the burning fuel, to prevent the heat of the fire from damaging the wall behind it, and to radiate heat from the fire into the room. The space behind a fireback should be of a solid material, though not a strong mix and not composed of loose rubble. The recommended mix is 1:2:4 lime, sand and broken brick; alternatively an insulating mix of 5 parts of vermiculite to 1 part of cement could be used. The temperature encountered by a fireback will be high: therefore the problem of expansion and subsequent contraction must be considered. The fireback should preferably be in two or more parts, because the lower half will become hotter than the upper half. Multi-piece firebacks also have the advantage of being easier to fit. It is also good practice to line the rear of a fireback with corrugated paper or a similar material, which will eventually smoulder away, leaving a small expansion gap at the back of the fireback.

Throat

The size and shape of the throat above the fireplace opening are of the utmost importance. A fireplace with an unrestricted outlet to the fire would create unpleasant draughts by drawing an unnecessary amount of air through the flue, and would reduce the efficiency of the fire by allowing too much heat to escape into the flue. A throat restriction of 100 mm will give reasonable efficiency without making chimney sweeping impossible (see Fig. 5.1.5).

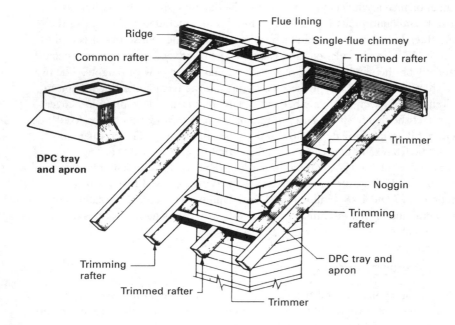

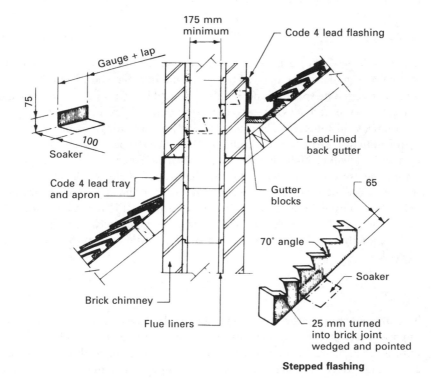

Figure 5.1.1   Chimney construction.

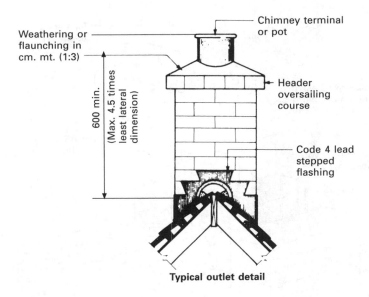

Weathering or flaunching in cm. mt. (1:3)

Chimney terminal or pot

600 min. (Max. 4.5 times least lateral dimension)

Header oversailing course

Code 4 lead stepped flashing

Typical outlet detail

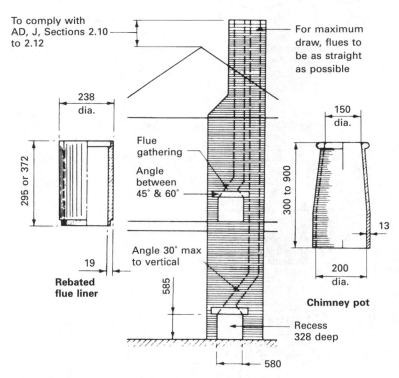

To comply with AD, J, Sections 2.10 to 2.12

For maximum draw, flues to be as straight as possible

238 dia.

150 dia.

295 or 372

Flue gathering

Angle between 45° & 60°

300 to 900

13

19

Angle 30° max to vertical

200 dia.

Rebated flue liner

585

Recess 328 deep

Chimney pot

580

Figure 5.1.2 Typical flue construction.

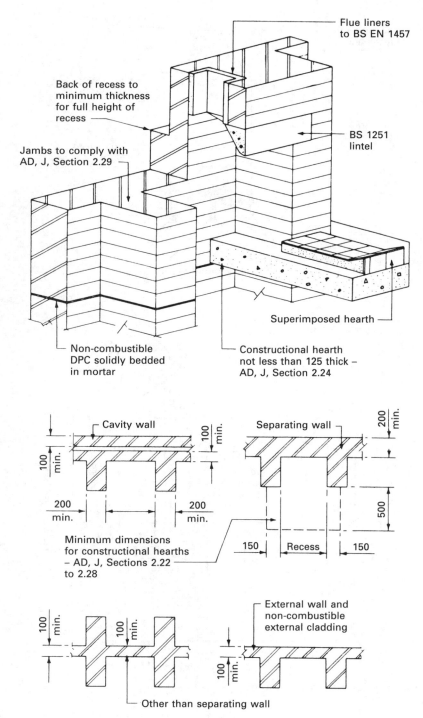

Figure 5.1.3   Fireplaces and Approved Document J.

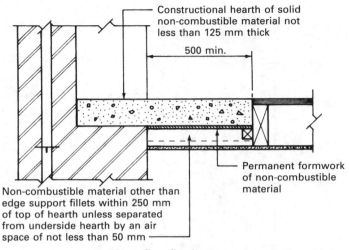

Constructional hearth of solid non-combustible material not less than 125 mm thick

500 min.

Permanent formwork of non-combustible material

Non-combustible material other than edge support fillets within 250 mm of top of hearth unless separated from underside hearth by an air space of not less than 50 mm

Upper floor fireplaces

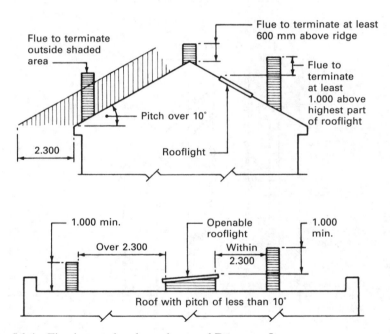

Flue to terminate outside shaded area

Flue to terminate at least 600 mm above ridge

Flue to terminate at least 1.000 above highest part of rooflight

Pitch over 10°

Rooflight

2.300

1.000 min.

Over 2.300

Openable rooflight

Within 2.300

1.000 min.

Roof with pitch of less than 10°

Figure 5.1.4   Fireplaces and outlets – Approved Document J.

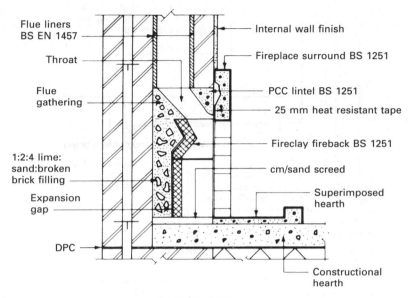

Flue liners BS EN 1457

Throat

Flue gathering

1:2:4 lime: sand:broken brick filling

Expansion gap

DPC

Internal wall finish

Fireplace surround BS 1251

PCC lintel BS 1251

25 mm heat resistant tape

Fireclay fireback BS 1251

cm/sand screed

Superimposed hearth

Constructional hearth

Open fireplaces

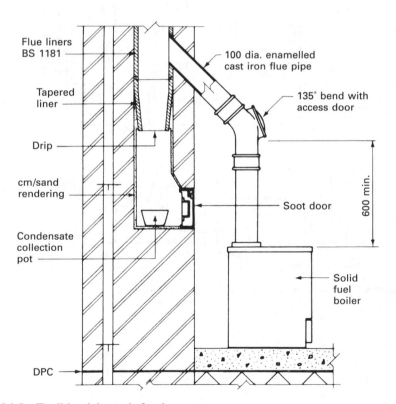

Flue liners BS 1181

Tapered liner

Drip

cm/sand rendering

Condensate collection pot

DPC

100 dia. enamelled cast iron flue pipe

135° bend with access door

Soot door

600 min.

Solid fuel boiler

Figure 5.1.5   Traditional domestic fireplaces.

Surrounds

This is the façade of a fireplace, and its main function is an attractive appearance. It can be of precast concrete with an applied finish such as tiles, or built *in-situ* from small brickettes or natural stones. If it is precast it is usually supplied in two pieces: the front and the hearth. The front is fixed by screws through lugs cast into the edges of the surround and placed against a 25 mm wide non-combustible cord or rope around the fireplace opening to allow for expansion and contraction. The hearth should be bedded evenly on the constructional hearth with at least 10 mm of 1:1:8 cement:lime:sand mix (see Fig. 5.1.5).

DEEP ASHPIT FIRE

This is a low-front fire with a fret or grate at or just below hearth level. A pit is constructed below the grate to house a large ashpan capable of holding several days' ash. The air for combustion is introduced through the ashpit and grate by means of a 75 mm diameter duct, at the end of which is an air control regulator. If the floor is of suspended construction the air duct would be terminated after passing through the fender wall, whereas with a solid floor the duct must pass under the floor and be terminated beyond the external wall, the end being suitably guarded against the entry of vermin.

BACK BOILER

This is an open fire of conventional design with part of the fireback replaced by a small boiler with a boiler flue and a control damper. These are rarely installed now, but were the main source of hot water supply for domestic premises prior to oil or gas central heating becoming more widespread in the 1960s.

OPEN FIRE CONVECTORS

These are designed to increase the efficiency of an open fire by passing warm air back into the room as well as the radiated heat from the burning fuel. The open convector is a self-contained unit consisting of a cast-iron box containing the grate and fireback forming a convection chamber. The air in the chamber heats up and flows into the room by convection currents moving the air up and out of the opening at the top of the unit.

ROOM HEATERS

These are similar appliances to open fire convectors, but they are designed to burn smokeless fuels and operate as a closed unit. In some models the strip glass front is in fact a door, and can therefore be opened to operate as an open fire. Room heaters are either free standing – that is, fixed in front of the surround with a plate reducing the size of the fireplace opening to the required flue pipe size – or inset into the fireplace recess, where the chimney flue aperture is reduced by a plate to receive the flue pipe of the appliance.

INDEPENDENT BOILERS

The function of these appliances is to heat water, whether it is for a central heating system or merely to provide hot water for domestic use. It is generally fixed in the kitchen because of the convenience to the plumbing pipe runs, and to utilise the background heat emitted. It must discharge into its own flue, because it is a slow combustion appliance, and the fumes emitted could, if joined to a common flue, cause a health hazard in other rooms of the building. Adequate access must be made for sweeping the flue and removing the fly ash that will accumulate with a smokeless fuel (see Fig. 5.1.5). Further details of boiler flues are considered in the next chapter.

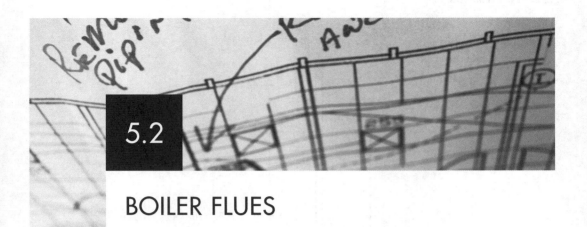

BOILER FLUES

Boilers are appliances designed to heat water by energy exchange from burning oil, gas or solid fuels. Applications of boilers to domestic systems of hot water circulation and distribution are provided in Chapter 11.1.

The term *boiler* is something of a misnomer, originating from Victorian times, when appliances were used to raise steam for heating and industrial processes. The modern domestic boiler in an open-vented system is designed to heat water to about 80 °C. At temperatures above this it is uneconomic and unsafe. They are either floor standing with a conventional open flue pipe, possibly using an existing chimney, or wall mounted with a room-sealed balanced flue. Wall mounting permits discreet positioning between or within kitchen units or other unobtrusive location. Floor-standing boilers are now less common for domestic use, as the technology has advanced sufficiently to provide more compact units for wall mounting with simple horizontal balanced flue.

Floor-standing boilers with conventional open flue incorporate an air opening from the room containing the boiler to aid combustion. The area of ventilation opening is considered in the previous chapter. Flue installation must allow for positioning of the terminal (outlet), flue length and size. The simplest installation for a gas boiler is a straight vertical pipe as shown in Fig. 5.2.1, with non-combustible sleeves provided where it passes through a floor and other combustible parts of the structure. The advantage of a simple straight flue is less surface exposure than a flue offset to the outer wall as in Fig. 5.2.2. Here there is more opportunity for condensation in the flue, resulting from flue gases cooling to the dew-point of water – approx. 60 °C.

◼◼◼ FLUE MATERIALS

Flue materials are detailed in the Building Regulations, Approved Document J. They can incorporate traditional brick chimneys, precast concrete flue blocks

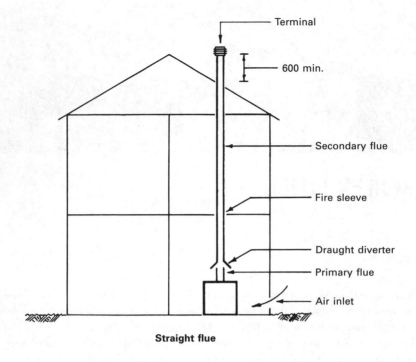

Straight flue

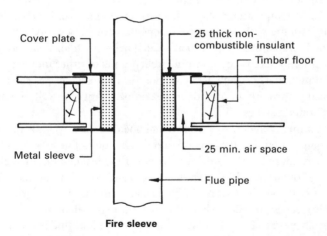

Fire sleeve

Figure 5.2.1   Internal open flue.

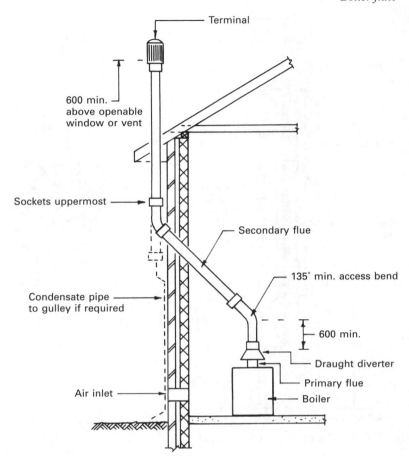

Terminal

600 min.
above openable
window or vent

Sockets uppermost

Secondary flue

135° min. access bend

Condensate pipe
to gulley if required

600 min.

Draught diverter

Primary flue

Boiler

Air inlet

Figure 5.2.2 Traditional open flue.

(gas appliances only), or flue pipes of various materials. Chimneys are unnecessarily large for domestic boiler purposes, but may be used with clay liners as defined in the preceding chapter. In older premises where linings were omitted, the chimney may receive a flexible stainless steel lining, inserted as shown in Fig. 5.2.3. These are not suitable for use with solid-fuel appliances as the products of combustion can deteriorate the lining.

Precast concrete flue blocks to BS 1289 are manufactured from high–alumina cement and dense aggregates, to resist chemical decomposition from acidity caused by flue gases and condensation. They are jointed with high–alumina cement mortar and incorporate a continuous void for the gases to escape. Dimensions coordinate with the inner leaf of blockwork, as shown in Fig. 5.2.4. The minimum cross-sectional flue area is 16 500 mm^2 for conveyance of flue gases from gas fires, convectors and other small appliances.

Connecting flue pipes are produced in a variety of materials to suit application. For solid-fuel and oil-burning appliances (up to 50 kW and 45 kW output

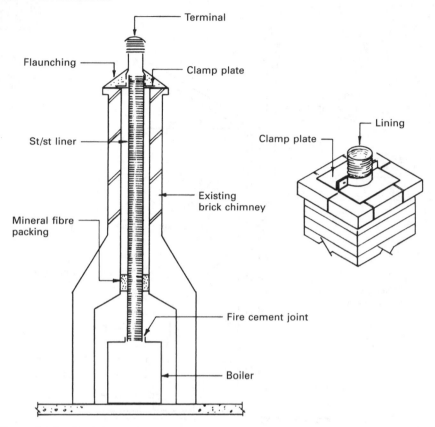

Figure 5.2.3   Flexible stainless steel flue lining.

respectively) with flue gas temperatures likely to exceed 250 °C, flue pipes may be of enamelled cast iron to BS 41, 3 mm minimum thickness mild steel to BS 1449, enamelled steel to BS 6999, or stainless steel as described in BS EN 10088-1. Gas appliances (up to 70 kW input) and oil-burning appliances (up to 45 kW output), with flue gas temperatures unlikely to exceed 250 °C, may use the preceding materials or lighter-weight connecting flue pipes of sheet metal to BS 715 or fibre cement to BS 7435.

Double-wall steel variations of annular construction (pipe within a pipe) are preferred, as the space between each pipe can be filled with insulation against heat transfer to the adjacent structure or the open air. The insulation will also reduce condensation within the flue.

Flue terminals or outlets should promote extraction of flue gases, resist potential down-draughts, and prevent the ingress of rain and snow. For solid-fuel and oil appliances, see the previous chapter (Fig. 5.1.4). For gas appliances flue terminals are shown in Fig. 5.2.5, with preferred location at or above the ridge of a pitched roof or at least 600 mm above the intersection with the roof pitch.

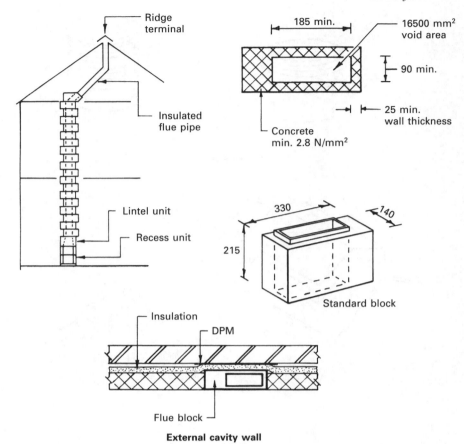

Figure 5.2.4 Precast concrete flue blocks.

Balanced flues are specifically suited to wall-mounted boilers. The air for combustion is drawn directly into the boiler through the terminal in the external wall: therefore no special provision for room ventilation is necessary. The burnt gases discharge through the combined flue/intake unit as shown in Fig. 5.2.6 to maintain a constant balanced air pressure at the gas burner, regardless of outside wind conditions. Some flues incorporate a fan for more efficient control of combustion gases. This also permits a longer horizontal flue where it is impractical to mount a boiler directly on an external wall.

Balanced flue location is important for efficient operation of the appliance, safe discharge of flue gases, and minimisation of heat damage to adjacent structure and components. Table 5.2.1 lists acceptable locations for the flues, and Fig. 5.2.7 illustrates these requirements.

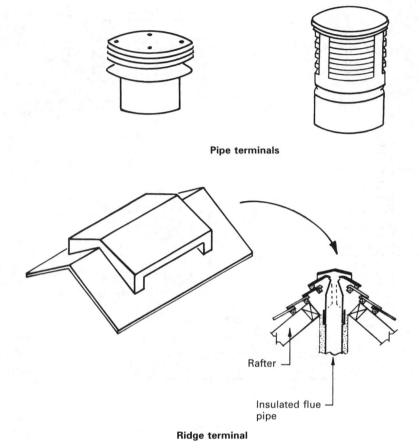

Pipe terminals

Ridge terminal

Figure 5.2.5   Flue terminals.

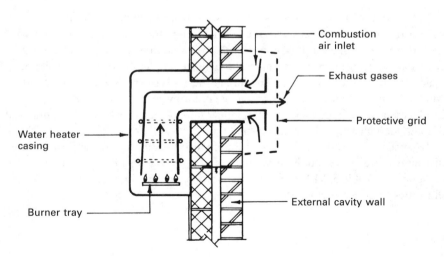

Figure 5.2.6   Room-sealed balanced-flue appliance.

Table 5.2.1   Minimum dimensions for location of balanced flue terminals

Position	Minimum distance (mm)	
	Natural draught	Fanned draught
Under an openable window or ventilator	300	300
Under rainwater goods or sanitation pipework	300	75
Under eaves	300	200
Under a balcony or a carport roof	600	200
From a window or door in the carport	1200	1200
Horizontally from vertical drain and soil pipes	300	150
Horizontally from internal or external corners	600	300
Above ground level or a projection, e.g. balcony	300	300
From an opposing surface	600	600
From an opposing terminal	600	1200
Vertically from another terminal	1500	1500
Horizontally from another terminal	300	300

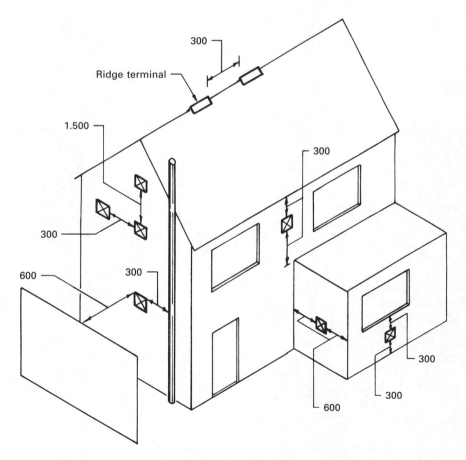

Figure 5.2.7   Balanced flue terminal positions – minimum dimensions (gas appliances).

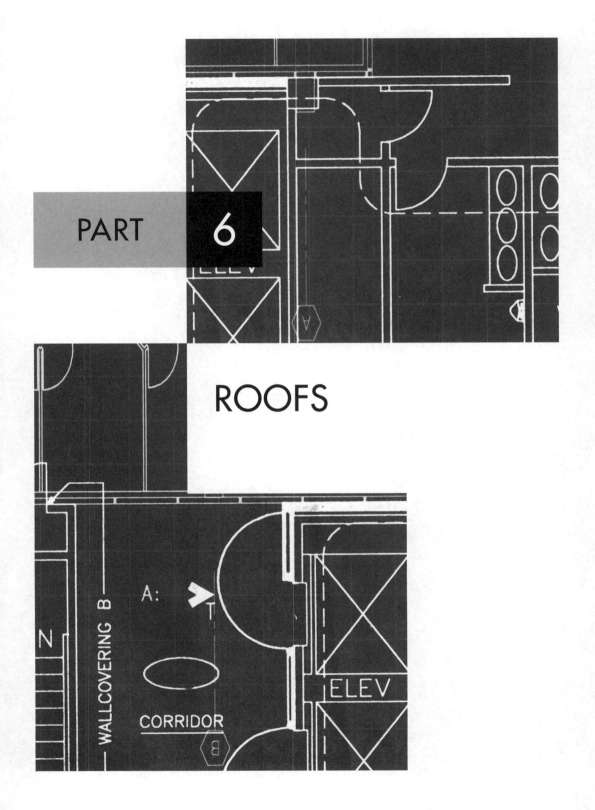

PART 6

ROOFS

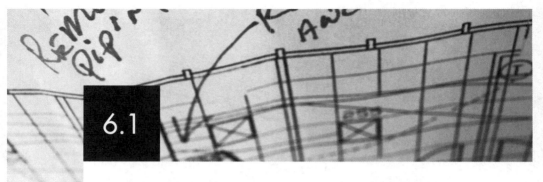

ROOFS: TIMBER, FLAT AND PITCHED

■ ■ ■ TIMBER FLAT ROOFS

A flat roof is essentially a low-pitched roof, and is defined in BS 6100: *Glossary of building and civil engineering terms* as a pitch of 10° or less to the horizontal. Generally the angle of pitch is governed by the type of finish that is to be applied to the roof.

The functions of any roof are:

- to keep out rain, wind, snow and dust;
- to prevent excessive heat loss in winter;
- to keep the interior of the building cool in summer;
- to accommodate all stresses encountered;
- to accept movement due to changes in temperature and moisture content;
- to provide lateral restraint and stability to adjacent walls;
- to resist penetration of fire and spread of flame from external sources.

The simplest form of roof construction to fulfil these functions is a timber flat roof covered with an impervious material to prevent rain penetration. This form of roof is suitable for spans up to 4000 mm; spans over this are usually covered with a reinforced concrete slab or a patent form of decking.

The disadvantages of timber flat roofs are as follows:

- They have limited capacity for insulation.
- They may contrast in style with other buildings in the vicinity and, if an extension, the building to which it is attached.
- Unless they are properly designed and constructed, pools of water will collect on the surface causing local variations in temperature. This results in deterioration of the covering and, consequently, high maintenance costs.
- They have little or no space to accommodate services.

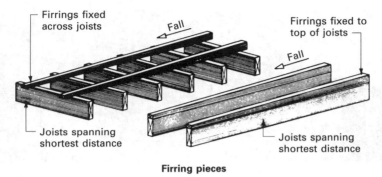

Firrings fixed across joists

Fall

Firrings fixed to top of joists

Fall

Joists spanning shortest distance

Joists spanning shortest distance

Firring pieces

Figure 6.1.1   Timber flat roof slope details.

CONSTRUCTION

The construction of a timber flat roof follows the same methods as those employed
for the construction of timber upper floors. Suitable joist sizes can be obtained
by design or by reference to tables in publications recommended in the Building
Regulations, Approved Document A. The spacing of roof joists is controlled by the
width of decking material to be used and/or the width of ceiling board on the
underside. Timber flat roofs are usually constructed to fall in one direction towards
a gutter or outlet. This can be achieved by sloping the joists to the required fall, but,
as this would give a sloping soffit on the underside, it is usual to fix wedge shaped
fillets called **firrings** to the top of the joists to provide the fall (see Fig. 6.1.1). The
materials used in timber flat roof construction are generally poor thermal insulators,
and therefore some form of non-structural material can be incorporated into the
roof if it has to comply with Part L of the Building Regulations.

Decking materials

Timber

This can be in the form of softwood boarding, chipboard, oriented strand board
or plywood. Plain-edge sawn softwood boards or tongued and grooved boards
are suitable for joists spaced at centres up to 450 mm. Exterior grade water- and
boil-proof (WBP) plywood is available in sheet form, which requires fixing on all
four edges. This means noggins will be required between joists to provide the
bearing for end fixings. Chipboard is also a suitable sheet material and is fixed in a
similar manner to plywood. This material can be susceptible to moisture movement:
therefore for roofing it should be specified 'moisture resistant to BS EN 312, type
C3, C4 or C5'. Flat roofs must have the void between ceiling and decking ventilated
to prevent condensation occurring. It is also advisable to use structural timbers that
have been treated against fungal and insect attack. In certain areas treatment to
prevent softwood infestation by the house longhorn beetle is a requirement under
the Building Regulations. Details can be found in the Approved Document to
support Regulation 7.

Compressed straw slabs

These are made from selected straw by a patent method of heat and pressure to a standard width of 1200 mm × 2400 mm length, the standard thickness being 58 mm, which gives sufficient strength for the slabs to span 600 mm. All edges of the slabs must be supported and fixed. Ventilation is of the utmost importance, and it is common practice to fix cross-bearers at right angles to and over the joists to give cross-ventilation. A bitumen scrim should be placed over the joints before the weathering membrane is applied.

Wood wool slabs

These are 610 mm wide slabs of various lengths, which can span up to 1200 mm. Thickness varies; for roof decking 51 mm is normally specified. The slabs are made of shredded wood fibres that have been chemically treated and are bound together with cement. The fixing and laying is similar to compressed straw slabs.

Insulating materials

There are many types of insulating material available, usually in the form of boards or quilts. Insulation boards laid over the decking create a **warm deck** roof, whereas quilted materials draped over the joists or placed between them, make a **cold deck**.

Boards

These originated from lightly compressed vegetable fibres, bonded with natural glues or resins. Compressed straw and wood wool slabs have been particularly successful as decking materials with the benefit of in-built insulation.

For insulation-only purposes, the most popular materials for roofing boards are high-density mineral wool with a tissue membrane bonded to one surface, expanded polystyrene, or polyurethane slabs. A warm deck roof has insulation board placed over the decking, but below the waterproof membrane. Conversely, an inverted warm deck has the insulation board above the waterproof membrane. The inverted warm deck insulation board must be unaffected by water and be capable of receiving a surface treatment of stone granules or ceramic pavings. With this technique, the waterproof membrane is protected from the stresses caused by exposure to weather extremes. Examples of the various forms of construction are shown in Fig. 6.1.2.

Quilts

These are made from mineral or glass wool that is loosely rolled, with the option of a kraft paper facing. The paper facing is useful as the wool is in fine shreds, which give rise to irritating scratches if handled. Quilts rely on the loose way in which the core is packed for their effectiveness, and therefore the best results are obtained when they are laid between joists.

A variety of loose fills are also available for placing between the joists and over the ceiling to act as thermal insulators. Existing flat roofs can have their resistance to heat loss improved by applying thermal insulation laminated plasterboard to the ceiling.

Weatherproof finishes

Suitable materials are asphalt, lead, copper, zinc, aluminium and built-up roofing felt; only the latter will be considered at this stage.

Built-up roofing felt

Most roofing felts consist essentially of a base sheet of glass fibre or polyester reinforcement, impregnated with hot bitumen during manufacture. This is coated on both sides with a stabilised weatherproof bitumen compound. The outer coating is dusted with sand while still hot and tacky. The underlayer may receive a thin layer of polythene to prevent the sheet from sticking to itself when rolled. After cooling the felt is cut to form rolls 1 m wide and 10 or 20 m long before being wrapped for dispatch.

BS 747: *Reinforced bitumen sheets for roofing. Specification* contains two types:

- Type 3: glass fibre reinforced;
- Type 5: polyester reinforced.

Both types have subdivisions relating to the surface finish. Other bituminous sheet materials are available that incorporate polymer modifications to the bitumen. These are styrene–butadiene–styrene (SBS) and atactic–polypropylene (APP). SBS improves the physical properties, and APP has a higher melting point, most suited to torch-on applications. Rag-fibre-reinforced bituminous felts are still available, but are no longer recommended for roofing habitable accommodation.

For flat roofs three layers of felt should be used, the first being laid at right angles to the fall, commencing at the eaves. If the decking is timber the first layer is secured with large flat-head felt nails, and the subsequent layers are bonded to it with a hot bitumen compound by a roll and pour method. This involves pouring molten bitumen on the decking or underlayer and unrolling the sheet over it. Torch-on is for specially made sheets which are heated to the underside to produce a wave of molten bitumen while the sheet is unrolled. If the decking is of a material other than timber, all three layers are bonded with hot bitumen compound. It is usually recommended that a vented first layer be used in case moisture is trapped during construction; this recommendation does not normally apply to roofs with a timber deck, as timber has the ability to 'breathe'. The minimum fall recommended for built-up roofing felt is 17 mm in 1000 mm or 1°.

In general the Building Regulations require a flat roof with a weatherproofing that has a surface finish of asbestos-based bituminous felt or is covered with a layer of stone chippings. The chippings protect the underlying felt, provide additional fire resistance, and give increased solar reflection. A typical application would be 12.5 mm stone chippings at approximately 50 kg to each 2.5 m^2 of roof area. Chippings of limestone, granite and light-coloured gravel would be suitable.

Vapour control layer/membrane

The problem of condensation should always be considered when constructing a flat roof. The insulation below the built-up roofing felt will not prevent condensation occurring, and as it is a permeable material water vapour will pass upwards through it and condense on the underside of the roofing felt. The drops of moisture so formed will soak into the insulating material, lowering its insulation value and possibly causing staining on the underside. To prevent this occurring a vapour control layer should be placed on the underside of the insulating material. Purpose-made plasterboard with a metallised polyester backing can be specified in this situation, to combine water vapour resistance with the ceiling lining. The inverted warm deck roof will not require a separate vapour control layer, as the waterproof membrane will provide this function. For typical timber flat roof details see Figs 6.1.1 and 6.1.2.

■■■ FLAT ROOF JOIST SIZING

Joist sizes can be determined by calculation or by reference to design tables. The calculation method for roof joists is in principle as shown in the example for floor joists in Chapter 4.3, with the exception of an imposed loading allowance for snow, i.e. $< 30°$ pitch, 1.5 kN/m^2; $> 30°$ pitch, 0.75 kN/m^2.

The sizing of flat roof timbers in accordance with Building Regulation A – Structure, is provided in the Approved Document *Span tables for solid timber members* published by the Timber Research and Development Association (TRADA). Table 6.1.1 gives some guidance for selection of commercial sawn timber sizes relative to span and loading.

■■■ TIMBER PITCHED ROOFS

The term **pitched roof** includes any roof whose angle of slope to the horizontal lies between 10° and 70°; below this range it would be called a flat roof and above 70° it would be classified as a wall.

The pitch is generally determined by the covering that is to be placed over the timber carcass, whereas the basic form is governed by the load and span. The terminology used in timber roof work and the basic members for various spans are shown in Figs 6.1.3 and 6.1.4.

The cost of constructing a pitched roof is greater than for a flat roof, but the pitch will create a useful void for locating cold water storage cisterns and for general storage. The timber used in roof work is structural softwood, the members being joined together with nails. The sloping components or rafters are used to transfer the covering, wind, rain and snow loads to the loadbearing walls on which they rest. The rafters are sometimes assisted in this function by struts and purlins in what is called a **purlin** or **double roof** (see Fig. 6.1.5). As with other forms of roof the spacing of the rafters and consequently the ceiling joists is determined by the module size of the ceiling boards that are to be fixed on the underside of the joists.

ROOF MEMBERS

- **Ridge**   This is the spine of a roof and is essentially a pitching plate for the rafters that are nailed to each other through the ridge board. The depth of ridge board is governed by the pitch of the roof: the steeper the pitch, the deeper will be the vertical or plumb cuts on the rafters abutting the ridge.

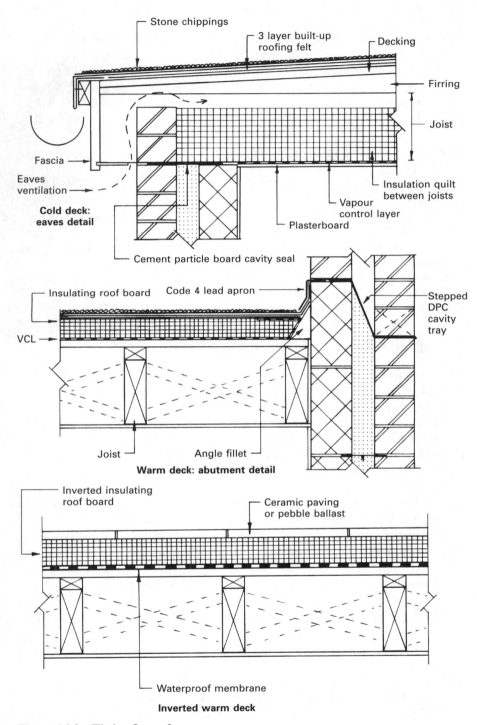

Figure 6.1.2   Timber flat roofs.

Table 6.1.1 Guide to span and loading potential for GS grade softwood timber flat roof joists (for timber grading categories, see Table 3.8.2).

	Dead weight of decking and ceiling, excluding self-weight of joist								
	< 50 kg/m^2 < 0.50 kN/m^2			50–75 0.50–0.75			75–100 0.75–1.00		
	Spacing of joists mm (c/c)								
Sawn size of joist (mm)	400	450	600	400	450	600	400	450	600
				Maximum clear span (m)					
38 × 125	1.80	1.79	1.74	1.74	1.71	1.65	1.68	1.65	1.57
38 × 150	2.35	2.33	2.27	2.27	2.25	2.18	2.21	2.18	2.09
38 × 175	2.88	2.85	2.77	2.77	2.74	2.64	2.68	2.64	2.53
38 × 200	3.47	3.43	3.29	3.33	3.28	3.16	3.21	3.16	3.02
38 × 225	4.08	4.03	3.71	3.90	3.84	3.56	3.75	3.66	3.43
50 × 125	2.06	2.05	2.00	2.00	1.98	1.93	1.95	1.93	1.86
50 × 150	2.68	2.65	2.59	2.59	2.58	2.47	2.51	2.47	2.38
50 × 175	3.27	3.25	3.14	3.14	3.10	2.99	3.04	2.99	2.86
50 × 200	3.93	3.86	3.61	3.76	3.70	3.47	3.62	3.56	3.35
50 × 225	4.60	4.47	4.07	4.38	4.30	3.91	4.21	4.13	3.78
63 × 100	1.67	1.66	1.63	1.63	1.61	1.57	1.59	1.93	1.86
63 × 125	2.31	2.29	2.24	2.24	2.21	2.15	2.17	2.15	2.07
63 × 150	2.98	2.95	2.87	2.87	2.84	2.74	2.78	2.74	2.63
63 × 175	3.62	3.59	3.41	3.48	3.43	3.28	3.36	3.30	3.16
63 × 200	4.34	4.29	3.90	4.15	4.08	3.75	3.99	3.92	3.62
63 × 225	5.00	4.82	4.39	4.82	4.64	4.22	4.62	4.48	4.08
75 × 125	2.50	2.48	2.42	2.42	2.40	2.32	2.35	2.32	2.24
75 × 150	3.23	3.19	3.11	3.11	3.07	2.96	3.00	2.96	2.84
75 × 175	3.91	3.87	3.61	3.75	3.69	3.47	3.61	3.55	3.35
75 × 200	4.66	4.53	4.13	4.45	4.36	3.97	4.28	4.20	3.84
75 × 225	5.28	5.09	4.65	5.09	4.90	4.47	4.92	4.74	4.32

- **Common rafters** The main loadbearing members of a roof; they span between a wall plate at eaves level and the ridge. Rafters have a tendency to thrust out the walls on which they rest, and this must be resisted by the walls and the ceiling joists. Rafters are notched over and nailed to a wall plate situated on top of a loadbearing wall; the depth of the notch should not exceed one-third the depth of the rafter.
- **Jack rafters** These fulfil the same function as common rafters but span from ridge to valley rafter or from hip rafter to wall plate.
- **Hip rafters** Similar to a ridge but forming the spine of an external angle and similar to a rafter spanning from ridge to wall plate.
- **Valley rafters** As hip rafters but forming an internal angle.
- **Wall plates** These provide the bearing and fixing medium for the various roof members, and distribute the loads evenly over the supporting walls; they are bedded in cement mortar on top of the loadbearing walls.
- **Dragon ties** Ties placed across the corners and over the wall plates to help provide resistance to the thrust of a hip rafter.

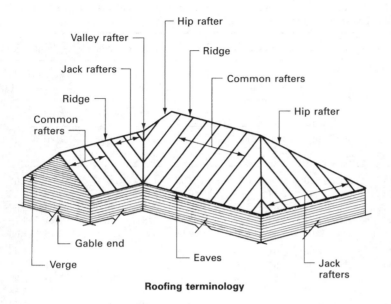

Hip rafter

Valley rafter

Ridge

Jack rafters

Common rafters

Ridge

Hip rafter

Common
rafters

Gable end

Eaves

Verge

Jack
rafters

Roofing terminology

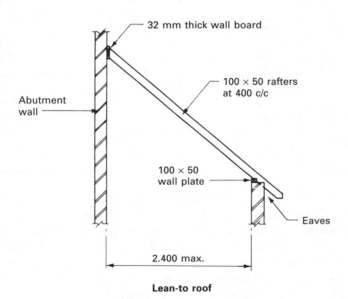

32 mm thick wall board

100 × 50 rafters
at 400 c/c

Abutment
wall

100 × 50
wall plate

Eaves

2.400 max.

Lean-to roof

Figure 6.1.3   Roofing terminology and lean-to roof.

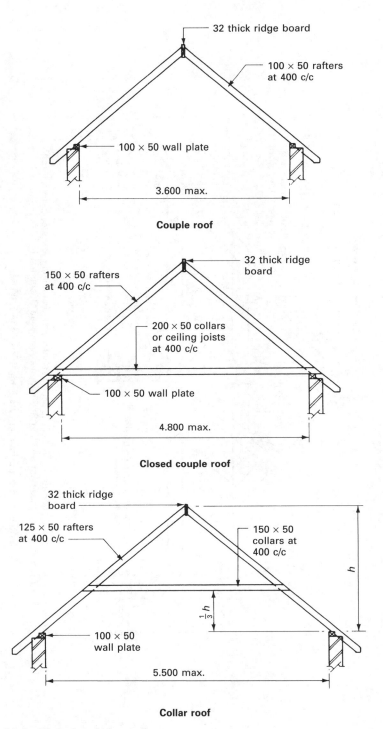

Figure 6.1.4 Pitched roofs for small spans.

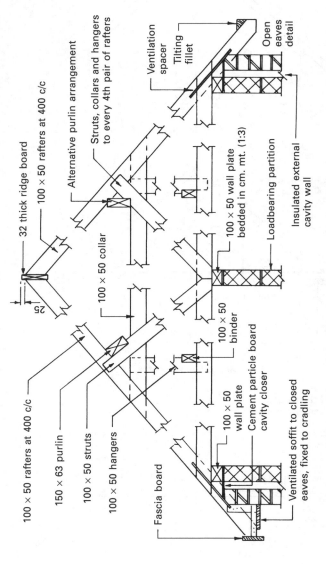

32 thick ridge board

100 × 50 rafters at 400 c/c

Alternative purlin arrangement

Struts, collars and hangers to every 4th pair of rafters

Ventilation spacer

Tilting fillet

Open eaves detail

100 × 50 collar

100 × 50 wall plate bedded in cm. mt. (1:3)

Loadbearing partition

Insulated external cavity wall

100 × 50 rafters at 400 c/c

150 × 63 purlin

100 × 50 struts

100 × 50 hangers

100 × 50 binder

100 × 50 wall plate

Cement particle board cavity closer

Fascia board

Ventilated soffit to closed eaves, fixed to cradling

Note: Mineral wool insulation (or equivalent) placed between and over ceiling joists or between rafters. Thickness in accordance with Building Regulations, Approved Document L.

Figure 6.1.5   Typical double or purlin roof details for spans up to 7200 mm.

■ **Ceiling joists** These fulfil the dual function of acting as ties to the feet of pairs of rafters and providing support for the ceiling boards on the underside and any cisterns housed within the roof void.

■ **Purlins** These act as beams, reducing the span of the rafters and enabling an economic section to be used. If the roof has a gable end they can be supported on a corbel or built in, but in a hipped roof they are mitred at the corners and act as a ring beam.

■ **Struts** These are compression members that transfer the load of a purlin to a suitable loadbearing support within the span of the roof.

■ **Collars** These are extra ties to give additional strength, and are placed at purlin level.

■ **Binders** These are beams used to give support to ceiling joists and counteract excessive deflections, and are used if the span of the ceiling joist exceeds 2400 mm.

■ **Hangers** Vertical members used to give support to the binders and allow an economic section to be used; they are included in the design if the span of the binder exceeds 3600 mm.

[*Note*: The arrangement of struts, collars and hangers occurs only on every fourth or fifth pair of rafters.]

PITCHED ROOF COMPONENT SIZING

The size of the principal timber members constituting a roof structure can be determined by calculation or from design tables as referred to under 'flat roof joist sizing'. Tables are extensive, to include the various roof components, grades of timber, different roof pitches and loadings. As an example, Table 6.1.2 is provided for guidance on selection of rafters for pitches between 30° and 45° with an imposed loading of 0.75 kN/m^2.

Table 6.1.2 Guide to span and loading potential for GS grade softwood timber rafters

	Dead weight tiles, battens, etc., excluding self-weight of rafter								
	< 50 kg/m^2 < 0.50 kN/m^2			50–75 0.50–0.75			75–125 0.75–1.25		
	Spacing of rafters (c/c)								
Sawn size of rafter (mm)	400	450	600	400	450	600	400	450	600
				Maximum clear span (m)					
38 × 100	2.28	2.23	2.10	2.10	2.05	1.91	1.96	1.91	1.76
38 × 125	3.07	2.95	2.69	2.87	2.77	2.52	2.65	2.56	2.35
38 × 150	3.67	3.53	3.22	3.44	3.31	3.01	3.26	3.14	2.85
50 × 100	2.69	2.59	2.36	2.53	2.43	2.21	2.38	2.30	2.09
50 × 125	3.35	3.23	2.94	3.15	3.03	2.76	2.98	2.87	2.61
50 × 150	4.00	3.86	3.52	3.76	3.62	3.30	3.57	3.44	3.13

Note: Clear span is measured between wallplate and purlin and between purlin and ridge board.

EAVES

The eaves of a roof is the lowest edge that overhangs the wall, thus giving the wall a degree of protection; it also provides the fixing medium for the rainwater gutter. The amount of projection from the wall of the eaves is a matter of choice but is generally in the region of 300–450 mm.

There are two basic types of eaves finish: **open eaves** and **closed eaves**. The former is less expensive, with rafters left exposed on the underside and treated with preservative. With both, the space between the rafters and the roof covering receives insulation over the top of the wall. A 50 mm air space must remain above the insulation for free air circulation through the roof in order to prevent condensation from occurring. This is easily achieved with a proprietary eaves ventilator secured between the rafters. A continuous triangular tilting fillet is fixed over the backs of the rafters to provide support for the bottom course of slates or tiles. A closed eaves is one in which the feet of the rafters are boxed in using a vertical fascia board, with the space between the fascia and the wall containing a ventilated soffit board. In a cheaper variant, the rafters are cut marginally beyond the wall face to leave space for ventilation and only a fascia board fixed to the rafter ends. This is called a **flush eaves**. Figures 6.1.5 and 6.1.6 indicate various rafter finishes, with provision for adequate through-ventilation.

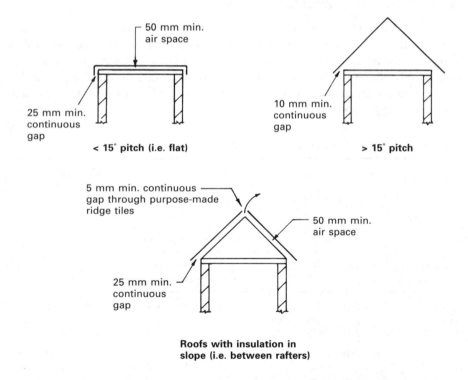

Note: Gaps need not be continuous, if staggered voids provide the equivalent area

Figure 6.1.6   Roof ventilation requirements.

TIMBER ROOF TRUSSES

These can be used on the larger spans in domestic work to give an area below the ceiling level free from loadbearing walls. Trusses are structurally designed frames based on the principles of triangulation, and serve to carry the purlins; they are spaced at 1800 mm centres with the space between being infilled with common rafters. It is essential that the members of a roof truss are rigidly connected together, as light sections are generally used. To make a suitable rigid joint, bolts and timber connectors are used. These are square- or circular-toothed plates; the teeth are pointed up and down and, when clamped between two members, bite into the surface, forming a strong connection and spreading the stresses over a greater surface area. A typical roof truss detail is shown in Fig. 6.1.7.

TRUSSED RAFTERS

This is another approach to the formation of a domestic timber roof giving a clear span; as with roof trusses it is based upon a triangulated frame, but in this case the members are butt jointed and secured with truss plates. All members in a trussed rafter are machined on all faces so that they are of identical thickness, ensuring a strong connection on both faces. The trussed rafters are placed at 600 mm centres and tied together over their backs with 38 mm × 25 mm tiling battens; no purlin or ridge is required. Stability is achieved from the tile battens and 100 mm × 25 mm diagonal wind braces from the bottom corner to a top corner. Also, two 100 mm × 25 mm longitudinal ties should run horizontally along the ceiling ties.

Truss manufacturers will design and prefabricate to the client's specification. This should include span, loading (type of tile), degree of exposure, pitch, spacing (if not 600 mm), and details of any special loadings such as water cistern location.

Truss or nail plates are generally of one or two forms:

- those in which holes are punched to take nails; suitable for site assembly using a nailing gun;
- those in which teeth are punched and bent from the plate; used in factory assembly using heavy presses.

In all cases truss plates are fixed to both faces of the butt joint.

Trussed rafters are also produced using gusset plates of plywood at the butt joints instead of truss plates; typical details of both forms are shown in Fig. 6.1.8.

LATERAL BRACING

Gable ladders and restraint straps secure pitched roofs to adjacent walls and, in turn, add stability to the wall. Figure 6.1.9 shows the application with straps at 2 m spacing. Straps may need to be more frequent, depending on building height and degree of exposure.

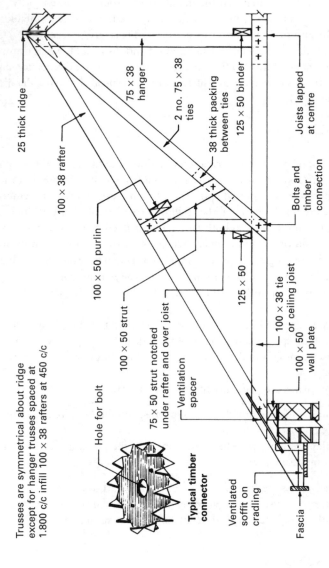

Trusses are symmetrical about ridge
except for hanger trusses spaced at
1.800 c/c infill 100 × 38 rafters at 450 c/c

25 thick ridge

100 × 38 rafter

100 × 50 purlin

75 × 38 hanger

2 no. 75 × 38 ties

38 thick packing between ties

125 × 50 binder

Joists lapped at centre

Bolts and timber connection

125 × 50

100 × 38 tie or ceiling joist

100 × 50 wall plate

100 × 50 strut

75 × 50 strut notched under rafter and over joist

Ventilation spacer

Hole for bolt

Typical timber connector

Ventilated soffit on cradling

Fascia

Figure 6.1.7 Typical truss details for spans up to 8000 mm.

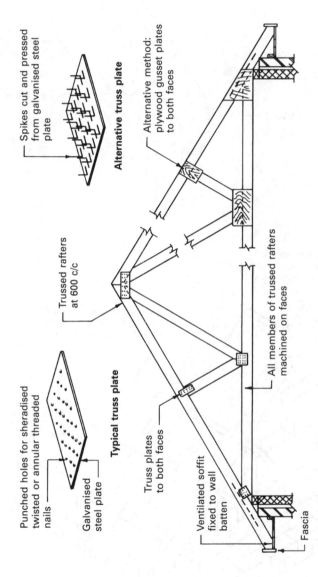

Spikes cut and pressed from galvanised steel plate

Alternative truss plate

Alternative method: plywood gusset plates to both faces

Trussed rafters at 600 c/c

All members of trussed rafters machined on faces

Punched holes for sheradised twisted or annular threaded nails

Galvanised steel plate

Typical truss plate

Truss plates to both faces

Ventilated soffit fixed to wall batten

Fascia

Figure 6.1.8   Typical trussed rafter details for spans up to 11 000 mm.

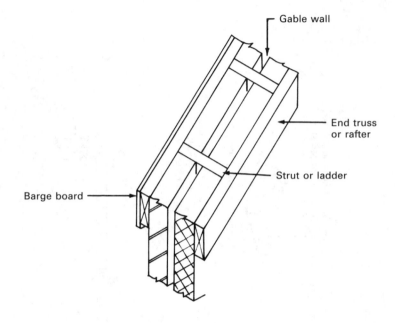

Gable wall

End truss
or rafter

Strut or ladder

Barge board

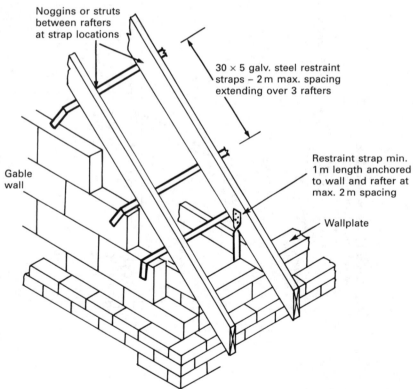

Noggins or struts
between rafters
at strap locations

30 × 5 galv. steel restraint
straps – 2 m max. spacing
extending over 3 rafters

Restraint strap min.
1 m length anchored
to wall and rafter at
max. 2 m spacing

Gable
wall

Wallplate

Figure 6.1.9   Lateral restraint to gable walls.

BUILDING REGULATIONS

Regulation 7

This requires that any building work shall be carried out with proper materials and in a workmanlike manner. The Approved Document supporting this regulation defines acceptable levels of performance for materials that include products, fittings, items of equipment, and backfilling for excavations. The aids for establishing fitness of materials suggested in the Approved Document are:

- past experience – such as a building in use;
- Agrément certificates or equivalent European Technical Approvals;
- British Standards and European Community national standards;
- independent certification schemes to the National Accreditation Council for Certification Bodies (NACCB) requirements;
- quality assurance schemes complying with BS 5750 and BS EN ISO 9000–9004;
- test and calculations for materials only using the National Measurement Accreditation Service (NAMAS) scheme for testing laboratories.

The Approved Document supporting Regulation 7 lists certain geographical areas where the softwood timber used for roof construction should be adequately treated with a suitable preservative to prevent infestation by the house longhorn beetle. This is a wise precaution whether it is recommended or not.

Regulation B4 (Section 14)

This requires the roof to offer adequate resistance to the spread of fire over the roof. Table 17 in Approved Document B gives limitations on roof coverings for dwelling houses by designations and distance from the boundary, the designations being defined in Table A5.

Regulation C2

This requires that the roof of a building shall adequately resist the passage of moisture to the inside of the building.

Regulation L1

This states that reasonable provision shall be made for the conservation of fuel and power in buildings. To satisfy this requirement Approved Document L requires that the roof construction is enhanced with suitable insulation material (see Chapter 8.2 for more detail).

Regulation A

The selection of suitable structural timber members for roofs with various loadings and spans can be determined from publications recommended in Approved Document A of the Building Regulations.

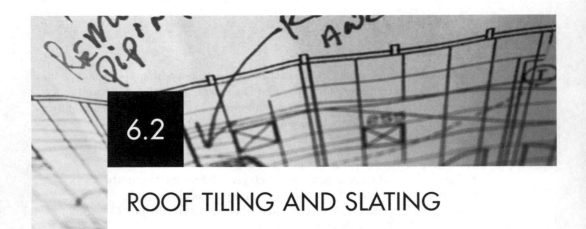

6.2

ROOF TILING AND SLATING

A roof can be defined as the upper covering of a building, and in this position
it is fully exposed to the rain, snow, wind, sun and general atmosphere: therefore
the covering to any roof structure must have good durability to meet these
conditions. Other factors to be taken into account are weight, maintenance
and cost.

Roofs are subjected to wind pressures, both positive and negative, the latter
causing uplift and suction, which can be overcome by firmly anchoring lightweight
coverings to the structure or by relying upon the deadweight of the covering
material. Domestic roofs are not usually designed for general access, and therefore
the chosen covering material should be either very durable or easily replaceable.
The total dead load of the covering will affect the type of support structure required
and ultimately the total load on the foundations, so therefore careful consideration
must be given to the medium selected for the roof covering.

In domestic work the roof covering usually takes the form of tiling or slating,
because these materials economically fulfil the above requirements and have
withstood the test of time.

◼◼◼ TILING

Tiles are manufactured from clay and concrete to a wide range of designs and
colours suitable for pitches from 15° to 45°, and work upon the principle of either
double or single lap. The vital factor for the efficient performance of any tile or slate
is the pitch, and it should be noted that the pitch of a tile is always less than the
pitch of the rafters owing to the overlapping technique.

Tiles are laid in overlapping courses and rely upon the water being shed off
the surface of one tile onto the exposed surface of the tile in the next course. The
problem of water entering by capillary action between the tiles is overcome by the
camber of the tile, by the method of laying, or by overlapping side joints. In all

methods of tiling a wide range of fittings are produced to enable all roof shapes to be adequately protected.

PLAIN TILING

This is a common method in this country and works on the double lap principle. The tiles can be of handmade or machine-pressed clay, the process of manufacture being similar to that of brickmaking. The handmade tile is used mainly where a rustic or distinctive roof character is required, because they have a wide variation in colour, texture and shape. They should not be laid on a roof of less than 45° pitch because they tend to absorb water, and if allowed to become saturated they may freeze, expand and spall or fracture in cold weather. Machine-pressed tiles are harder, denser and more uniform in shape than handmade varieties and can be laid to a minimum pitch of 35°.

A suitable substitute for plain clay tiles is the concrete plain tile. These are produced in a range of colours to the same size specifications as the clay tiles and with the same range of fittings. The main advantage of concrete tiles is their lower costs; the main disadvantage is the extra weight. Plain clay tile details are specified in BS EN 1304 and BS EN 538. Concrete tiles have separate designation in BS EN 490 and 491.

Plain tiling, in common with other forms of tiling, provides an effective barrier to rain and snow penetration, but wind is able to penetrate into the building through the gaps between the tiling units: therefore a barrier in the form of boarding or sheeting is placed over the roof carcass before the battens on which the tiles are to be hung are fixed.

The rule for plain tiling is that there must always be at least two thicknesses of tiles covering any part of the roof, and bonded so that no 'vertical' joint is immediately over a 'vertical' joint in the course below. To enable this rule to be maintained shorter-length tiles are required at the eaves and the ridge; each alternate course is commenced with a wider tile of one-and-a-half tile widths. The apex or ridge is capped with a special tile bedded in cement mortar over the general tile surface. The hips can be covered with a ridge tile, in which case the plain tiling is laid underneath and mitred on top of the hip; alternatively a special bonnet tile can be used where the plain tiles bond with the edges of the bonnet tiles. Valleys can be formed by using special tiles or mitred plain tiles, or by forming an open gutter with a durable material such as lead. The verge at the gable end can be formed by bedding plain tiles face down on the gable wall as an undercloak and bedding the plain tiles in cement mortar on the upper surface of the undercloak. The verge tiling should overhang its support by at least 50 mm. Abutments are made watertight by dressing a flashing over the upper surface of tiling, between which is sandwiched a soaker. The soaker in effect forms a gutter. An alternative method is to form a cement fillet on top of the tiled surface, but this method sometimes fails by the cement shrinking away from the surface of the wall.

The support or fixing battens are of softwood extended over and fixed to at least three rafters, the spacing or gauge being determined by the lap given to the tiles thus:

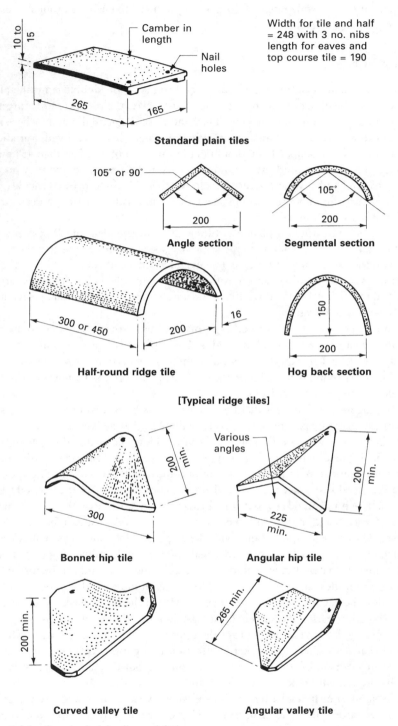

10 to 15

Camber in length

Nail holes

Width for tile and half = 248 with 3 no. nibs length for eaves and top course tile = 190

265

165

Standard plain tiles

105° or 90°

200

Angle section

105°

200

Segmental section

300 or 450

200

16

Half-round ridge tile

150

200

Hog back section

[Typical ridge tiles]

200 min.

300

Bonnet hip tile

Various angles

200 min.

225 min.

Angular hip tile

200 min.

Curved valley tile

265 min.

Angular valley tile

Figure 6.2.1   Standard plain tiles and fittings.

$$\text{Gauge} = \frac{\text{length of tile} - \text{lap}}{2}$$

$$\text{Gauge} = \frac{265 - 65}{2} = 100 \text{ mm.}$$

Plain tiles are fixed with two galvanised nails to each tile in every fourth or fifth course. Details of plain tiles, fittings and methods of laying are shown in Figs 6.2.1 to 6.2.5.

SINGLE-LAP TILING

Single-lap tiles are laid with overlapping side joints to a minimum pitch of 35° and are not bonded like the butt-jointed single-lap plain tiles; this gives an overall reduction in weight as fewer tiles are used. A common form of single-lap tile is the **pantile**, which has opposite corners mitred to overcome the problem of four tile thicknesses at the corners (see Fig. 6.2.6). The pantile is a larger unit than the plain tile and is best employed on large roofs with gabled ends, because the formation of hips and valleys is difficult and expensive. Other forms of single-lap tiling are Roman tiling, Spanish tiling and interlocking tiling. The latter types are produced in both concrete and clay and have one or two grooves in the overlapping edge to give greater resistance to wind penetration; they can generally be laid as low as 15° pitch (see Fig. 6.2.6).

■■■ SLATING

Slate is a naturally dense material, which can be split into thin sheets and used to provide a suitable covering to a pitched roof. Slates are laid to the same basic principles as double-lap roofing tiles except that every slate should be twice nailed. Slates come mainly from Wales, Cornwall and the Lake District and are cut to a wide variety of sizes – the Westmorland slates are harder to cut and are usually supplied in random sizes. Slates can be laid to a minimum pitch of 25° and are fixed by head nailing or centre nailing. Centre nailing is used to overcome the problem of vibration caused by the wind and tending to snap the slate at the fixing if nailed at the head; it is used mainly on the long slates and pitches below 35°.

The gauge of the battens is calculated thus:

$$\text{Head-nailed gauge} = \frac{\text{length of slate} - (\text{lap} + 25 \text{ mm})}{2}$$

$$= \frac{400 - (75 + 25)}{2} = 150 \text{ mm}$$

$$\text{Centre-nailed gauge} = \frac{\text{length of slate} - \text{lap}}{2}$$

$$= \frac{400 - 76}{2} = 162 \text{ mm.}$$

Roofing slates are covered by BS EN 12326-1, which gives details of standard sizes, thicknesses and quality. Typical details of slating are shown in Fig. 6.2.7.

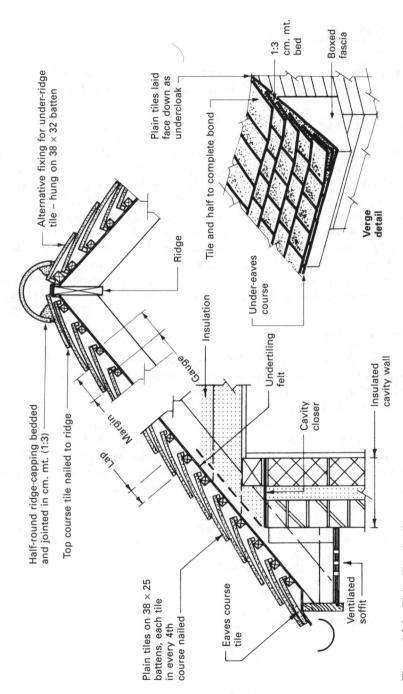

Alternative fixing for under-ridge tile – hung on 38 × 32 batten

Plain tiles laid face down as undercloak

1:3 cm. mt. bed

Boxed fascia

Tile and half to complete bond

Verge detail

Ridge

Under-eaves course

Half-round ridge-capping bedded and jointed in cm. mt. (1:3)

Top course tile nailed to ridge

Insulation

Undertiling felt

Gauge

Margin

Lap

Cavity closer

Insulated cavity wall

Plain tiles on 38 × 25 battens, each tile in every 4th course nailed

Eaves course tile

Ventilated soffit

Figure 6.2.2   Plain tiling details.

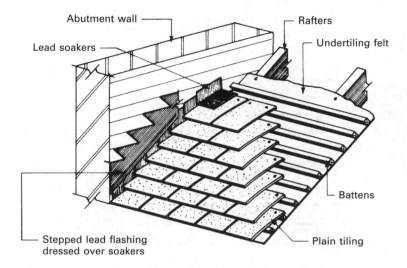

Abutment wall

Lead soakers

Rafters

Undertiling felt

Battens

Stepped lead flashing dressed over soakers

Plain tiling

Typical abutment detail (see also Fig. 5.1.1)

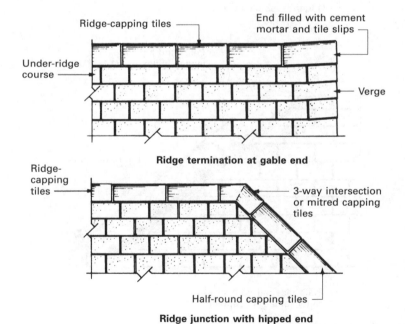

Ridge-capping tiles

End filled with cement mortar and tile slips

Under-ridge course

Verge

Ridge termination at gable end

Ridge-capping tiles

3-way intersection or mitred capping tiles

Half-round capping tiles

Ridge junction with hipped end

Figure 6.2.3 Abutment and ridge details.

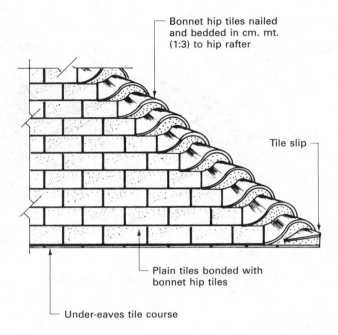

Bonnet hip tiles nailed
and bedded in cm. mt.
(1:3) to hip rafter

Tile slip

Plain tiles bonded with
bonnet hip tiles

Under-eaves tile course

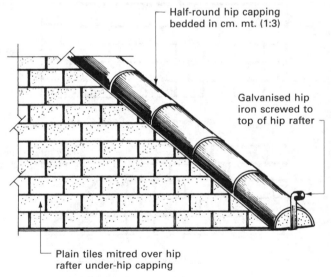

Half-round hip capping
bedded in cm. mt. (1:3)

Galvanised hip
iron screwed to
top of hip rafter

Plain tiles mitred over hip
rafter under-hip capping

Figure 6.2.4 Hip treatments.

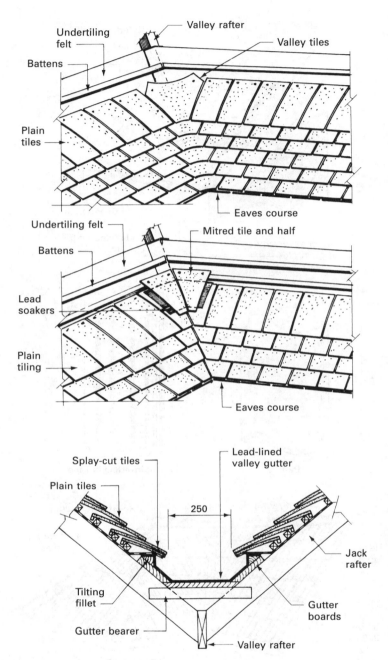

Figure 6.2.5 Valley treatments.

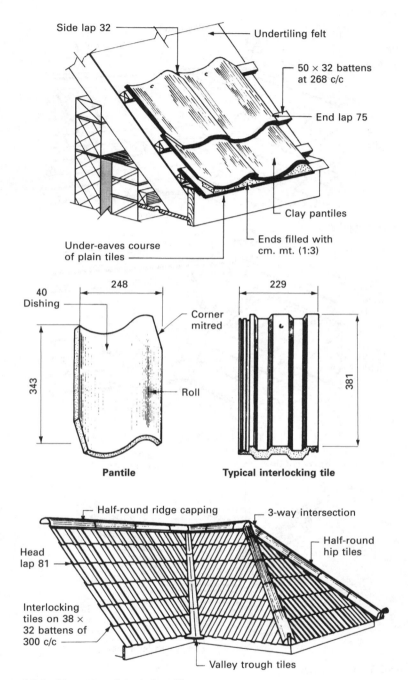

Figure 6.2.6   Examples of single-lap tiling.

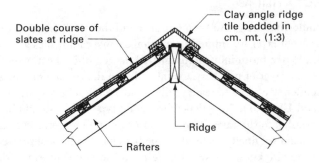

Typical ridge detail

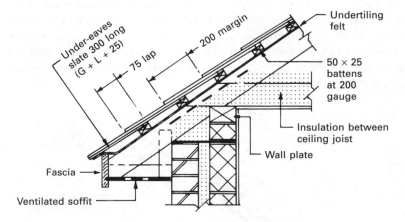

Head-nailed slating using 500 × 250 slates

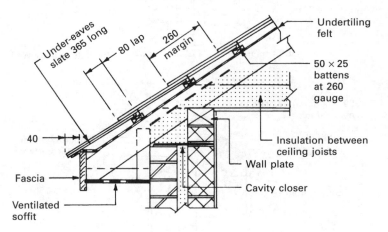

Centre-nailed slating using 600 × 300 slates

Figure 6.2.7   Typical slating details.

▦ ▦ ■ UNDERTILING FELTS

The object of undertiling felt is to keep out dust and the wind. It also provides a secondary waterproof layer for limited protection against rain if tiles are damaged. It consists basically of a bituminous impregnated matted fibre sheet, which can be reinforced with a layer of jute hessian embedded in the coating on one side to overcome the tendency of felts to tear readily. Undertiling felts are supplied in rolls 1 m wide and 10 or 20 m long depending upon type. They should be laid over the rafters and parallel to the eaves with 150 mm laps and temporarily fixed with large-head felt nails until finally secured by the battens. Traditional first-generation undertiling felts are also known as **sarking felts** and should conform to the requirements of BS 747. They should also be permeable to water vapour, to relieve any possibility of condensation in the roof space. Subsequently there has been a second generation of lightweight tiling underlays produced from reinforced plastics. As for traditional bituminous felt, there is a risk that condensation can form on the underside. Third-generation underlays are of triple-ply construction, comprising a waterproof and vapour-permeable core between layers of non-woven, spun-bonded polypropylene. In effect, they permit internal moisture to pass through but remain watertight to external conditions.

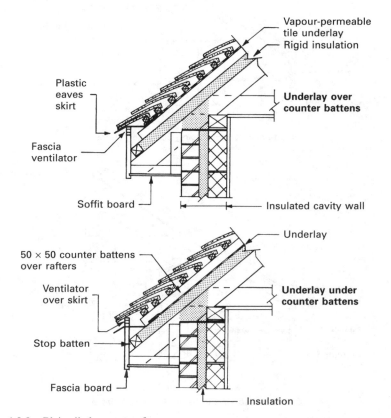

Figure 6.2.8 Plain tiled warm roof.

■■■ COUNTER BATTENS

If a roof is boarded before the roofing felt is applied the tiling battens will provide ledges on which dirt and damp can collect. To overcome this problem counter battens are fixed to the boarding over the rafter positions to form a cavity between the tiling battens and the boarding. Boarding a roof over the rafters is a method of providing a wind barrier and adding to the thermal insulation properties, but it is seldom used in new work because of the high cost.

Use of counter battening is common where pitched roofs are of warm construction, i.e. the insulation is above the roof slope to create a habitable room in the roof void. See Fig. 6.2.8. Cold roof construction with insulation at ceiling joist level is shown in Figs 6.2.2 and 6.2.7.

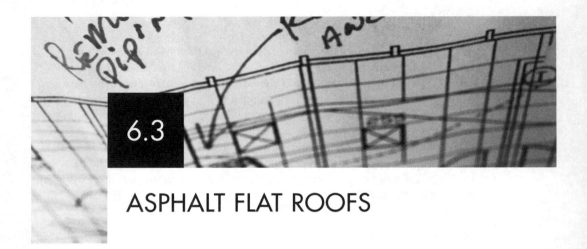

6.3

ASPHALT FLAT ROOFS

Flat roofs are often considered to be a simple form of construction, but unless correctly designed and constructed they can be an endless source of trouble. Flat roofs can have the advantage of providing an extra area to a building for the purposes of recreation and an additional viewpoint. Mastic asphalt provides an ideal covering material, especially where foot traffic is required.

Mastic asphalt consists of an aggregate with a bituminous binder, which is cast into blocks ready for reheating on site (see Chapter 7.8). The blocks are heated in cauldrons or cookers to a temperature of over 200 °C and are then transported in a liquid state in buckets for application to the roof deck by hand spreading. Once the melted asphalt has been removed from the source of heat it will cool and solidify rapidly: therefore the distance between the cauldron and the point of application should be kept to a minimum. Mastic asphalt can be applied to most types of rigid substructure and proprietary structural deckings. To prevent slight movements occurring between the substructure and the finish an isolating membrane should be laid over the base before the mastic asphalt is applied.

Roof work will often entail laying mastic asphalt to horizontal, sloping and vertical surfaces, and these are defined as follows:

- horizontal surfaces – up to 10° pitch;
- sloping surfaces – between 10° and 45° pitch;
- vertical surfaces – over 45° pitch.

The thickness and number of coats required will depend on two factors:

- surface type;
- substructure or base.

Horizontal surfaces with any form of rigid base should have a two-coat application of mastic asphalt laid breaking the joint and built up to a minimum total thickness of 20 mm. An isolating membrane of black sheathing felt complying with BS 747 Type 4A(i) should be used as an underlay laid dry over insulation boards with 50 mm lapped joints.

Vertical and sloping surfaces other than those with a timber base require a three-coat application built up to a 20 mm total thickness without an isolating membrane.

Timber substructures with vertical and sloping surfaces should have a three-coat 20 mm mastic asphalt finish applied to expanded metal lathing complying with BS 1369 fixed at 150 mm centres over an isolating membrane.

If the mastic asphalt surface is intended for foot traffic the total thickness of the two-coat application should be increased to a minimum of 25 mm.

Roofs with a mastic asphalt finish can be laid to falls so that the run-off of water is rapid and efficient. Puddles of water left on the roof surface will create small areas of different surface temperatures, which could give rise to local differential thermal movements and cause cracking of the protective covering. Depressions will also provide points at which dust and rubbish can collect: it is therefore essential that suitable falls be provided to direct the water to an eaves gutter or specific outlet points. The falls should be formed on the base or supporting structure to a gradient of not less than 1 in 80 (see typical details in Figs 6.3.1 and 6.3.2).

Alternatively the roof can be designed to act as a reservoir by a technique sometimes called **ponding**. The principle is to retain on the roof surface a 'pond' of water some 150 mm deep by having the surface completely flat, with high skirtings and outlets positioned 150 mm above the roof level. The main advantage is that differential temperatures are reduced to a minimum; the disadvantages are the need for a stronger supporting structure to carry the increased load, a three-coat 30 mm thick covering of mastic asphalt, and the need to flood the roof in dry hot weather to prevent the pond from completely evaporating away.

Resistance to heat loss can be achieved by incorporating within the structure a dry insulation such as cork slabs, wood fibre boards, glass or mineral fibre boards, rigid polyisocyanurate (Pi) foam board or rigid lightweight expanded polystyrene (EPS) board. This may be deployed to create warm or cold deck roof insulating principles as described in Chapter 6.1. If fixed over the roof structure it will reduce the temperature variations within the roof to a minimum and hence the risk of unacceptable thermal movements. A lightweight concrete screed will also prove a useful insulative supplement and a means of providing the roof slope. Suitable aggregates are furnace clinker, foamed blast furnace slag, pumice, expanded clay, sintered pulverised fuel ash, exfoliated vermiculite and expanded perlite. The thickness should not be less than 40 mm, and it may be necessary to apply to the screed a 1:4 cement:sand topping to provide the necessary surface finish. Mix ratios of 1:8–10 are generally recommended for screeds of approximately 1100 kg/m^3 density using foamed slag, sintered pulverised fuel ash and expanded clay, whereas a 1:5 mix is recommended for exfoliated vermiculite and expanded perlite aggregates giving a low density of under 640 kg/m^3. When mixing screeds containing porous aggregates, high water/cement ratios are required to give workable mixes, and therefore the screeds take a long time to dry out.

Moisture in the form of a vapour will tend to rise within the building and condense on the underside of the covering or within the thickness of the insulating

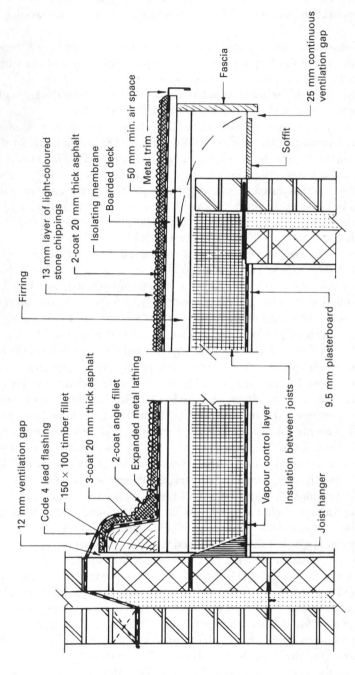

Figure 6.3.1   Timber flat roof with mastic asphalt covering.

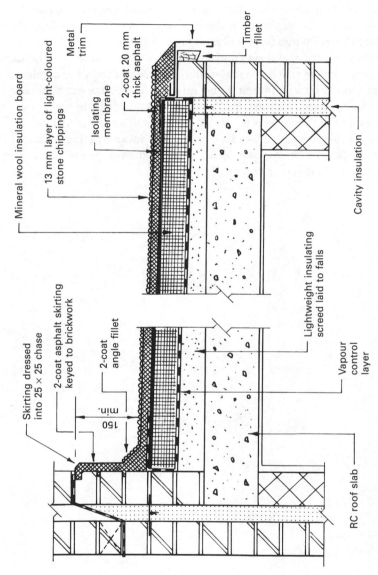

Figure 6.3.2   Reinforced concrete flat roof with mastic asphalt covering.

material. If an insulating material becomes damp its efficiency will decrease, and if composed of organic material it can decompose. To overcome this problem a vapour control layer of a suitable impermeable material such as polythene sheet or aluminium foil should be placed beneath the insulating layer. Care must be taken to see that all laps are complete and sealed and that the edges of the insulating material are similarly protected against the infiltration of moisture vapour. This is achieved by turning the vapour control layer around the edges of insulation board and bonding about 300 mm (min. 150 mm) to the top face of board.

The surface of a flat roof, being fully exposed, will gain heat by solar radiation and if insulated will be raised to a temperature in excess of the air or ambient temperature, as the transfer of heat to the inside of the building has been reduced by the insulating layer. The usual method employed to reduce the amount of solar heat gain of the covering is to cover the upper surface of the roof with light-coloured chippings to act as a reflective finish. Suitable aggregates are white spar, calcined flint, white limestone or any light-coloured granite of 12 mm size and embedded in a bitumen compound.

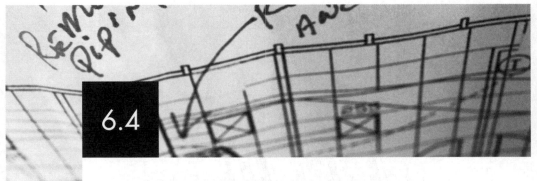

6.4

LEAD-COVERED FLAT ROOFS

Lead as a building material has been used extensively for over 5000 years, and is obtained mainly from the mineral galena, of which Australia, Canada, Mexico and the USA are the main producers. The raw material is mined, refined to a high degree of purity and then cast into bars or pigs, which can be used to produce lead sheet, pipe and extruded products.

Lead is a durable and dense material (11 340 kg/m^3) of low strength but is very malleable and can be worked cold into complicated shapes without fracture. In common with other non-ferrous metals, lead oxidises on exposure to the atmosphere and forms a thin protective film or coating over its surface. When in contact with other metals there is seldom any corrosion by electrolysis, and therefore fixing is usually carried out by using durable copper nails.

For application to flat roofs, milled lead sheet should comply with the recommendations of BS EN 12588. The sheet is supplied in rolls of widths varying between 150 mm and 2400 mm, in 3 m and 6 m standard lengths. For easy identification, lead sheet carries a colour guide for each code number:

BS Code No.	Thickness (mm)	Weight (kg/m^2)	Colour
3	1.32	14.97	green
4	1.80	20.41	blue
5	2.24	25.40	red
6	2.65	30.10	black
7	3.15	35.72	white
8	3.55	40.26	orange

The code number is derived from the former imperial notation of 5 lb/ft^2 = No. 5 lead.

The thickness or code number of lead sheet for any particular situation will depend upon the protection required against mechanical damage and the shape

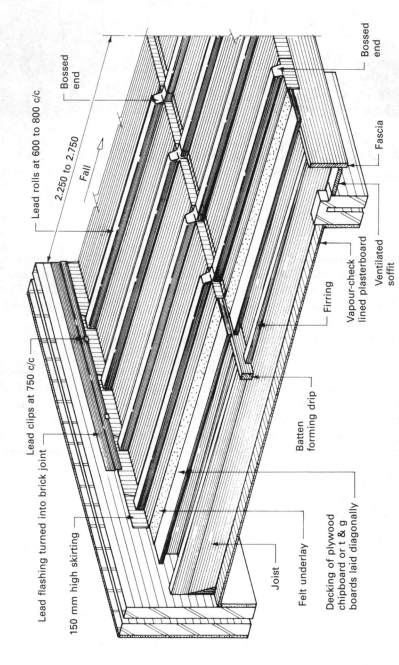

Bossed
end

Bossed
end

Lead rolls at 600 to 800 c/c

2.250 to 2.750

Fall

Fascia

Lead flashing turned into brick joint

Lead clips at 750 c/c

150 mm high skirting

Firring

Vapour-check
lined plasterboard

Ventilated
soffit

Batten
forming drip

Joist

Felt underlay

Decking of plywood
chipboard or t & g
boards laid diagonally

Figure 6.4.1 Typical layout of lead flat roof.

required. The following thicknesses can therefore be considered as a general guide for flat roofs:

- small areas without foot traffic, No. 4 or 5;
- small areas with foot traffic, No. 5, 6 or 7;
- large areas with or without foot traffic, No. 5, 6 or 7;
- flashings, No. 4 or 5;
- aprons, No. 4 or 5.

Milled lead sheet may be used as a covering over timber or similar deckings and over smooth screeded surfaces. In all cases an underlay of felt or stout building paper should be used to reduce frictional resistances, decrease surface irregularities and, in the case of a screeded surface, isolate the lead from any free lime present that might cause corrosion. Provision must also be made for the expansion and contraction of the metal covering. This can be achieved by limiting the area and/or length of the sheets being used. The following table indicates the maximum recommended area and length for any one piece of lead:

BS Code No.	*Max. length between drips* (m)	*Max. area* (m^2)
3	Only suitable for soakers	–
4	1.50	1.13
5	2.00	1.60
6	2.25	1.91
7	2.50	2.25
8	3.00	3.00

Joints that can accommodate the anticipated thermal movements are in the form of rolls running parallel to the fall and drips at right angles to the fall positioned so that they can be cut economically from a standard sheet: for layout and construction details see Figs 6.4.1 and 6.4.2.

[*Note*: Current legislation will require these details to incorporate cavity insulation in the walls with an insulative concrete block inner leaf. Insulation must also be included within the roof construction, as shown in Chapter 6.1.]

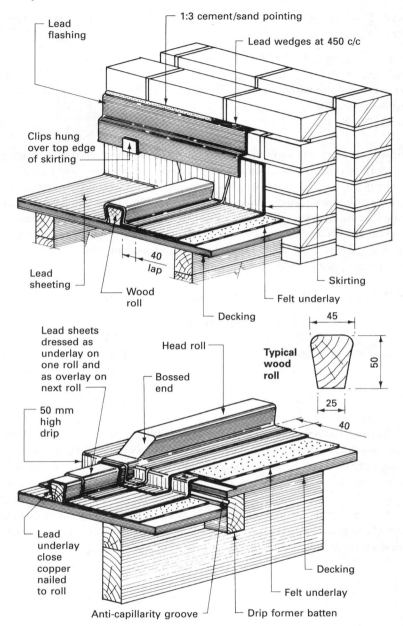

Lead flashing

1:3 cement/sand pointing

Lead wedges at 450 c/c

Clips hung over top edge of skirting

Lead sheeting

Wood roll

40 lap

Decking

Felt underlay

Skirting

Lead sheets dressed as underlay on one roll and as overlay on next roll

Head roll

Bossed end

Typical wood roll

45

50

25

40

50 mm high drip

Lead underlay close copper nailed to roll

Anti-capillarity groove

Drip former batten

Felt underlay

Decking

Figure 6.4.2   Lead flat roof details.

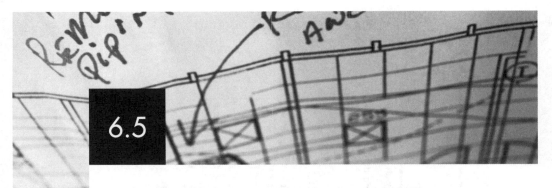

6.5

COPPER-COVERED FLAT ROOFS

Copper, like lead, has been used as a building material for many centuries. It is obtained from ore deposits, which have to be mined, crushed, and ground to a fine powder. The copper dust, which is about 4% of the actual rock mined, is separated from the waste materials, generally by a system of flotation, and transferred to the smelting furnace and then cast into cakes of blister copper or thick slabs called **anodes**. The metal is now refined and formed into copper wire, strip, sheets and castings. Copper is also used to form alloys that are used in making components for the building industry. Common alloys are copper and zinc (forming brass) and copper and tin (forming bronze).

Copper is a dense material (8900 kg/m^3) that is highly ductile and malleable, and can be cold-worked into the required shape or profile. The metal hardens with cold-working, but its original dead soft temper can be restored by the application of heat with a blowlamp or oxyacetylene torch and quenching with water or by natural air cooling. If the dead soft temper is not maintained the hardened copper will be difficult to work and may fracture. On exposure to the atmosphere copper forms on its upper surface a natural protective film or patina, which varies in colour from green to black, making the copper beneath virtually inert.

For covering flat roofs rolled copper complying with the recommendations of BS 2870 is generally specified. Rolled copper is available in three forms:

- **Sheet** Flat material of exact length, over 0.15 mm up to and including 10.0 mm thick and over 450 mm in width.
- **Strip** Material over 0.15 mm up to and including 10.0 mm thick and any width, and generally not cut to length. It is usually supplied in coils but can be obtained flat or folded.
- **Foil** Material 0.15 mm thick and under, of any width supplied flat or in a coil; because of its thickness foil has no practical application in the context of roof coverings.

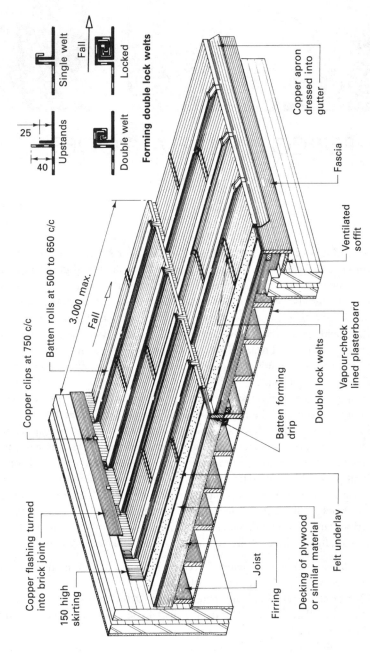

Single welt

Upstands

25

40

Locked

Double welt

Forming double lock welts

Fall

Copper apron
dressed into
gutter

Fascia

Ventilated
soffit

Batten rolls at 500 to 650 c/c

Copper clips at 750 c/c

3,000 max.

Fall

Vapour-check
lined plasterboard

Double lock welts

Batten forming
drip

Copper flashing turned
into brick joint

150 high
skirting

Joist

Firring

Decking of plywood
or similar material

Felt underlay

Figure 6.5.1　Typical layout of copper flat roof.

In general copper strip is used for flashings and damp-proof courses, whereas sheet or strip can be considered for general roof covering application; widths of 600 and 750 mm, according to thickness, are used with a standard length of 1.800 m.

The copper sheet or strip is laid in bays between rolls or standing seams to allow for the expansion of the covering. Standing seams are not recommended for pitches under 5° because they may be flattened by foot traffic and become a source of moisture penetration due to a capillary action. The recommended maximum bay widths for the common sheet thicknesses used are:

- 0.60 mm thick – maximum bay width 600 mm;
- 0.70 mm thick – maximum bay width 750 mm.

Transverse joints in the form of double lock welts should be used on all flat roofs. The welts should be sealed with a mastic or linseed oil before being folded together. Cross-welts may be staggered or continuous across the roof when used in conjunction with batten rolls (see Fig. 6.5.1).

An exception is the relatively new concept of long strip copper roofing. It uses special sheet lengths of up to 10 m in standard widths, without the need for cross-welts. Panels are machined together with standing welts/seams, which include sliding clips to allow for longitudinal movement. The process is less labour intensive than traditional, and therefore more economical to apply, and the roof appearance is smoother.

The substructure supporting the covering needs to be permanent, even, smooth, and laid to the correct fall of not less than 1:60 (40 mm in 2.400 m). With a concrete structural roof a lightweight screed is normally laid to provide the correct surface and fall; if the screed has a sulphate content it will require sealing coats of bitumen to prevent any moisture present from forming dilute acids which might react with the copper covering. The 50 mm × 25 mm dovetail battens inserted in the screed to provide the fixing medium for the batten rolls should be impregnated with a suitable wood preservative. Timber flat roofs should be wind tight and free from 'spring', and may be covered with 25 mm nominal tongued and grooved boarding laid with the fall or diagonally, or any other decking material that fulfils the general requirements given above.

On flat roofs, drips at least 65 mm high are required at centres not exceeding 3.000 m to increase the flow of rainwater across the roof to the gutter or outlet. To lessen the wear on the copper covering as it expands and contracts a separating layer of felt (BS 747 Type 4A(ii)) should be incorporated into the design. The copper sheet should be secured with copper wire nails not less than 2.6 mm thick and at least 25 mm long; the batten rolls should be secured with well countersunk brass screws. Typical constructional details are shown in Fig. 6.5.2.

[*Note*: For habitable accommodation current Building Regulations require the external walls to include a lightweight concrete block inner leaf and cavity insulation. The roof will also need insulating, as shown in Chapter 6.1.]

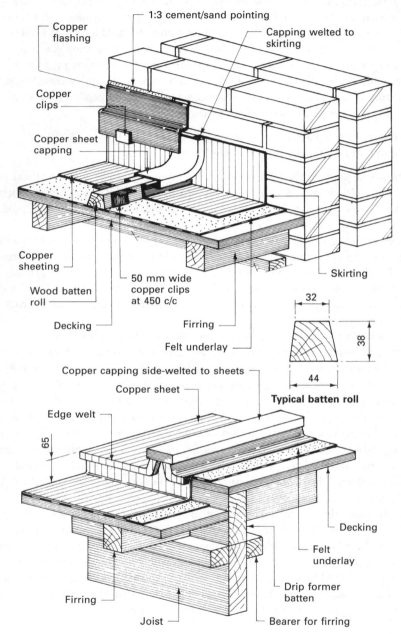

Copper flashing

1:3 cement/sand pointing

Capping welted to skirting

Copper clips

Copper sheet capping

Copper sheeting

Wood batten roll

Decking

50 mm wide copper clips at 450 c/c

Firring

Felt underlay

Skirting

32

38

44

Typical batten roll

Copper capping side-welted to sheets

Copper sheet

Edge welt

65

Decking

Felt underlay

Drip former batten

Firring

Joist

Bearer for firring

Figure 6.5.2   Copper flat roof details.

ROOFLIGHTS IN PITCHED ROOFS

Rooflights can be included in the design of a pitched roof to provide daylight and ventilation to rooms within the roof space, or to supplement the daylight from windows in the walls of medium- and large-span single-storey buildings.

In domestic work a rooflight generally takes one of two forms – the dormer window or skylight. A **dormer window** has a vertical sash, and therefore projects from the main roof; the cheeks or sides can be clad with a sheet material such as lead or tile hanging, and the roof can be pitched or flat of traditional construction (see Fig. 6.6.1).

A **skylight** is fixed in a trimmed opening and follows approximately the pitch of the roof. It can be constructed as an opening or dead light (see Fig. 6.6.2). In common with all rooflights in pitched roofs, making the junctions water- and weathertight present the greatest problems, and careful attention must be given to the detail and workmanship involved in the construction of dormer windows and rooflights.

Roofs of the type used on medium-span industrial buildings with coverings such as corrugated fibre cement sheeting supported by purlins and steel roof trusses require a different treatment. The amount of useful daylight entering the interior of such a building from windows fixed in the external walls will depend upon:

- the size of windows;
- the height of window above the floor level;
- the span of building.

Generally the maximum distance that useful daylight will penetrate is approximately 10.000 m; over this distance artificial lights or rooflights will be required during the daylight period. Three methods are available for the provision of rooflights in profiled sheeted roofs.

Special rooflight units of fibre cement consisting of an upstand kerb surmounted by either a fixed or opening glazed sash can be fixed instead of a standard profiled

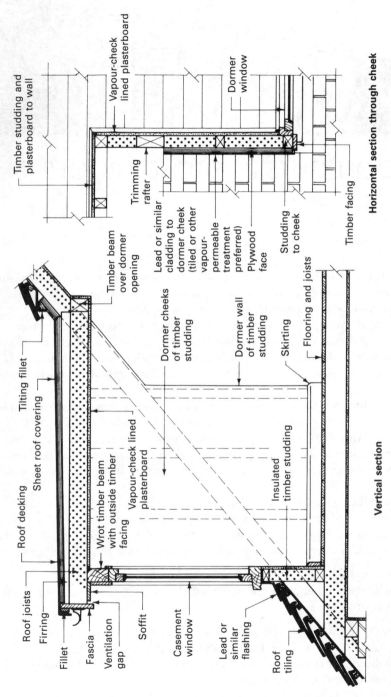

Horizontal section through cheek

Timber studding and plasterboard to wall

Vapour-check lined plasterboard

Dormer window

Trimming rafter

Lead or similar cladding to dormer cheek (tiled or other vapour-permeable treatment preferred)

Plywood face

Studding to cheek

Timber facing

Timber beam over dormer opening

Dormer cheeks of timber studding

Dormer wall of timber studding

Skirting

Flooring and joists

Tilting fillet

Sheet roof covering

Wrot timber beam with outside timber facing

Vapour-check lined plasterboard

Insulated timber studding

Roof decking

Roof joists

Firring

Fillet

Fascia

Ventilation gap

Soffit

Casement window

Lead or similar flashing

Roof tiling

Vertical section

Figure 6.6.1   Typical flat roof dormer window details.

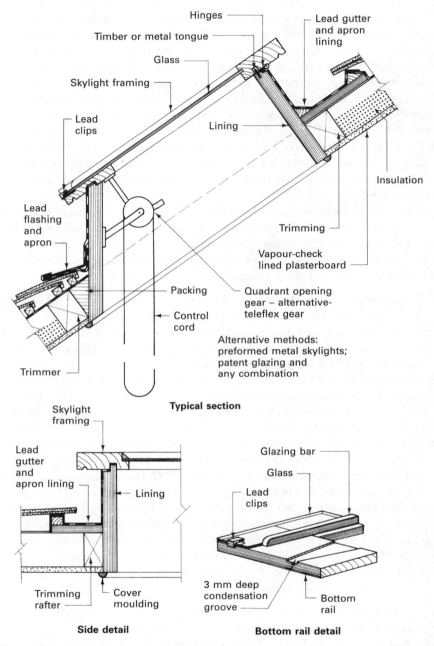

Figure 6.6.2 Typical timber opening skylight.

sheet. These units are useful where the design calls for a series of small isolated glazed rooflights to supplement the natural daylight. An alternative is to use translucent profiled sheets that are of the same size and profile as the main roof covering. In selecting the type of sheet to be used the requirements of Part B of the Building Regulations must be considered. Approved Document B4 (Section 14) deals specifically with the fire risks of roof coverings, and refers to the designations defined in BS 476-3. These designations consist of two letters: the first letter represents the time of penetration when subjected to external fire, and the second letter is the distance of spread of flame along the external surface. Each group of designations has four letters, A–D, and in both cases the letter A has the highest resistance. Specimens used in the BS 476 test are tested for use on either a flat surface or a sloping surface, and therefore the material designation is preceded by either EXT. F or EXT. S.

Most of the translucent profiled sheets have a high light transmission, are light in weight, and can be fixed in the same manner as the general roof covering. It is advisable to weather-seal all lapping edges of profiled rooflights with silicon or mastic sealant to accommodate the variations in thickness and expansion rate of the adjacent materials. Typical examples are:

- **Polyester glass fibre sheets**   Made from polyester resins reinforced with glass fibre and nylon to the recommendations of BS 4154. These sheets can be of natural colour or tinted, and are made to suit most corrugated fibre cement and metal profiles. Typical designations are EXT. S.AA for self-extinguishing sheets and EXT. S.AB for general-purpose sheets.
- **Wire-reinforced PVC sheets**   Made from unplasticised PVC reinforced with a fine wire mesh to give a high resistance to shattering by impact. Designation is EXT. S.AA, and they can therefore be used for all roofing applications. Profiles are generally limited to Categories A and B defined in BS EN 494.
- **PVC sheets**   Made from heavy gauge clear unplasticised rigid PVC to the recommendations of BS 4203, they are classified as self-extinguishing when tested in accordance with method 508A of BS 2782: Part O, Annex C, and may be used on the roof of a building provided that part of the roof is at least 6.000 m from any boundary. If that part of the roof is less than 6.000 m from any boundary and covers a garage, conservatory or outhouse with a floor area of less than 40 m^2, or is on an open balcony, carport or the like, or a detached swimming pool, PVC sheets can be used without restriction. Table 6.3 of Approved Document B defines the use of these sheets for the roof covering of a canopy over a balcony, veranda, open carport, covered way or detached swimming pool.

As an alternative to profiled rooflights in isolated areas, continuous rooflights can be incorporated into a corrugated or similar roof covering by using flat-wired glass and patent glazing bars. The bars are fixed to the purlins and spaced at 600 mm centres to carry either single or double glazing. The bars are available as a steel bar sheathed in lead or PVC or as an aluminium alloy extrusion (see Fig. 6.6.3). Many sections with different glass-securing techniques are manufactured under the patents granted to the producers, but all have the same basic principles. The bar is generally an inverted 'T' section, the flange providing the bearing for the glass and the stem

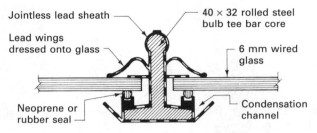

Jointless lead sheath — — 40 × 32 rolled steel bulb tee bar core

Lead wings dressed onto glass — — 6 mm wired glass

Neoprene or rubber seal — — Condensation channel

Crittall-Hope lead-clothed steel bar

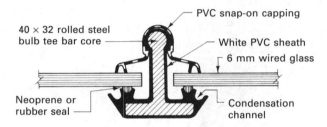

40 × 32 rolled steel bulb tee bar core — — PVC snap-on capping

— White PVC sheath

— 6 mm wired glass

Neoprene or rubber seal — — Condensation channel

Crittall-Hope polyclad bar

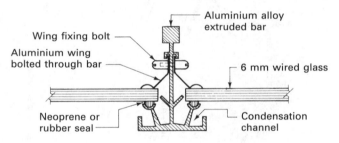

Wing fixing bolt — — Aluminium alloy extruded bar

Aluminium wing bolted through bar — — 6 mm wired glass

Neoprene or rubber seal — — Condensation channel

British Challenge aluminium bar

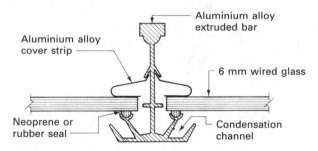

Aluminium alloy cover strip — — Aluminium alloy extruded bar

— 6 mm wired glass

Neoprene or rubber seal — — Condensation channel

Heywood Williams 'Aluminex' bar

Figure 6.6.3   Typical patent glazing bar sections.

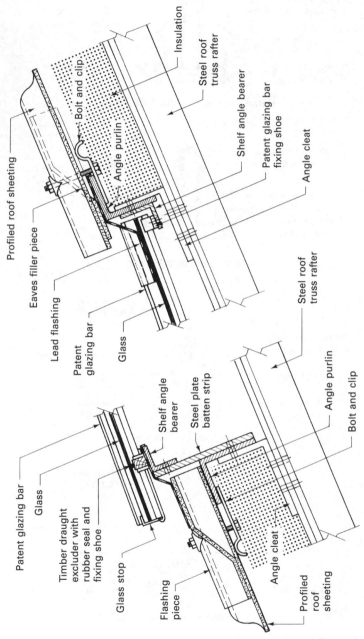

Figure 6.6.4   Patent glazing and profiled roof covering connection details.

depth giving the bar its spanning properties. Other standard components are fixing shoes, glass weathering springs or clips, and glass stops at the bottom end of the bar (see Fig. 6.6.3).

As the glass and the glazing bar are straight they cannot simply replace a standard profiled sheet; they must be fixed below the general covering at the upper end and above the covering at the lower end to enable the rainwater to discharge onto the general roof surface. Great care must be taken with this detail and with the quality of workmanship on site if a durable and satisfactory junction is to be made. Typical details are shown in Fig. 6.6.4.

The total amount of glazing to be used in any situation involves design appreciation beyond the scope of this volume, but a common rule of thumb method is to use an area of glazing equal to 10% of the total roof area. The glass specified is usually a wired glass of suitable thickness for the area of pane being used. Wired glass is selected so that it will give the best protection should an outbreak of fire occur; the splinters caused by the heat cracking the glass will adhere to the wire mesh and not shatter onto the floor below. A successful alternative is laminated glass composed of two or more sheets of float glass with a plastic interlayer. Polyvinylbutyral (PVB) is the most common interlayer, which is bonded to the glass laminates by heat and pressure. If broken, the glass remains stuck to the PVB to provide the benefit of safety and security. It also remains intact following impact from bullets or explosive blasts.

As with timber skylights, provision should be made to collect the condensation that can occur on the underside of the glazing to prevent the annoyance of droplets of water falling to the floor below. Most patent glazing bars for single glazing have condensation channels attached to the edges of the flange, which directs the collected condensation to the upper surface of the roof below the glazing line (see Figs 6.6.3 and 6.6.4). Contemporary frames can be produced from uPVC or extruded aluminium. The latter is often coated in polyester powder for colour enhancement and protection. A double-glazed example is shown in Fig. 6.6.5. This applies the principle of tension between the glazing frame and glass for security of fit, with the ability to respond to thermal movement without loss of seal.

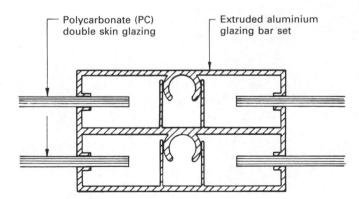

Figure 6.6.5   'Glidevale' patent double–glazing unit.

Polycarbonate (PC) can be an economic option in preference to glass, particularly for rooflight applications. It has the benefit of a double skin if required and, although produced in flat sheet format, can be moulded to domed profiles. Colour variants are possible, in addition to transparent, translucent and opaque composition. Dimensions, types and characteristics are defined in BS EN ISO 11963.

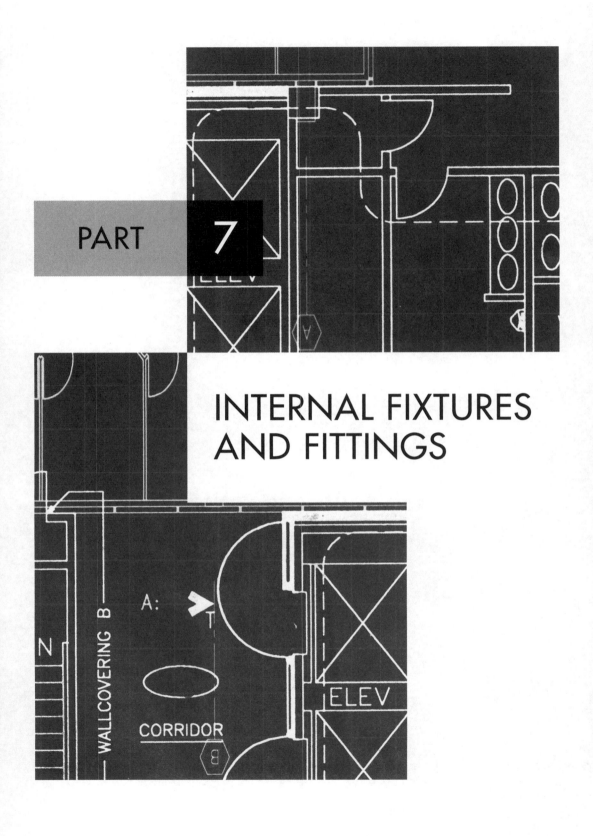

PART 7

INTERNAL FIXTURES
AND FITTINGS

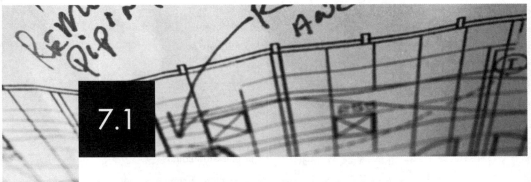

DOORS, DOOR FRAMES AND LININGS

■ ■ ■ DOORS

A door is a screen used to seal an opening into a building or between rooms within a building. It can be made of timber, glass, metal or plastic or any combination of these materials. Doors can be designed to swing from one edge, slide, slide and fold, or roll to close an opening. The doors to be considered in this book are those made of timber and those made of timber and glass that are hung so that they swing from one edge. All doors may be classified by their position in a building, by their function, or by their method of construction.

EXTERNAL DOORS

These are used to close the access to the interior of a building and provide a measure of security. They need to be weather resistant, as in general they are exposed to the elements: this resistance is provided by the thickness, stability and durability of the construction and materials used, together with protective coatings of paint or polish. The external walls of a building are designed to give the interior of a building a degree of thermal and sound insulation; doors in such walls should therefore be constructed, as far as is practicable, to maintain the insulation properties of the external enclosure.

The standard sizes for external timber doors are 1981 mm high × 762 or 838 mm wide × 45 mm thick, which is a metric conversion of the old Imperial door sizes. Metric doors are produced so that, together with the frame, they fit into a modular coordinated opening size, and are usually supplied as door sets with the door already attached or hung in the frame.

INTERNAL DOORS

These are used to close the access through internal walls and partitions and to the inside of cupboards. As with external doors the aim of the design should be to

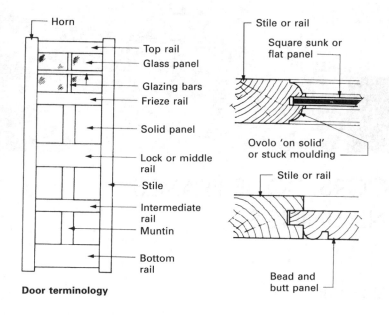

Door terminology

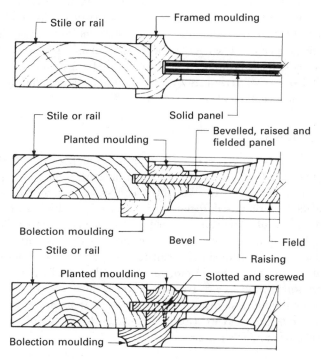

Figure 7.1.1   Purpose-made doors and mouldings.

maintain the properties of the wall in which they are housed. Generally, internal doors are thinner than their external counterparts as weather protection is no longer a requirement. Standard sizes are similar to those of external doors but with a wider range of widths to cater for narrow cupboard openings.

Purpose-made doors

The design and construction of these doors are usually the same as those of manufacturers' standard doors but to non-standard sizes, shapes or designs. Most door manufacturers produce a range of non-standard doors, which are often ornate and are used mainly for the front elevation doors of domestic buildings. Purpose-made doors are also used in buildings such as banks, civic buildings, shops, theatres and hotels to blend with or emphasise the external façade design or internal decor (see Fig. 7.1.1).

Methods of construction

The British Standard Code of Practice 151 covers all aspects of door construction. Unfortunately it is no longer published, but access to old copies will provide a useful reference for all doors of timber construction and their frames. British Standard 459 is published especially for matchboarded wooden doors for use externally.

Standard doors are used extensively because they are mass produced to known requirements, are readily available from stock and are cheaper than purpose-made doors.

Panelled and glazed wood doors

The Code of Practice gives a wide variety of types, all of which are based upon the one-, two-, three- or four-panel format. They are constructed of timber, which should be in accordance with the recommendations of BS 1186 and BS EN 942 with plywood or glass panels. External doors with panels of plywood should be constructed using an external-quality plywood (see Fig. 7.1.2).

The joints used in framing the doors can be a dowelled joint or a mortise and tenon joint. The dowelled joint is considered superior to the mortise and tenon joint and is cheaper when used in the mass production of standard doors. Bottom and lock rails have three dowels, top rails have two dowels, and intermediate rails have a single dowel connection (see Fig. 7.1.3). The plywood panels are framed into grooves with closely fitting sides with a movement allowance within the depth of the groove of 2 mm. The mouldings at the rail intersections are scribed, whereas the loose glazing beads are mitred. Weatherboards for use on external doors can be supplied to fit onto the bottom rail of the door, which can also be rebated to close over a water bar (see Fig. 7.1.2).

A fire door specification of FD 30 (see section headed: Flush fire doors) can be attained for 45 mm panelled doors by lining the existing panel with 6 mm glass-reinforced plasterboard or by replacing the panels with a double layer of 6 mm glass-reinforced plasterboard.

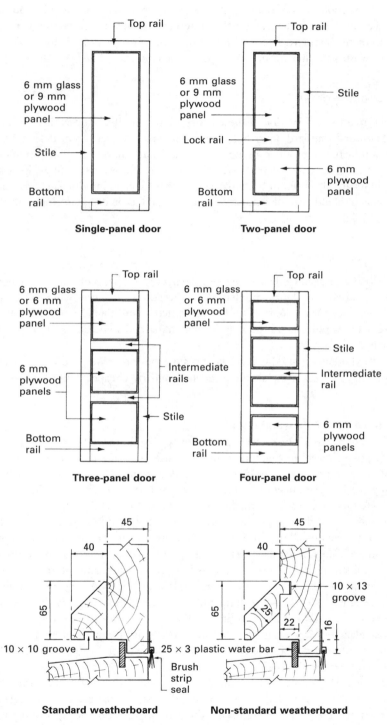

Figure 7.1.2   Standard panelled doors and weatherboards.

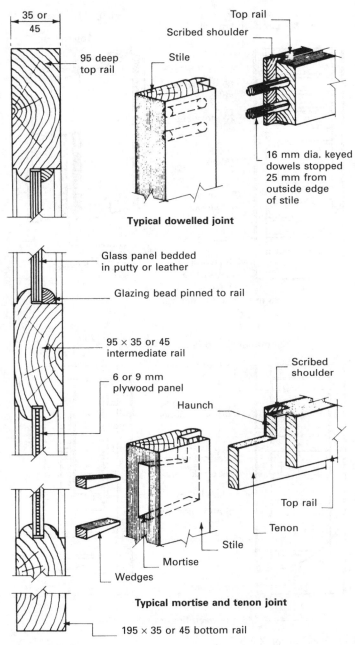

35 or 45

95 deep top rail

Top rail

Scribed shoulder

Stile

16 mm dia. keyed dowels stopped 25 mm from outside edge of stile

Typical dowelled joint

Glass panel bedded in putty or leather

Glazing bead pinned to rail

95 × 35 or 45 intermediate rail

6 or 9 mm plywood panel

Scribed shoulder

Haunch

Top rail

Tenon

Stile

Mortise

Wedges

Typical mortise and tenon joint

195 × 35 or 45 bottom rail

Figure 7.1.3   Panelled door details.

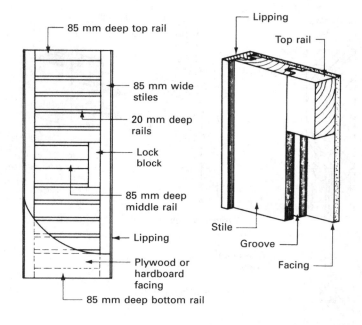

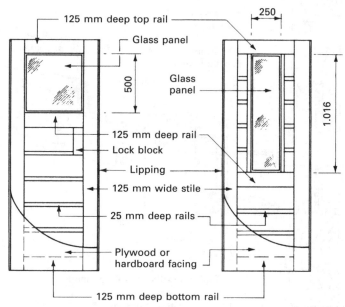

Figure 7.1.4   Skeleton core flush doors.

Flush doors

This type of door has the advantage of a plain face that is easy to clean and decorate; it is also free of the mouldings that collect dust. Flush doors can be faced with hardboard, plywood or a plastic laminate, and by using a thin sheet veneer of good-quality timber the appearance of high-class joinery can be created.

The method of constructing flush doors is left to the manufacturer, which provides complete freedom for design. Therefore the forms of flush door construction are many and varied, but basically they can be considered as either skeleton core doors or solid core doors. The former consists of an outer frame with small-section intermediate members over which is fixed the facing material. The facing has a tendency to deflect between the core members, and this can be very noticeable on the surface, especially if the facing is coated with gloss paint. Solid core doors are designed to overcome this problem and at the same time improve the sound insulation properties of flush doors. They also provide a higher degree of fire resistance and security. Solid doors of suitably faced block or lamin board are available for internal and external use. Another method of construction is to infill the voids created by a skeleton core with a lightweight material such as foamed plastic, which will give support to the facings but will not add appreciably to the weight of the door.

The facings of flush doors are very vulnerable to damage at the edges: therefore a lipping of solid material should be fixed to at least the vertical edges (good-quality doors have lippings on all four edges).

Small glazed observation panels can be incorporated in flush doors when the glass panel is secured by loose fixing beads (see Figs 7.1.4 and 7.1.5).

Flush fire doors

These doors provide an effective barrier to the passage of fire for the time designated by a reference coding. They can be assessed in terms of stability and integrity in the presence of fire tests. For example, a coding of 30/30 indicates the time in minutes for which a door is 'stable' under simulated fire conditions and sufficiently 'integral' to resist the penetration of flame and hot gases. To achieve this they must be used in conjunction with the correct frame.

However, the term 'stability' is now considered less appropriate, and in accordance with BS 476-22: *Methods for determination of the fire resistance of non-loadbearing elements of construction*, integrity and insulation are sufficient criteria for grading (see BS 476-20 for test procedures). Furthermore, BS 5588: *Fire precautions in the design, construction and use of buildings* and the Building Regulations Approved Document B (see Appendix B, Table B1) designate fire doors by integrity performance only.

The performance of fire doors is simplified by expressing the category with the initials FD and the integrity in minutes numerically. For example, FD 60 indicates a fire door with 60 minute integrity, and if the coding is followed by the letter S it indicates the door or frame has a facility to resist the passage of smoke, i.e. FD 60S. Standard ratings are 20, 30, 60 and 90, but other ratings are available to order. A timber frame is unlikely to satisfy British Standard tests for a 120 rating.

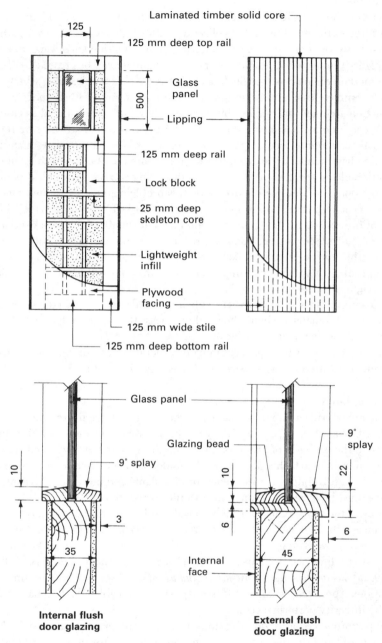

Figure 7.1.5   Solid core doors.

Resistance to smoke penetration can be achieved by fitting a brush strip seal to the frame. These are primarily for draught-proofing, but will be effective in providing a barrier to cold smoke. A more effective barrier to smoke and flame is an intumescent strip fitted to the door edge or integral with the frame. The latter is usually the most practical, being secured in a frame recess with a rubber-based or PVA adhesive. At temperatures of about 150 °C the seal expands to prevent passage of fire and smoke for the designated period. The seal will not prevent door movement for access and escape of personnel.

Notes

1. Doorsets/hardware and glass must be selected with compatibility to the door specification (see BS 476-31 for conformity of doorsets).
2. A colour-coded plastic plug is fitted into the hanging style of the door to indicate fire resistance. For example:

   - blue core/white background:
     - FD 20 without intumescent strip in frame;
     - FD 30 with intumescent strip in frame;
   - red core/blue background:
     - FD 60 with two intumescent strips in frame.

   A full list of coloured core and background codes can be found in BS 8214, the *Code of practice for fire door assemblies* (see Table 1).
3. Half-hour fire-rated doors are hung using one pair of hinges, whereas one-hour doors require one-and-a-half pairs of hinges (see Fig. 7.1.6).
4. All fire doors should have an automatic self-closing device, except cupboards and accesses to service ducts that are normally kept locked.

Matchboarded doors

These doors can be used as external or internal doors and, as a standard door, take one of two forms, a ledged and braced door or a framed ledged and braced door, the latter being a stronger and more attractive version.

The face of the door is made from tongue-and-grooved boarding that has edge chamfers to one or both faces; these form a vee joint between consecutive boards. Three horizontal members called **ledges** clamp the boards together, and in this form a non-standard door, called a **ledged and battened door**, has been made. It is simple and cheap to construct but has the disadvantage of being able to drop at the closing edge, thus pulling the door out of square; the only resistance offered is that of the nails holding the boards to the ledges. The use of this type of door is limited to buildings such as sheds and outhouses and to small units such as trapdoors (see Fig. 7.1.7). In the standard door, braces are added to resist the tendency to drop out of square; the braces are fixed between the ledges so that they are parallel to one another and slope downwards towards the hanging edge (see Fig. 7.1.7).

In the second standard type a mortise and tenoned frame surrounds the matchboarded panel, giving the door added strength and rigidity (see Fig. 7.1.7). If wide doors of this form are required, the angle of the braces becomes too

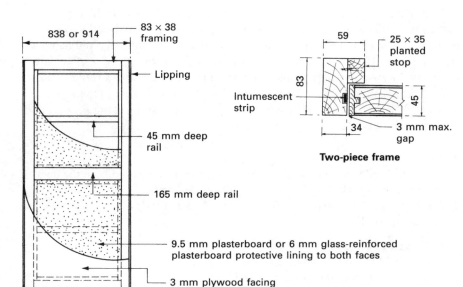

Half-hour fire door and frame (FD 30)

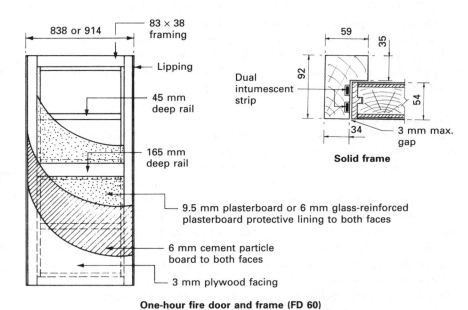

One-hour fire door and frame (FD 60)

Figure 7.1.6   Fire doors and frames.

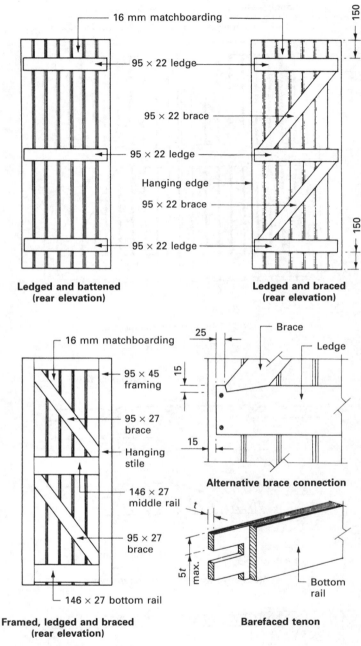

Figure 7.1.7   Matchboarded doors.

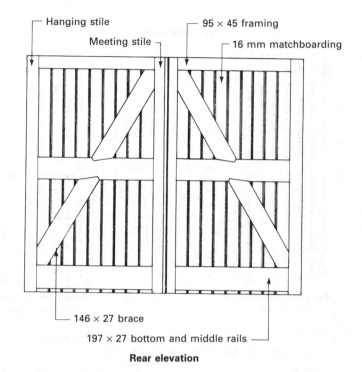

Hanging stile

Meeting stile

95 × 45 framing

16 mm matchboarding

146 × 27 brace

197 × 27 bottom and middle rails

Rear elevation

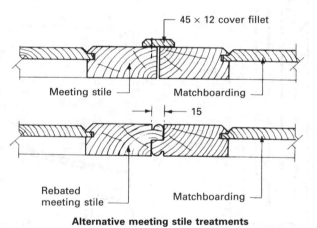

45 × 12 cover fillet

Meeting stile

Matchboarding

15

Rebated meeting stile

Matchboarding

Alternative meeting stile treatments

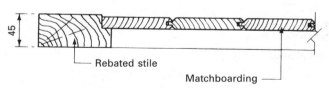

45

Rebated stile

Matchboarding

Alternative stile treatment

Figure 7.1.8   Matchboarded double doors.

low to be of value as an effective restraint, and the brace must therefore be framed as a diagonal between the top and bottom rails. Wide doors of this design are not covered by the British Standard but are often used in pairs as garage doors or as wide entrance doors to workshops and similar buildings (see Fig. 7.1.8).

The operation of fixing a door to its frame or lining is termed **hanging**, and entails: removing the protective horns from the top and bottom of the stiles; planing the stiles to reduce the door to a suitable width; cutting and planing the top and bottom to the desired height; marking out and fitting the butts or hinges that attach the door to the frame; and fitting any locks and door furniture that is required. The hinges should be positioned 225 mm from the top and bottom of the door; where one-and-a-half pairs of hinges are specified for heavy doors, the third hinge is positioned midway between the bottom and top hinges.

A door, irrespective of the soundness of its construction, will deteriorate if improperly treated during transportation and storage, and after hanging. It should receive a wood priming coat of paint before or immediately after delivery, and be stored in the dry and in a flat position so that it does not twist; it should also receive the finishing coats of paint as soon as practicable after hanging.

■ ■ ■ FRAMES AND LININGS

A door frame or lining is attached to the opening in which a door is to be fitted; it provides a surround for the door and is the member to which a door is fixed or hung. Door sets consisting of a storey-height frame with a solid or glazed panel over the door head are also available; these come supplied with the door ready hung on lift-off hinges (see Fig. 7.1.9).

TIMBER DOOR FRAMES

These are made from rectangular section timber in which a rebate is formed or to which a planted door stop is fixed to provide the housing for the door. Generally a door frame is approximately twice as wide as its thickness plus the stop. Frames are used for most external doors, heavy doors, doors situated in thin non-loadbearing partitions and internal fire doors.

A timber door frame consists of three or four members: the head, two posts or jambs, and a sill or threshold. The members can be joined together by wedged mortise and tenon joints, combed joints, or mortise and tenon joints pinned with a metal star-shaped dowel or a round timber dowel. All joints should have a coating of adhesive (see Fig. 7.1.10).

Door frames that do not have a sill are fitted with mild steel dowels driven into the base of the jambs and cast into the floor slab or, alternatively, grouted into preformed pockets as a means of securing the feet of the frame to the floor. If the frame is in an exposed position it is advisable to sit the feet of the jambs on a damp-proof pad such as lead or bituminous felt, to prevent moisture soaking into the frame and creating the conditions for fungal attack.

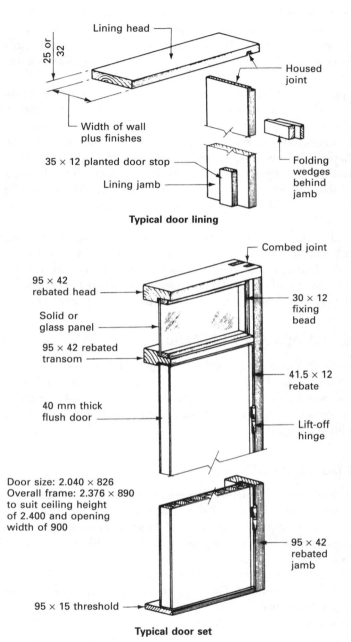

25 or 32

Lining head

Housed joint

Width of wall plus finishes

35 × 12 planted door stop

Lining jamb

Folding wedges behind jamb

Typical door lining

Combed joint

95 × 42 rebated head

30 × 12 fixing bead

Solid or glass panel

95 × 42 rebated transom

41.5 × 12 rebate

40 mm thick flush door

Lift-off hinge

Door size: 2.040 × 826
Overall frame: 2.376 × 890
to suit ceiling height
of 2.400 and opening
width of 900

95 × 42 rebated jamb

95 × 15 threshold

Typical door set

Figure 7.1.9   Door linings and door sets.

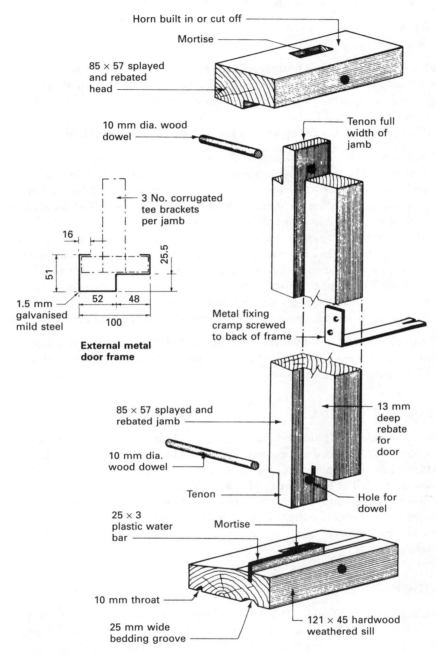

Horn built in or cut off

Mortise

85 × 57 splayed
and rebated
head

10 mm dia. wood
dowel

Tenon full
width of
jamb

3 No. corrugated
tee brackets
per jamb

16

25.5

51

52   48

1.5 mm
galvanised
mild steel

100

**External metal
door frame**

Metal fixing
cramp screwed
to back of frame

85 × 57 splayed and
rebated jamb

13 mm
deep
rebate
for
door

10 mm dia.
wood dowel

Tenon

Hole for
dowel

25 × 3
plastic water
bar

Mortise

10 mm throat

25 mm wide
bedding groove

121 × 45 hardwood
weathered sill

Figure 7.1.10 Door frames.

Door frames fitted with a sill are designed for one of two conditions:

- doors opening out;
- doors opening in.

In both cases the sill must be designed to prevent the entry of rain and wind under the bottom edge of the door. Doors opening out close onto a rebate in the sill, whereas doors opening in have a rebated bottom rail, and close over a water bar set into the sill (see Fig. 7.1.10).

Timber door frames can be fixed to a wall by the following methods:

- Built into the brick or block wall as the work proceeds by using L-shaped ties or cramps. The ties are made from galvanised mild steel with one end turned up 50 mm, with two holes for wood screws, the other end being 100–225 mm long and fish-tailed for building into brick or block bed joints. The ties are fixed to the back of the frame for building in at 450 mm centres.
- Fixed into a brick opening at a late stage in the contract to prevent damage to the frame during the construction period. This is a more expensive method and is usually employed only when high-class joinery using good-quality timber is involved. The frames are fixed to timber plugs inserted into the reveals with wood screws, whose heads are sunk below the upper surface of the frame; this is made good by inserting over the screw heads plugs or pellets of matching timber.

Timber door frames of softwood are usually finished with several applications of paint, whereas frames of hardwood are either polished or oiled. Frames with a factory coating of plastic are also available.

METAL DOOR FRAMES

These are made from mild steel pressed into one of three standard profiles, and are suitable for both internal and external positions. The hinges and striking plates are welded on during manufacture, and the whole frame receives a rust-proof treatment before delivery. The frames are fixed in a similar manner to timber frames using a tie or tee bracket, which fits into the back of the frame profile and is built into the bed joints of the wall (see Fig. 7.1.10). The advantage of this type of frame is that they will not shrink or warp, but they are more expensive than their timber counterparts.

DOOR LININGS

These are made from timber board 25 or 32 mm thick and as wide as the wall plus any wall finishes. They are usually specified only for internal doors. Door linings are not built in but are fixed into an opening by nailing or screwing directly into block walls, or into plugs in the case of brick walls. Timber packing pieces or folding wedges are used to straighten and plumb up the sides or jambs of the lining (see Fig. 7.1.9).

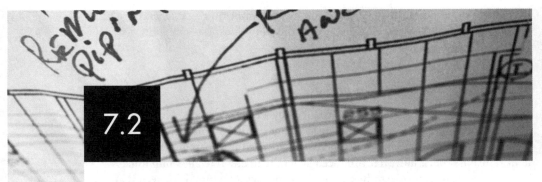

GLASS AND GLAZING

■■■ GLASS

Glass is made mainly from soda, lime, silica and other minor ingredients such as magnesia and alumina, to produce a material suitable for general window glazing. The materials are heated in a furnace to a temperature range of 1490–1550 °C, where they fuse together in a molten state; they are then formed into sheets by a process of drawing, floating or rolling. BS 952-1: *Glass for glazing* classifies glass for use in buildings by composition, dimensions, mass and available sizes.

DRAWN CLEAR SHEET GLASS

There are two principal methods of producing drawn clear sheet glass. The first is by vertical drawing from a pool of molten glass that, when 1 m or so above the pool level, is rigid enough to be engaged by a series of asbestos-faced rollers that continue to draw the ribbon of glass up a tower some 10 m high, after which the ribbon is cut into sheets and washed in a dilute acid to remove surface deposits. In the second method the glass is initially drawn in the vertical plane but is turned over a roller so that it is drawn in the horizontal direction for some 60 m and passes into an annealing furnace, at the cold end of which it is cut into sheets.

Clear sheet glass is a transparent glass with 85% light transmission, with a fire-finished surface, but because the two surfaces are never perfectly flat nor parallel there is always some distortion of vision and reflection.

The only thicknesses produced in the UK are 3 and 4 mm; 5 and 6 mm are available as an import.

FLOAT GLASS

This is gradually superseding the sheet glass process, and provides a glass with undistorted vision. It is formed by floating a continuous ribbon of molten glass over a bath of liquid metal at a controlled rate and temperature. A general glazing quality and a selected quality are produced in thicknesses ranging from 3 to 25 mm.

ROLLED AND ROUGH CAST GLASS

This is a term applied to a flat glass produced by a rolling process. Generally the glass produced in this manner is translucent, and transmits light with varying degrees of diffusion so that vision is not clear. A wired transparent glass with 80% light transmission is, however, produced generally in one thickness of 6 mm. The glass is made translucent by rolling onto one face a texture or pattern that will give 70–85% light transmission. Rough cast glass has an irregular texture to one side. Wired rough cast glass comes in two forms: Georgian wired, which has a 12 mm square mesh electrically welded wire reinforcement; and hexagonally wired, which is reinforced with hexagonal wire of approximately 20 mm mesh. Rough cast glass is produced in 5, 6 and 10 mm thicknesses and is made for safety and fire-resistant glazing purposes.

SPECIAL GLASS PRODUCTS

Numerous special glasses are available. Some of the most significant include:

- toughened glass;
- laminated glass;
- fire-resisting glass.

Toughened glass

Processed glass is subjected to heat and then cooled by jets of cold air. The effect is to create compressive stresses at the surface with balancing tensile stresses in the centre. The result is a glass of a strength 4–5 times greater than that of annealed glass. It can, however, still fracture under extreme bending loads or severe impact from a sharp implement. When broken it shatters into small blunt-edged fragments, which reduce the risk of personal injury. Because of its great strength and potential for large glazed areas it provides considerable scope for freedom of design.

Laminated glass

As the name suggests, laminated glass is built up in layers. It is composed of an outer and inner layer of annealed float glass with a heat and pressure-sealed intermediate bonding layer of polyvinylbutyral (PVB). Performance can be varied by changing the number and thickness of glass layers to provide a wide range of applications and purposes. These include rooflights, internal doors and partitions, noise and solar controls, as well as its suitability for security and safety situations due to its inherent resistance to bullets and explosives.

Table 7.2.1 Fire test performance of Pilkington Pyrodur and Pyrostop glass

Glass type	Thickness (mm)	Integrity (min.)	Insulation (min.)
Pyrodur	10	30	N/A
	13	60	N/A
Pyrostop	15	60	30
	21	60	60
	44	90	90
	50	120	120

Fire-resisting glass

Fire resistance can be measured in terms of **integrity** and **insulation**: see also Chapter 7.1. BS 476–20 defines integrity as 'the ability of a specimen of a separating element to contain a fire to specified criteria of collapse, freedom from holes, cracks and fissures and sustained flaming on the unexposed face'. Insulation is also defined in the same BS as 'the ability of a specimen of a separating element to restrict the temperature rise of the unexposed face to below specified levels'. 'Specified levels' means an average of no more than 140 °C and, in any specific position, 180 °C.

Wired glass, composed of an electrically welded mesh embedded centrally in rough glass during the rolling process, is well established as an effective means of resisting fire. It is frequently used in doors for security reasons and its ability to remain integral in fire for up to 2 hours. Although the glass fractures in heat, the close proximity of mesh retains the unit in one piece. Clear fire-resisting glass is produced by Pilkington UK Ltd under the trade names of Pyrodur and Pyrostop. Pyrodur is suitable for use internally or externally in doors and screens. Standard thicknesses are 10 and 13 mm, composed of three or four glass layers respectively. Interlayers are of intumescent material with an ultraviolet bonding laminate of polyvinylbutyral to ensure impact resistance performance. It is primarily intended to satisfy the Building Regulations' requirements for integrity, as shown in Table 7.2.1, although it has insulation properties for up to 22 minutes. Pyrostop is produced in two grades corresponding to internal or external application. For internal use it is made up of four glass layers with three intumescent interlayers. External specification has an additional glass layer and an ultraviolet filter interlayer positioned to the outside to protect the intumescent material against degradation. It is also produced in double-glazed format with an 8 mm air gap achieved with steel spacer bars. Overall thicknesses are 15, 21, 44 and 50 mm, to attain integrity and insulation for up to 2 hours.

▓▓■ GLAZING

The securing of glass in prepared openings such as doors, windows and partitions is termed **glazing**. It is logical that, as the area of glass pane increases, so must its thickness: similarly, position, wind load and building usage must be taken into account.

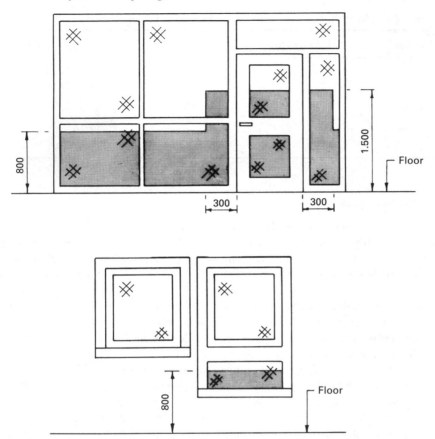

Figure 7.2.1   Zones of safe glazing – shown shaded.

The Building Regulations effect suitable standards for the safe use and application of glass through Approved Document N. Position is very important, and critical locations are shown in Fig. 7.2.1. These are defined within 800 mm of the finished floor level and 1500 mm if in a door or adjacent side panel. Glass in critical locations must be classified as safe, which means it satisfies one of the following:

- It conforms to the 'safe break' characteristics defined in BS 6206: *Specification for impact performance requirements for flat safety glass and safety plastics for use in buildings.*
- Annealed glass within the limitations of Table 7.2.2. Small panes isolated or in groups separated by glazing bars: maximum pane width is 250 mm, maximum pane area is 0.5 m², nominal thickness minimum 6 mm.
- A robust glass substitute material such as polycarbonate. It is permanently protected by a screen, as shown in Fig. 7.2.2.

The effect of wind loading on glazed area and glass thickness can be determined by reference to BS 6262: *Code of practice for glazing for buildings.* Tables and graphs

Table 7.2.2 Maximum dimensions of annealed glass panels

Annealed glass thickness (mm)	Maximum width (m)	Maximum height (m)
8	1.10	1.10
10	2.25	2.25
12	4.50	3.00
15	Any	Any

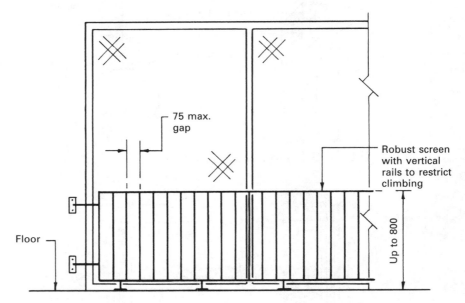

Figure 7.2.2 Screen protection to glazing.

permutate data for topography, ground roughness, life and glass factors to provide the recommended minimum glass thickness.

GLAZING WITHOUT BEADS

This is a suitable method for general domestic window and door panes. The glass is bedded in a compound, secured with sprigs, pegs or clips, and fronted with a weathered surface putty. Putty is a glazing compound that will require a protective coating of paint as soon as practicable after glazing. Two kinds of putty are in general use:

- **Linseed oil putty**   For use with primed wood members; made from linseed oil and whiting, usually to the recommendations of BS 544.
- **Metal casement putty**   For use with metal or non-absorbent wood members; made from refined vegetable drying oils and finely ground chalk.

[*Note*: A general-purpose putty is available that will satisfy application for both wood and metal glazing.]

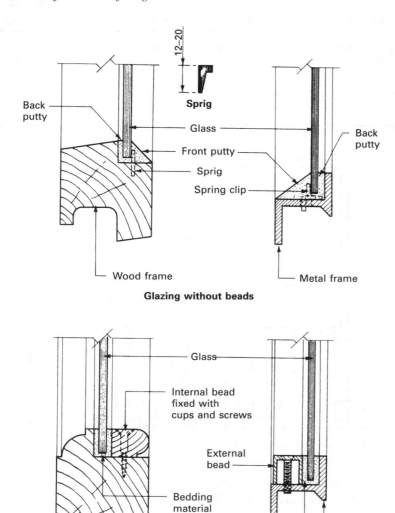

Figure 7.2.3 Glazing details.

The glass pane should be cut to allow a minimum clearance of 2 mm all round for both wood and metal frames. Sufficient putty is applied to the rebate to give at least 2 mm of back putty when the glass is pressed into the rebate, any surplus putty being stripped off level or at an angle above the rebate. The glass should be secured with sprigs or clips at not more than 440 mm centres and finished off on the front edge with a weathered putty fillet so that the top edge of the fillet is at or just below the sightline (see Fig. 7.2.3).

Compounds and sealants

Putty has the disadvantage of setting after a period of time, which may be unsuitable for large glazed areas subject to thermal movement. Therefore several non-setting compounds and sealants are available. Rubberised compounds are usually in two parts for mixing on site and applying by hand or with an extruder gun. One-part sealants of polysulphide, silicone and urethane cure by chemical reaction with exposure to the atmosphere. They form a firm, resilient seal with a degree of flexibility. One-part sealants of butyl or acrylic do not cure, but remain soft and pliable. These are preferred for bedding glass and beads only, as they can deteriorate if exposed. Two-part polysulphide- or polyurethane-based sealants are mixed for application to cure into rubbery material within 14 days. These have excellent adhesion.

GLAZING WITH BEADS

For domestic work, glazing with beads is generally applied to good-class joinery. The beads should be secured with either panel pins or screws; for hardwoods it is usual to use cups and screws. The glass is bedded in a compound or a suitable glazing felt, mainly to prevent damage by vibration to the glass. Beads are usually mitred at the corners to give continuity of any moulding. Beads for metal windows are usually supplied with the surround or frame, and fixing of glass should follow the manufacturer's instructions (see Fig. 7.2.3).

DOUBLE GLAZING

Double glazing is now standard practice in window installation to satisfy UK Building Regulation requirements for energy conservation in external walls. Factory-made units contain two parallel glass panes sealed with an air gap between the glass. This not only reduces heat loss considerably but will also improve sound insulation and restrict surface condensation. Typical manufactured units comprise two panes with a dry air gap of between 3 and 20 mm. The greater the air space, the better the thermal insulation, although care must be observed when applied at extremes of atmospheric conditions as explosion or implosion may occur with wider-gapped units. Generally, the insulative effect is to reduce the thermal transmittance coefficient (U value) by about half, from approximately 5.6 W/m^2 K for single glazing to approximately 2.8 W/m^2 K, depending on air gap and exposure. Some applications are shown in Fig. 7.2.4.

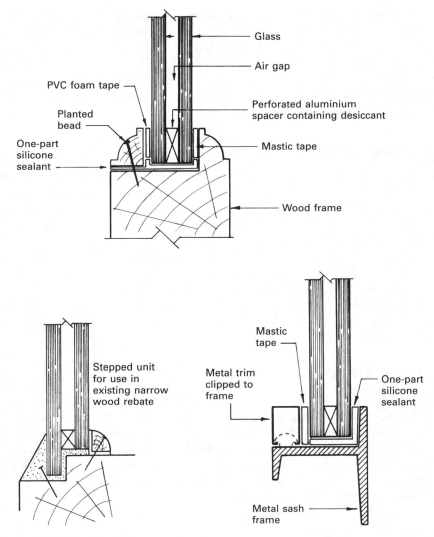

Figure 7.2.4   Double glazing.

LOW-EMISSIVITY GLASS

The use of low-emissivity or 'Low E' glass in double-glazed units can considerably enhance window insulation. Units are manufactured with a standard pane of float glass to the outside and a microscopic-metal-coated outer surface to the inner pane of glass, as shown in Fig. 7.2.5.

The effect is to permit short-wave radiation such as sunlight to pass through the glazed unit, while reflecting long-wave radiation such as internal heat from radiators and fires back into the room. Using argon gas, which has lower thermal conductivity than air, between the panes will also reduce heat losses. Table 7.2.3 compares the insulation properties of glazed systems.

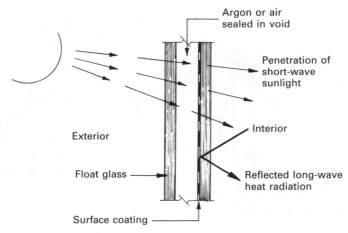

Figure 7.2.5 'Low E' double glazing.

Table 7.2.3 Typical U values based on 6 mm thick glass and a 12 mm void in double glazing.

Glazing system	Void	U value (W/m² K)
Single glazing	–	5.6
Double glazing:		
Float glass × 2	air	2.8
Float glass × 2	argon	2.7
Float glass + 'Low E'	air	1.9
Float glass + 'Low E'	argon	1.6

▨ ▧ ◼ GLASS BLOCK WALLING

Glass blocks are produced in a variety of sizes, colour tones and surface finishes. They can be used for non-loadbearing partitioning, particularly in offices where a degree of separation is required while still retaining access for light. They are also acceptable as a light source through external walling, provided they are within a structural support frame.

Standard units are shown in Fig. 7.2.6. These are manufactured in two halves, rather like square ash trays, of 10 mm thick glass and then heat-fused together to make a hollow block. The periphery is sprayed with a white adhesive textured finish to bond with mortar. Blocks can be staggered or stretcher-bonded like traditional brickwork. For aesthetic reasons they are usually laid with continuous vertical and horizontal joints, i.e. stack bonded with steel wire reinforced bed joints.

EXTERNAL APPLICATION

Blocks are laid and tied to a structural support frame of maximum area 9 m²
and 3 m in any direction, as shown in Fig. 7.2.7. The lowest course is laid on a bituminous bed and mortar. The mortar to this and subsequent courses comprises

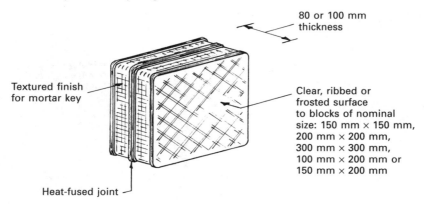

Figure 7.2.6   Standard glass block.

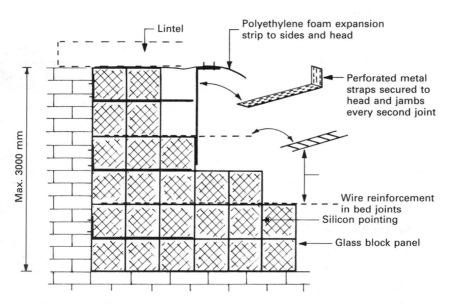

Figure 7.2.7   Typical glass block framing and laying.

1 part white Portland cement, 0.5 part lime and 4 parts white quartzite or silver sand, measured by volume. A waterproofing agent is added to the mix. Pointing can be with a weather-resistant silicon sealant. To accommodate expansion and contraction, a polyethylene foam strip is located around the periphery of blocks at the sides and head of the support frame. 9 gauge (3.7 mm) steel reinforcement in ladder pattern is placed in every horizontal course with 300 mm blocks, every other course with 200 mm blocks and every third course with 150 mm blocks (sizes nominal).

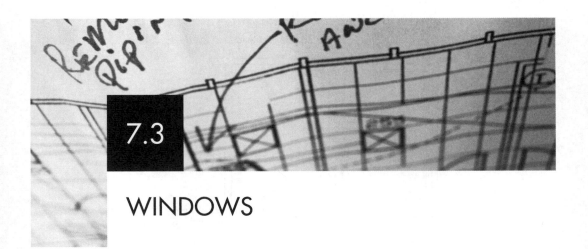

7.3

WINDOWS

The primary function of a window is to provide a means for admission of natural daylight to the interior of a building. A window can also serve as a means of providing the necessary ventilation of dwellings, as required under Building Regulation F1, by including opening lights in the window design.

Windows, like doors, can be made from a variety of materials, or a combination of these materials, such as timber, metal and plastic. They can also be designed to operate in various ways by arranging the sashes to slide, pivot or swing, or hang as a casement to one of the frame members.

■ ■ ■ BUILDING REGULATIONS

Approved Document F details the requirements for ventilating habitable rooms, kitchens, utility rooms, bathrooms and sanitary accommodation in domestic buildings. A habitable room is defined as a room used for dwelling purposes but not a kitchen. A habitable room must have ventilation openings unless it is adequately ventilated by mechanical means.

Passive stack ventilation (PSV) is an acceptable alternative in some rooms (see Section 1, Approved Document F). This is a simple system of ducts used to convey air by natural movement to external terminals. Non-domestic buildings have different requirements to suit various purposes, with criteria applied to occupied rooms instead of habitable rooms and greater emphasis on mechanical ventilation and floor areas (see Section 2, Approved Document F).

A ventilation opening includes any permanent or closable means of ventilation that opens directly to the external air. Included in this category are openable parts of windows, louvres, air bricks, window trickle ventilators and doors that open directly to the external air. The three general objectives of providing a means of ventilation are:

■ to extract moisture from rooms where it is produced in significant quantities, such as kitchens and bathrooms;

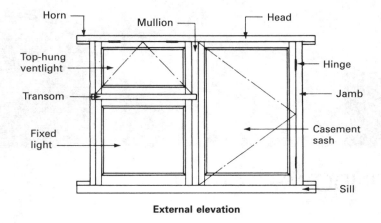

External elevation

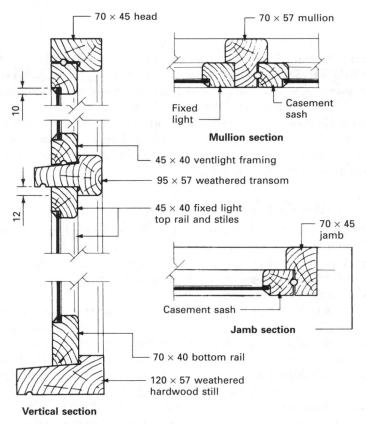

Figure 7.3.1   Traditional timber casement window.

- to provide for occasional rapid ventilation where moisture presence is likely to produce condensation;
- to achieve a background ventilation that is adequate without affecting the comfort of the room.

Guidance in the Approved Document allows for a variety of means to achieve these objectives. These range from simple trickle-type ventilators (see Fig. 7.3.3) to opening sashes, mechanical extract fans and fully ducted air ventilation systems. The minimum height for ventilators is 1.700 m from finished floor level.

Unless a mechanical system of air movement is used throughout a dwelling, background ventilation through trickle vents of minimum equivalent area of 6000 mm^2 is required in all habitable rooms. Additionally, these rooms should have rapid or purge ventilation by opening sash or an extract fan with a capability of at least 4 room air changes per hour. Intermittent use of mechanical fan extract ventilation is also required for the following specific situations:

- kitchen – minimum 60 litres per second; 30 l/s if adjacent to hob;
- utility room – 30 l/s;
- bathroom – 15 l/s;
- sanitary accommodation – 6 l/s.

If any of the above situations is without a window with an opening sash, a combined light/extract fan switch should be fitted with an over-run facility for the fan. A gap of at least 10 mm should be provided between the door and floor finish.

■ ■ ■ TRADITIONAL CASEMENT WINDOWS

Figure 7.3.1 shows a typical arrangement and details of this type of window. A wide range of designs can be produced by using various combinations of the members, the only limiting factor being the size of glass pane relevant to its thickness.

The general arrangement of the framing is important; heads and sills always extend across the full width of the frame and in many cases have projecting horns for building into the wall. The jambs and mullions span between the head and sill; these are joined to them by a wedged or pinned mortise and tenon joint. This arrangement gives maximum strength, as the vertical members will act as struts; it will also give a simple assembly process.

The traditional casement window frame has deep rebates to accommodate the full thickness of the **sash**, which is the term used for the framing of the opening ventilator. If fixed glazing or lights are required it is necessary to have a sash frame surround to the glass, because the depth of rebate in the window frame is too great for direct glazing to the frame.

STANDARD WOOD CASEMENT WINDOWS

BS 644: *Timber windows. Factory assembled windows of various types. Specification* provides details of the quality, construction and design of a wide range of wood casement windows. The frames, sashes and ventlights are made from standard sections of softwood timbers arranged to give a variety in design and size. The

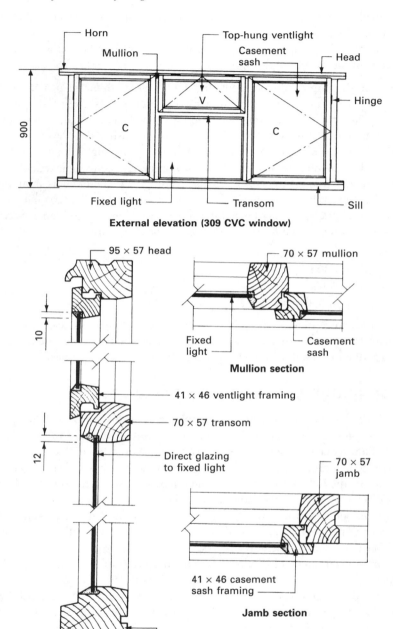

Horn

Mullion

Top-hung ventlight

Casement sash

Head

Hinge

900

C

V

C

Fixed light

Transom

Sill

External elevation (309 CVC window)

95 × 57 head

70 × 57 mullion

10

Fixed light

Casement sash

Mullion section

41 × 46 ventlight framing

70 × 57 transom

12

Direct glazing to fixed light

70 × 57 jamb

41 × 46 casement sash framing

Jamb section

95 × 70 softwood or hardwood sill

Vertical section

Figure 7.3.2　Typical modified BS casement window (see Figure 7.3.3 for double–glazed vertical section).

sashes and ventlights are designed so that their edges rebate over the external face of the frame to form a double barrier to the entry of wind and rain. The general construction is similar to that described for traditional casement windows, and the fixing of the frame into walls follows that described for door frames.

Most joinery manufacturers produce a range of modified standard casement windows following the basic principles set out in BS 644 but with improved head, sill and sash sections. The range produced is based on brick modules to avoid unnecessary cutting and disturbance to the masonry appearance. Window types are identified by a notation of figures and letters, for example

212 CV

where   2 = width divided into two units
       12 = 1200 mm height
        C = casement
        V = ventlight

For typical details see Figs 7.3.1–7.3.3.

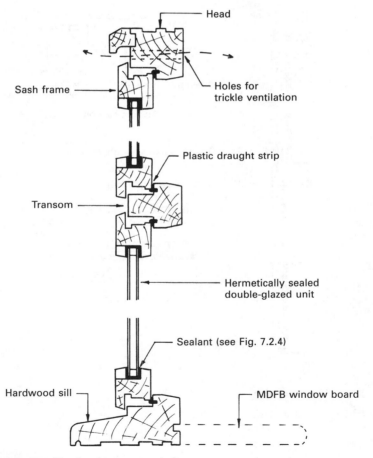

Figure 7.3.3 Double-glazed casement window.

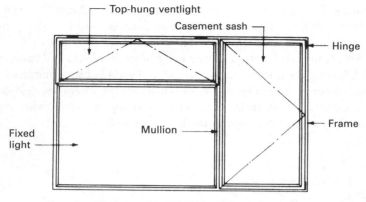

External elevation (18 FCT 11 RH window)

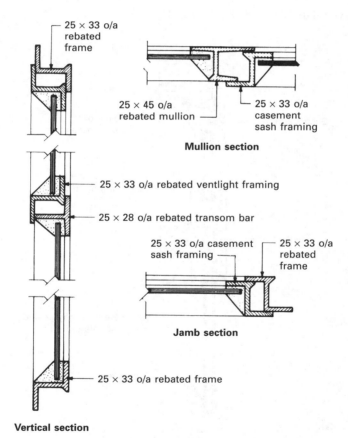

Figure 7.3.4   Typical steel window details.

STEEL CASEMENT WINDOWS

These windows are produced to conform to the recommendations of BS 6510, which gives details of construction, sections, sizes, composites and hardware. The standard range covers fixed lights, hung casements, pivot casements and doors. They are produced in dimensions to suit both imperial and metric applications, the latter conforming to the basic space first preference of 300 mm, giving the following range (in mm):

500; 600; 800; 900; 1200; 1500; 1800.

Frame heights are based upon basic spaces for the preferred head and sill heights for public sector housing, giving the following sizes (in mm):

200; 500; 700; 900; 1100; 1300; 1500.

Steel windows, like wood windows, are identified by a notation of numbers and letters:

- Prefix number: × 100 = basic space length.
- Code letters: F = fixed light.
  C = side-hung casement opening out.
  V = top-hung casement opening out and extending full width of frame.
  T = top-hung casement opening out and extending less than full width of frame.
  B = bottom casement opening inwards.
  S = fixed sublight.
- Suffix number: × 100 = basic space height.
- Suffix code: RH = right-hand casement as viewed from outside.
  LH = left-hand casement as viewed from outside.

The basic range of steel windows can be coupled together to form composite frames by using transom and mullion coupling sections without increasing the basic space module of 100 mm. The actual size of a steel frame can be obtained by deducting the margin allowance of 6 mm from the basic space size.

All the frames are made from basic rolled steel sections, which are mitred and welded at the corners to form right angles; the internal bars are tenoned and riveted to the outer frame and to each other. The completed frame receives a hot–dip galvanised protective finish after manufacture and before delivery.

Steel windows can be fixed into an opening by a number of methods, such as direct fixing to the structure or by using a wood surround that is built into the reveals and secured with fixing ties or cramps. The wood surround will add 100 or 50 mm to the basic space size in each direction using either a nominal 75 mm × 75 mm or 50 mm × 75 mm timber section. Typical details of steel windows, couplings and fixings are shown in Figs 7.3.4 and 7.3.5.

The main advantage of steel windows is the larger glass area obtained for any basic space size, because of the smaller frame sections used. The main disadvantage is the condensation that can form on the frames because of the high thermal conductivity of the metal members.

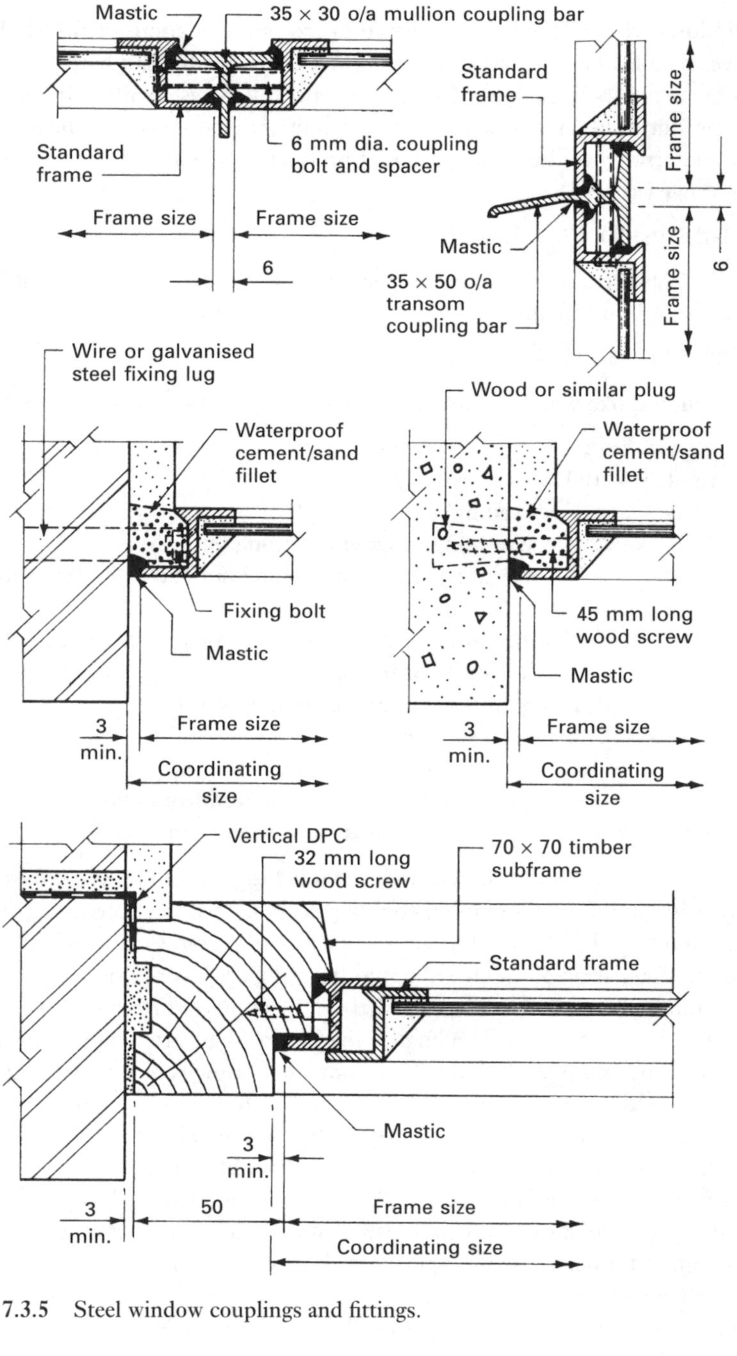

Figure 7.3.5　Steel window couplings and fittings.

ALUMINIUM HOLLOW PROFILE CASEMENT WINDOWS

Extruded aluminium profiles can be designed for use in window and door framing. The main advantages over other materials in these applications are high strength and durability relative to light weight, and very little maintenance except for occasional cleaning. The aluminium is protected from surface oxidisation and corrosive atmospheres with a coloured polyester powder coating. Standard colours are white or dark brown, but other options are available to order.

Metals are very poor thermal insulators, but the overall thermal performance of aluminium profiled casements can be considerably improved by specifying a closed cell foam infill to the hollow sections. Cold or thermal bridging and associated condensation is prevented by incorporating a thermal barrier or break between internal and external components. The thermal break is produced from a high strength two–part polyurethane resin.

Typical section details are shown in Fig. 7.3.6.

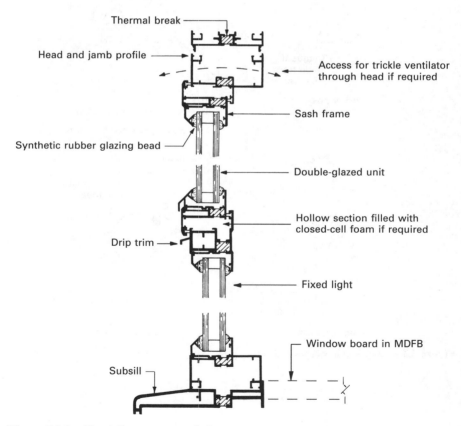

Figure 7.3.6 Aluminium casement window.

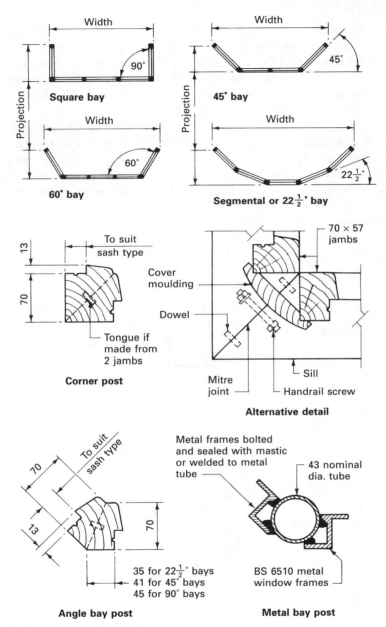

Figure 7.3.7   Bay window types and corner posts.

UPVC HOLLOW PROFILE CASEMENT WINDOWS

uPVC (or PVC-U) window components are produced in similar sectional profiles to that shown for hollow section aluminium windows. Plastic is an excellent thermal insulating material, so there is no need to incorporate a thermal break. uPVC has lower strength characteristics than aluminium: therefore manufacturers include box sections of aluminium or galvanised steel within the profiles of larger framing members.

uPVC windows are very popular in both new build and the refurbishment housing market. Manufacturers' designs can emulate most traditional features associated with timber casements, with the added advantage of excellent thermal insulation, improved security and a maintenance-free finish. Pigmenting in the plastic provides for a variety of colour options, generally white, light or dark brown.

■■■ BAY WINDOWS

Any window which projects in front of the main wall line is considered to be a bay window; various names are, however, given to various plan layouts (see Fig. 7.3.7). Bay windows can be constructed of timber, and/or metal and designed with casement or sliding sashes; the main difference in detail is the corner post, which can be made from the solid, jointed or masked in the case of timber and tubular for metal windows (see Fig. 7.3.7).

The bay window can be applied to one floor only or continued over several storeys. Any roof treatment can be used to cover in the projection and weather seal it to the main wall (see Fig. 7.3.8). No minimum headroom heights for bay windows or habitable rooms are given in the Building Regulations but 2.000 in bay windows and 2.300 in rooms would be considered reasonable. A bay window which occurs only on upper storeys is generally called an oriel window.

■■■ DOUBLE-HUNG SASH WINDOWS

These windows are sometimes called **vertical sliding sash windows** and consist of two sashes sliding vertically over one another. They are costly to construct, but are considered to be more stable than side-hung sashes and have a better control over the size of ventilation opening, thus reducing the possibility of draughts.

In timber, two methods of suspension are possible:

■ weight-balanced type;
■ spring-balanced type.

The former is the older method, in which the counterbalance weights suspended by cords are housed in a boxed framed jamb or mullion. It has been generally superseded by the metal spring balance, which uses a solid frame and needs less maintenance (see Figs 7.3.9 and 7.3.10).

Double-hung sashes in metal are supported and controlled by spring balances or by friction devices, but the basic principles remain the same.

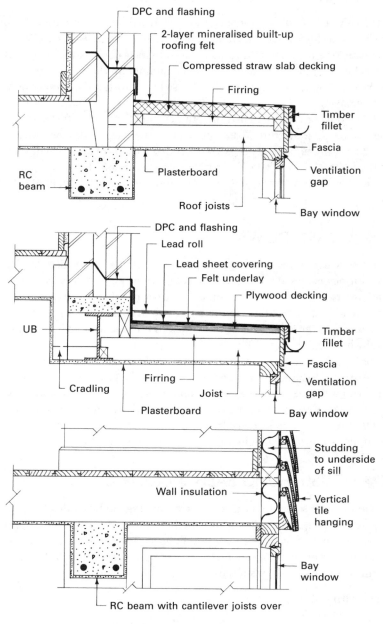

Figure 7.3.8   Typical existing bay window roofs.

Note: Current practice is to insulate flat roofs (see Chapter 6.1) and stud framing (see Chapter 3.7)

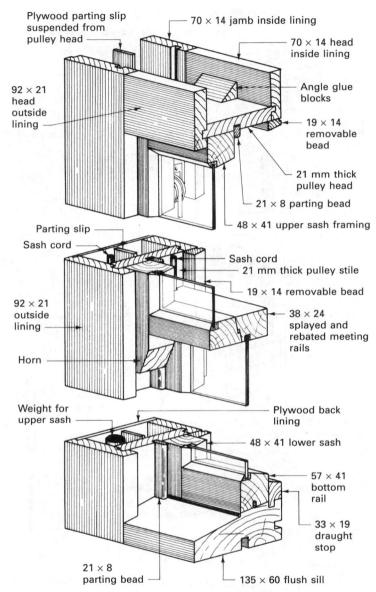

Figure 7.3.9 Traditional double-hung weight-balanced sliding sash windows.

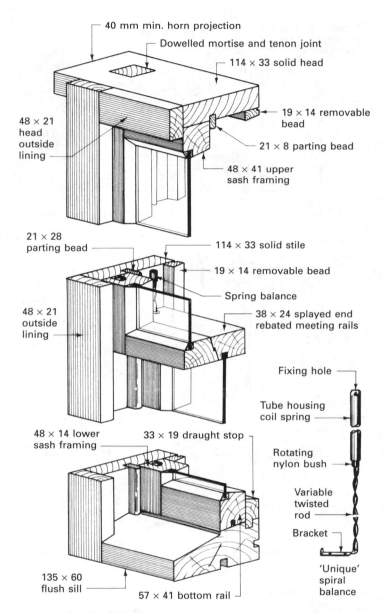

40 mm min. horn projection

Dowelled mortise and tenon joint

114 × 33 solid head

48 × 21 head outside lining

19 × 14 removable bead

21 × 8 parting bead

48 × 41 upper sash framing

21 × 28 parting bead

114 × 33 solid stile

19 × 14 removable bead

Spring balance

48 × 21 outside lining

38 × 24 splayed end rebated meeting rails

Fixing hole

Tube housing coil spring

Rotating nylon bush

48 × 14 lower sash framing

33 × 19 draught stop

Variable twisted rod

Bracket

135 × 60 flush sill

57 × 41 bottom rail

'Unique' spiral balance

Note: If 114 × 60 solid stiles are used, balances can be housed in grooves within the stile thickness

Figure 7.3.10   Double-hung spring-balanced sash windows.

■ ■ ■ PIVOT WINDOWS

The basic construction of the frame and sash is similar to that of a standard casement frame and sash. The sash can be arranged to pivot horizontally or vertically on friction pivots housed in the jambs or in the sill and head. These windows give good adjustment for ventilation purposes, and in the main both faces of the glazing can be cleaned from the inside of the building.

7.4

TIMBER STAIRS

A stair is a means of providing access from one floor level to another. Modern stairs with their handrails are designed with the main emphasis on simplicity, trying to avoid the elaborate and costly features used in the past.

The total rise of a stair – that is, the distance from floor finish to floor finish in any one storey height – is fixed by the storey heights and floor finishes being used in the building design: therefore the stair designer has only the total going or total horizontal distance with which to vary the stair layout. It is good practice to keep door openings at least 450 mm away from the head or the foot of a stairway, and to allow at least the stair width as circulation space at the head or foot of the stairway.

Stairs can be designed as one straight flight between floor levels, which is the simplest and cheapest layout; alternatively they can be designed to turn corners by the introduction of quarter space (90°) or half space (180°) intermediate landings. Stairs that change direction of travel use tapered steps or are based on geometrical curves in plan. Irrespective of the plan layout, the principles of stair construction remain constant, and are best illustrated by studying the construction of simple straight-flight stairs in this volume.

Terminology
- **Stairwell**   The space in which the stairs and landings are housed.
- **Stairs**   The actual means of ascent or descent from one level to another.
- **Tread**   The upper surface of a step on which the foot is placed.
- **Nosing**   The exposed edge of a tread, usually projecting with a square, rounded or splayed edge.
- **Riser**   The vertical member between two consecutive treads.
- **Step**   Riser plus tread.
- **Going**   The horizontal distance between two consecutive risers or, as defined in Approved Document K, the horizontal dimensions from front to back of a tread less any overlap with the next tread above.

- **Rise**   The vertical height between two consecutive treads.
- **Flight**   A series of steps without a landing.
- **Newel**   Post forming the junction of flights of stairs with landings or carrying the lower end of strings.
- **Strings**   The members receiving the ends of steps, which are generally housed to the string and secured by wedges, called wall or outer strings according to their position.
- **Handrail**   Protecting member usually parallel to the string and spanning between newels. This could be attached to a wall above and parallel to a wall string.
- **Baluster**   The vertical infill member between a string and handrail.
- **Pitch line**   A line connecting the nosings of all treads in any one flight.

Building Regulations Part K

Approved Document K defines three categories of stairs:

1. **Private**: intended for use solely in connection with one dwelling.
2. **Institutional** and assembly: serves one place in which a substantial number of people will gather.
3. **Other**: serves in buildings other than dwellings, institutional and assembly.

The practical limitations applicable to each of these three categories are shown in Table 7.4.1.

Table 7.4.1   Stairs (Approved Document K)

Category	Rise (mm)		Going (mm)
Private	155–220 or 165–200	with	245–260 or 223–300
Institutional and assembly*	135–180	with	280–340
Other*	150–190	with	250–320

* Subject to requirements for access for disabled people. See Building Regulations, Approved Document M, Section 3.51 (max. rise 170 mm, min. going 250 mm).

▪▪▪ CONSTRUCTION AND DESIGN

It is essential to keep the dimensions of the treads and risers constant throughout any flight of steps to reduce the risk of accidents by changing the rhythm of movement up or down the stairway. The height of the individual step rise is calculated by dividing the total rise by the chosen number of risers. The individual step going is chosen to suit the floor area available so that it, together with the rise, meets the requirements of the Building Regulations (see Fig. 7.4.1). It is important to note that in any one flight there will be one more riser than treads, as the last tread is in fact the landing.

Stairs are constructed by joining the steps into the spanning members or strings by using housing joints, glueing and wedging the steps into position to form a complete and rigid unit. Small angle blocks can be glued at the junction of tread

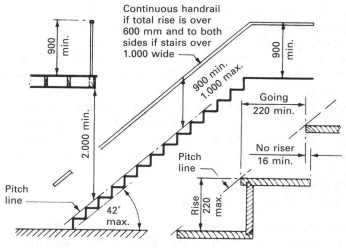

Sum of going + twice rise = 550 min. to 700 max.
in any flight all risers of equal height and all
goings of equal width

Private stairways

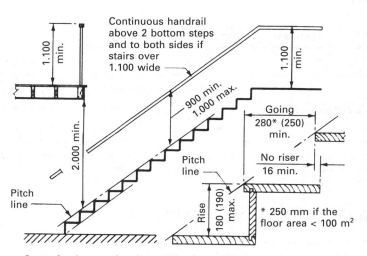

Sum of going + twice rise = 550 min. to 700 max.
in any flight all risers of equal height and all
goings of equal width

Institutional and assembly and (other) stairways

Figure 7.4.1   Timber stairs and Approved Document K.

and riser in a step to reduce the risk of slight movement giving rise to the annoyance of creaking. The flight can be given extra rigidity by using triangular brackets placed under the steps on the centreline of the flight. A central beam or carriage piece with rough brackets as a support is used only on wide stairs over 1200 mm, especially where they are intended for use as a common stairway (see Fig. 7.4.2).

Stairs can be designed to be fixed to a wall with one outer string, fixed between walls, or free standing; the majority have one wall string and one outer string. The wall string is fixed directly to the wall along its entire length, or is fixed to timber battens plugged to the wall, the top of the string being cut and hooked over the trimming member of the stairwell. The outer string is supported at both ends by a newel post: in the case of the bottom newel this rests on the floor; in the case of the upper newel, it is notched over and fixed to the stairwell trimming member. If the upper newel is extended to the ground floor to give extra support, it is called a **storey newel post**. The newel posts also serve as the termination point for handrails that span between them and are then infilled with balusters, balustrade or a solid panel to complete the protection to the sides of the stairway (see Fig. 7.4.3).

If the headroom distance is critical it is possible to construct a bulkhead arrangement over the stairs, as shown in Fig. 7.4.4; this may give the increase in headroom required to comply with the Building Regulations. The raised floor in the room over the stairs can be used as a shelf or form the floor of a hanging cupboard.

LAYOUT ARRANGEMENTS

A stair flight incorporating landings or tapered steps will enable the designer to economise with the space required to accommodate the stairs. Landings can be quarter space, giving a 90° turn, or half space, giving a 180° turn: for typical arrangements see Fig. 7.4.5. The construction of the landing is similar to that of a timber upper floor except that, with the reduced span, joist depths can be reduced (see Fig. 7.4.5). The landing can be incorporated in any position up the flight, and if sited near the head may well provide sufficient headroom to enable a cupboard or cloakroom to be constructed below the stairs. A dog–leg or string over string stair is economical in width, as it will occupy a width less than two flights, but this form has the disadvantage of a discontinuous handrail because this abuts to the underside of the return or upper flight.

TAPERED STEPS

Prior to the introduction of the Building Regulations tapered steps or winders were frequently used by designers to use space economically, because three treads occupied the area required for the conventional quarter space landing, which is counted as one tread. These steps had the following disadvantages:

- hazard to the aged and very young, because of the very small tread length at or near the newel post;
- difficult to carpet, requiring many folds or wasteful cutting;
- difficult to negotiate with furniture owing to a rapid rise on the turn;
- expensive to construct.

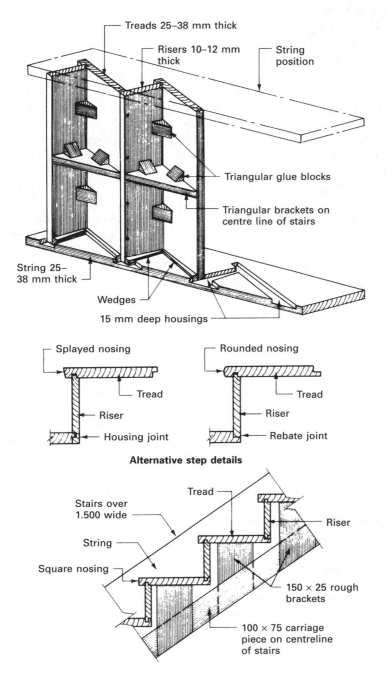

Treads 25–38 mm thick

Risers 10–12 mm thick

String position

Triangular glue blocks

Triangular brackets on centre line of stairs

String 25–38 mm thick

Wedges

15 mm deep housings

Splayed nosing

Rounded nosing

Tread

Tread

Riser

Riser

Housing joint

Rebate joint

Alternative step details

Tread

Stairs over 1.500 wide

Riser

String

Square nosing

150 × 25 rough brackets

100 × 75 carriage piece on centreline of stairs

Figure 7.4.2   Stair construction details.

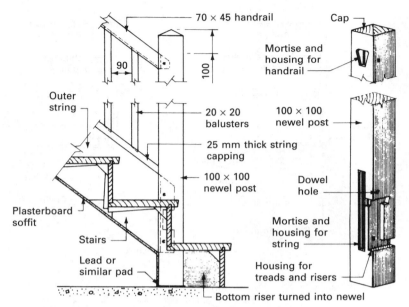

70 × 45 handrail

Cap

Mortise and housing for handrail

90

100

Outer string

20 × 20 balusters

100 × 100 newel post

25 mm thick string capping

100 × 100 newel post

Dowel hole

Mortise and housing for string

Plasterboard soffit

Stairs

Lead or similar pad

Housing for treads and risers

Bottom riser turned into newel

Typical detail at bottom newel

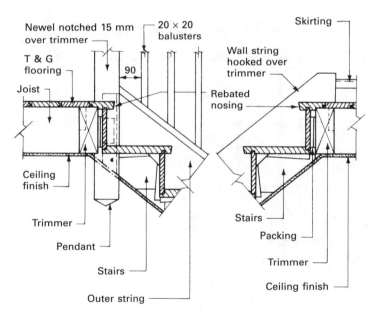

Newel notched 15 mm over trimmer

20 × 20 balusters

Skirting

T & G flooring

90

Wall string hooked over trimmer

Joist

Rebated nosing

Ceiling finish

Stairs

Trimmer

Packing

Pendant

Trimmer

Stairs

Ceiling finish

Outer string

Typical details at landing

Figure 7.4.3   Stair support and fixing details.

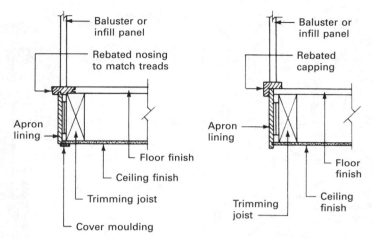

Typical stairwell finishes

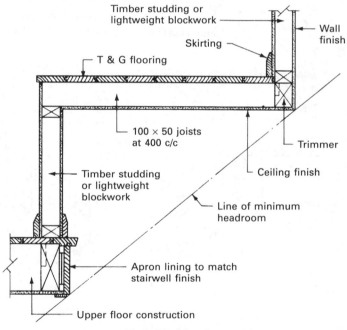

Typical bulkhead over stairs

Figure 7.4.4   Stairwell finishes and bulkhead details.

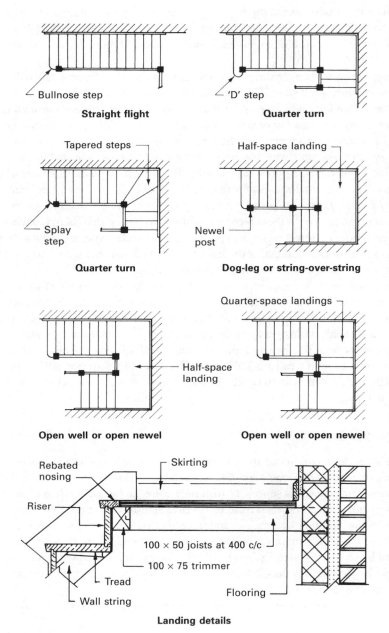

Figure 7.4.5   Typical stair layouts.

With the introduction of the Building Regulations special attention has been given to the inclusion of tapered steps in Approved Document K, which makes the use of tapered steps less of an economic proposition and more difficult to design (see Fig. 7.4.6).

A popular space-saving variation is the application of continuous tapered steps to create a spiral stair. The example illustrated in Fig. 7.4.7 can be in hard or softwood timber, but contemporary design versions are produced in stainless steel or in any combination of steel, timber and plastic. With spiral stairs every tread has a continuous taper, and the arrangement of these treads to produce a stairway should comply with the safety in use requirements indicated in the Building Regulations. Approved Document K1 of these regulations gives a specific reference for spiral stairs as conforming to BS 5395-2: *Stairs, ladders and walkways. Code of practice for the design of helical and spiral stairs.* Joinery manufacturers can be expected to adhere to these recommendations to ensure that there is a market for their product. However, before specifying a particular type, the details should be checked with the local authority building control department. Any structural alteration to the floor and landing will also require approval.

Spiral stair units are generally produced as a kit for site assembly of simple fit components about a central column or continuous newel post. The column is centralised over a lower-floor secured pattress, from which are placed alternate spacers and treads. Balusters fit to the periphery of the treads and provide support to the helical spire handrail. On plan, overall diameters from outside of tread to outside of tread are 1200 and 1400 mm at 14 treads per 360° turn, and 1600 and 2200 mm at 16 treads per 360° turn. Ascent may be clockwise or anti-clockwise.

OPEN-TREAD STAIRS

These are a contemporary form of stairs used in homes, shops and offices based on the simple form of access stair, which has been used for many years in industrial premises. The concept is simplicity with the elimination of elaborate nosings, cappings and risers. The open tread or riserless stair must comply fully with Part K of the Building Regulations and in particular Approved Document K, which recommends a minimum tread overlap of 16 mm. Where they are likely to be used by children under 5 years, i.e. dwellings, pre-school nurseries, etc., the opening between treads must contain a barrier sufficient to prevent a 100 mm sphere passing through.

Four basic types of open-tread stairs can be produced, as described below.

Closed string

This will terminate at the floor and landing levels and be fixed as for traditional stairs. The treads are tightly housed into the strings, which are tied together with long steel tie bars under the first, last and every fourth tread. The nuts and washers can be housed into the strings and covered with timber inserts (see Fig. 7.4.8).

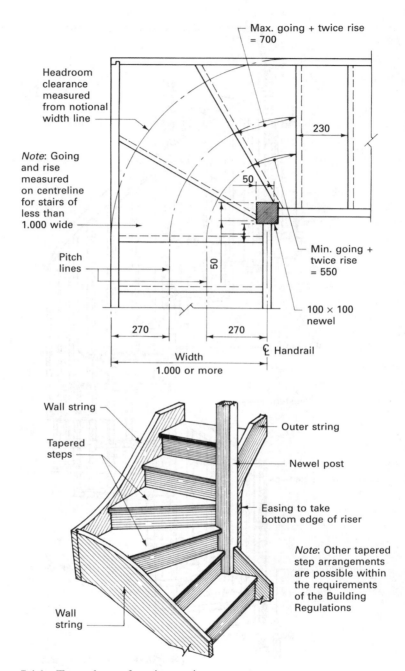

Max. going + twice rise = 700

Headroom clearance measured from notional width line

230

Note: Going and rise measured on centreline for stairs of less than 1.000 wide

50

Min. going + twice rise = 550

Pitch lines

50

100 × 100 newel

270

270

℄ Handrail

Width
1.000 or more

Wall string

Outer string

Tapered steps

Newel post

Easing to take bottom edge of riser

Note: Other tapered step arrangements are possible within the requirements of the Building Regulations

Wall string

Figure 7.4.6 Tapered steps for private stairways.

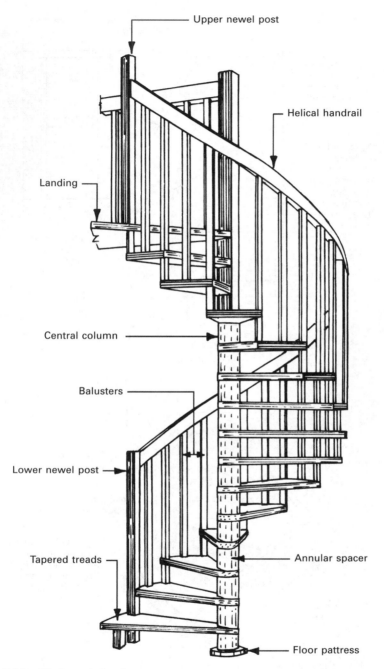

Figure 7.4.7   Timber spiral stair.

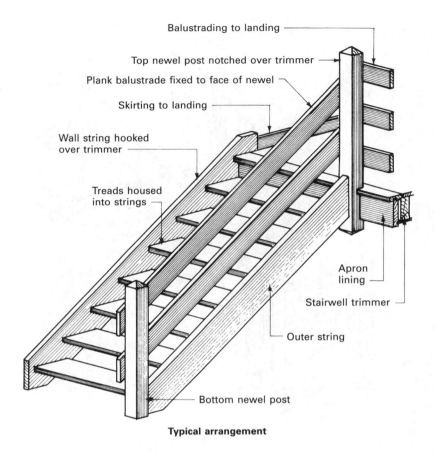

Balustrading to landing

Top newel post notched over trimmer

Plank balustrade fixed to face of newel

Skirting to landing

Wall string hooked over trimmer

Treads housed into strings

Apron lining

Stairwell trimmer

Outer string

Bottom newel post

Typical arrangement

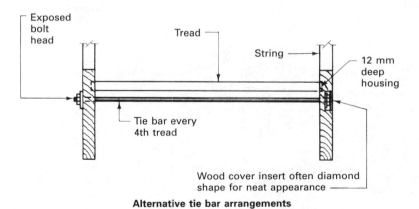

Exposed bolt head

Tread

String

12 mm deep housing

Tie bar every 4th tread

Wood cover insert often diamond shape for neat appearance

Alternative tie bar arrangements

Figure 7.4.8   Closed string open tread stairs.

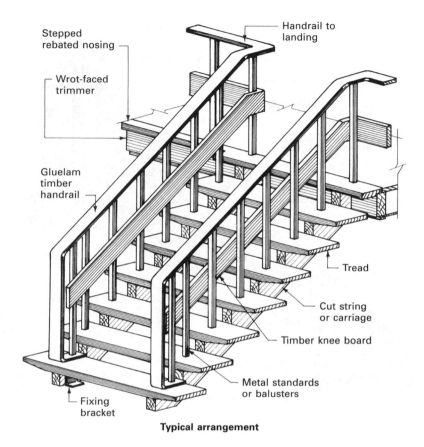

Stepped rebated nosing

Wrot-faced trimmer

Gluelam timber handrail

Handrail to landing

Tread

Cut string or carriage

Timber knee board

Metal standards or balusters

Fixing bracket

Typical arrangement

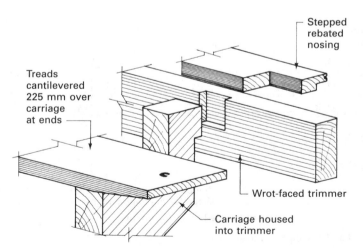

Stepped rebated nosing

Treads cantilevered 225 mm over carriage at ends

Wrot-faced trimmer

Carriage housed into trimmer

Figure 7.4.9   Cut string open tread stairs.

Cut strings or carriages

These are used to support cantilever treads, and can be worked from the solid or of laminated construction. The upper end of the carriage can be housed into the stairwell trimming member, with possible additional support from metal brackets. The foot of the carriage is housed in a purpose-made metal shoe or fixed with metal angle brackets (see Fig. 7.4.9).

Mono-carriage

Sometimes called a spine beam, this employs a single central carriage with double cantilever treads. The carriage, which is by necessity large, is of laminated construction and very often of a tapered section to reduce the apparently bulky appearance. The foot of the carriage is secured with a purpose-made metal shoe in conjunction with timber connectors (see Fig. 7.4.10).

Alternating tread stairs

These make economic use of space and are frequently applied to loft conversions. They have a pitch of about 60° and paddle-shaped treads, which are both controversial issues in the interests of user safety. However, the Building Regulations, Approved Document K, accepts 'familiarity and regular use' as a reasonable safety argument provided the stair accesses only one habitable room. Additional requirements include a non-slip surface, handrails both sides, a minimum going of 220 mm with a maximum rise of 220 mm, and a gap between treads of no more than 100 mm if likely to be used by children under 5 years (see Fig. 7.4.11).

Treads

These must be of adequate thickness, because there are no risers to give extra support; usual thicknesses are 38 and 50 mm. To give a lighter appearance it is possible to taper the underside of the treads at their cantilever ends for a distance of 225–250 mm. This distance is based on the fact that the average person will ascend or descend a stairway at a distance of about 250 mm in from the handrail.

Balustrading

Together with the handrail, balustrading provides both the visual and the practical safety barrier to the side of the stair. Children present special design problems, because they can and will explore any gap big enough to crawl through. BS 5395 for wood stairs recommends that the infill under handrails should have no openings that would permit the passage of a sphere 90 mm in diameter, a slight reduction on the 100 mm stated in Approved Document K1, Section 1.29, to the Building Regulations. Many variations of balustrading are possible, ranging from simple newels with planks to elaborate metalwork of open design (see Figs 7.4.8, 7.4.9 and 7.4.10).

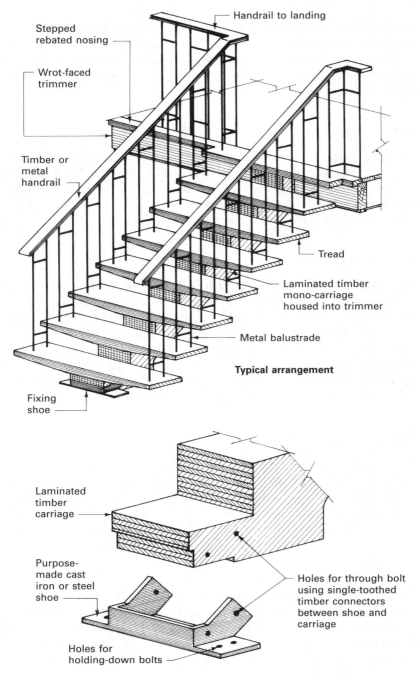

Handrail to landing

Stepped
rebated nosing

Wrot-faced
trimmer

Timber or
metal
handrail

Tread

Laminated timber
mono-carriage
housed into trimmer

Metal balustrade

Typical arrangement

Fixing
shoe

Laminated
timber
carriage

Purpose-
made cast
iron or steel
shoe

Holes for through bolt
using single-toothed
timber connectors
between shoe and
carriage

Holes for
holding-down bolts

Figure 7.4.10   Mono–carriage open tread stairs.

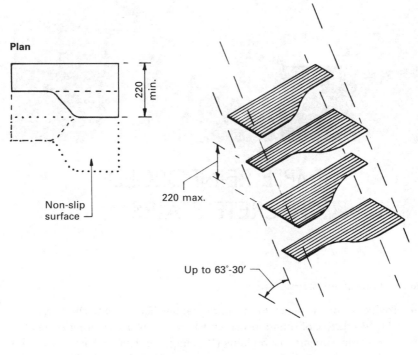

Figure 7.4.11 Alternating tread stairs.

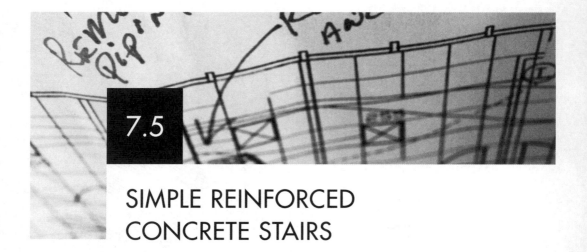

7.5

SIMPLE REINFORCED CONCRETE STAIRS

The functions of any stairway are:

- to provide for the movement of people from one floor level to another;
- to provide a degree of sound insulation where part of a separating element between compartments in a building (Building Regulations, Parts E1 and E3);
- to provide a suitable means of escape in event of fire (Building Regulations, Part B1).

A timber stairway will maintain its strength and integrity for a reasonable period during an outbreak of fire. However, timber is combustible and will contribute to the fire spread, thus increasing the hazards that could be encountered along an escape route. It is for this reason that application of timber stairs is limited by the Building Regulations to certain domestic and some other small building types. Stairs in this category are not normally in a protected shaft, and therefore not part of a specifically planned escape route.

Stairs, other than the exceptions given above, must therefore be constructed of non-combustible materials, but combustible materials are allowed to be used as finishes to the upper surface of the stairway or landing. Reinforced concrete stairs are non-combustible, strong and hard wearing. They may be constructed *in-situ* or precast in sections ready for immediate installation and use when delivered to site. The general use of cranes on building sites has meant that many of the large flight arrangements, which in the past would have been cast *in-situ*, can now be precast under factory-controlled conditions.

Many variations of plan layout and spanning direction are possible, but this study will be confined to the simple straight flight spanning from floor to floor or from floor to intermediate landing. The designer will treat the stairs as being an inclined slab spanning simply between the supports, the steps being treated as triangular loadings evenly distributed over the length. Where intermediate landings are included in the design the basic plan is similar to the open well or newel timber stairs. Difficulty is sometimes experienced with the intersection of the upper and lower flight soffits with the landing. One method of overcoming this problem is to have, in plan, the top and bottom steps out of line so that the soffit intersections form a straight line on the underside of the landing (see Fig. 7.5.1). The calculations to determine the rise, going and number of steps are the same as those used for timber stairs; it should be noted that the number of risers in each flight must not exceed 16. To achieve a greater tread length without increasing the actual horizontal going it is common to use a splayed riser face giving a 25 mm increase to the tread length.

The concrete specification should be very strong, with a mix ratio of 1:1.5:3–10 mm aggregate or 25–30 N/mm^2 compressive strength. Water/cement ratio is no greater than 0.5, otherwise concrete will flow over the formwork riser boards. Concrete cover to reinforcement is at least 15 mm or the bar diameter, whichever is the greater. This cover is sufficient for one-hour fire resistance in most buildings, but will need to be increased for buildings in certain purpose groups: see Table A2 in Appendix A of Approved Document B to the Building Regulations. The thickness of concrete required is dependent on the loading and span, but is not generally less than 100 mm or more than 150 mm measured across the waist, which is the distance from the soffit to the intersection of tread and riser (see Fig. 7.5.1).

Mild steel or high-yield steel bars can be used to reinforce concrete stairs, the bars being lapped to starter bars at the ground floor and taken into the landing or floor support slab. The number, diameter and spacing of the main and distribution reinforcement must always be calculated for each stairway.

Handrails and balustrading must be constructed of a non-combustible material, continuous and to both sides if the width of the stairs exceeds 1.0 m. The overall height of the handrail up the stairs should be between 900 and 1000 mm measured vertically and have a height above the floor of 1.1 m minimum. The capping can be of a combustible material such as plastic provided that it is fixed to or over a non-combustible core. Methods of securing balustrades and typical handrail details are shown in Fig. 7.5.2. Flights of over 1.8 m width in public buildings should have a centrally dividing handrail. Widths of stairs designated for fire escape purposes should be determined by reference to tables and calculations in Section 5 of the Building Regulations, Approved Document B1.

A wide variety of finishes can be applied to the tread surface of the stairs. If the appearance is not of paramount importance, such as in a warehouse, a natural finish could be used, but it would be advisable to trowel into the surface some carborundum dust to provide a hard-wearing non-slip surface. Alternatively, rubber or carborundum insert strips could be fixed or cast into the leading edges of the treads. Finishes such as PVC tiles, rubber tiles and carpet mats are applied and fixed in the same manner as for floors. The soffits can be left as struck from the formwork and decorated, or finished with a coat of spray plaster or a skim coat of finishing plaster.

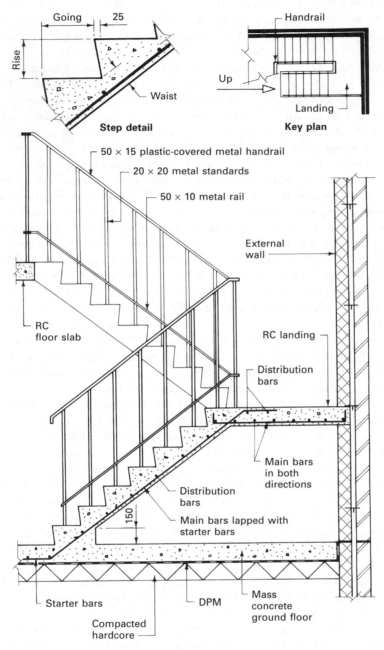

Figure 7.5.1 Simple reinforced concrete *in-situ* stairs.

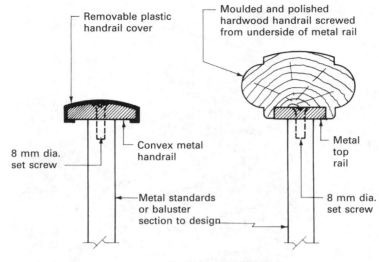

Removable plastic
handrail cover

8 mm dia.
set screw

Convex metal
handrail

Metal standards
or baluster
section to design

Moulded and polished
hardwood handrail screwed
from underside of metal rail

Metal
top
rail

8 mm dia.
set screw

Typical handrails

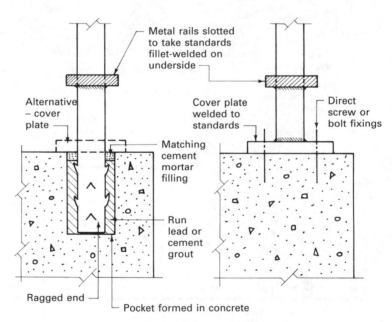

Metal rails slotted
to take standards
fillet-welded on
underside

Alternative
– cover
plate

Matching
cement
mortar
filling

Run
lead or
cement
grout

Ragged end

Pocket formed in concrete

Cover plate
welded to
standards

Direct
screw or
bolt fixings

Typical fixing methods

Figure 7.5.2   Handrails and balustrades.

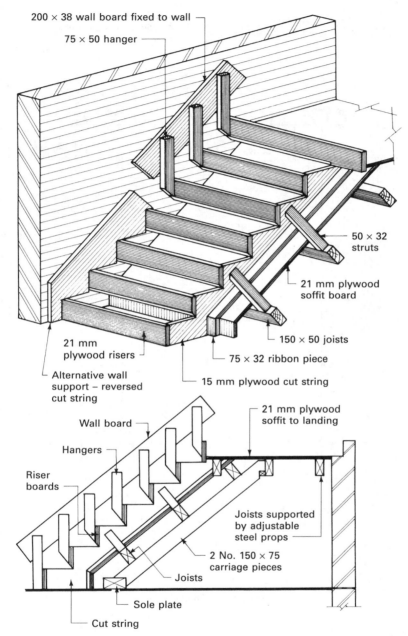

200 × 38 wall board fixed to wall

75 × 50 hanger

50 × 32 struts

21 mm plywood soffit board

21 mm plywood risers

Alternative wall support – reversed cut string

150 × 50 joists

75 × 32 ribbon piece

15 mm plywood cut string

21 mm plywood soffit to landing

Wall board

Hangers

Riser boards

Joists supported by adjustable steel props

2 No. 150 × 75 carriage pieces

Joists

Sole plate

Cut string

Figure 7.5.3 Typical framework to reinforced concrete *in-situ* stairs.

▨ ▤ ■ FORMWORK

The basic requirements are the same as for formwork to a framed structure. The stair profile is built off an adequately supported soffit of sheet material by using a cut string. Riser boards are used to form the leading face of the steps; these should have a splayed bottom edge to enable complete trowelling of the tread surfaces and to ensure that air is not trapped under the bottom edge of the riser board, thus causing voids. If the stair abuts a vertical surface two methods can be considered to provide the abutment support for the riser boards: a reverse-cut string or a wall board with hangers (see Fig. 7.5.3). Wide stairs can have a reverse-cut string as a central support to the riser boards to keep the thickness of these within an acceptable coat limit.

7.6

SIMPLE PRECAST CONCRETE STAIRS

Precast concrete stairs can be designed and constructed to satisfy a number of different requirements. They can be a simple inclined slab, a cranked slab, open riser stairs, or constructed from a series of precast steps built into, and if required cantilevered from, a structural wall.

The design considerations for the simple straight flight are the same as those for *in-situ* stairs of comparable span, width and loading conditions. The fixing and support, however, require a different approach. Bearings for the ends of the flights must be provided at the floor or landing levels in the form of a haunch, rebate or bracket, and continuity of reinforcement can be achieved by leaving projecting bars and slots in the floor into which they can be grouted (see Fig. 7.6.1).

Ideally the delivery of precast stairs should be arranged so that they can be lifted, positioned and fixed direct from the delivery vehicle, thus avoiding double handling. Precast components are usually designed for 2 conditions:

- lifting and transporting;
- final fixed condition.

It is essential that the flights are lifted from the correct lifting points, which may be in the form of loops or hooks projecting from or recessed into the concrete member, if damage by introducing unacceptable stresses during lifting is to be avoided.

Balustrade and handrail requirements, and the various methods of fixing, are as described for *in-situ* reinforced concrete stairs. Any tread finish that is acceptable for an *in-situ* stair will also be suitable for the precast alternative.

The use of precast concrete steps to form a stairway is limited to situations such as short flights between changes in floor level and external stairs to basements and areas. They rely on the loadbearing wall for support and, if cantilevered, on the downward load of the wall to provide the necessary reaction. The support wall has to provide this necessary load and strength, and at the same time it has to be bonded or cut around the stooled end(s) of the steps. It is for these reasons that the application of precast concrete steps is restricted.

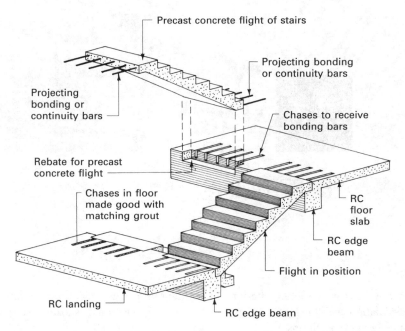

Precast concrete flight of stairs

Projecting bonding
or continuity bars

Projecting
bonding or
continuity bars

Chases to receive
bonding bars

Rebate for precast
concrete flight

Chases in floor
made good with
matching grout

RC
floor
slab

RC edge
beam

Flight in position

RC landing

RC edge beam

Simple precast concrete stairs

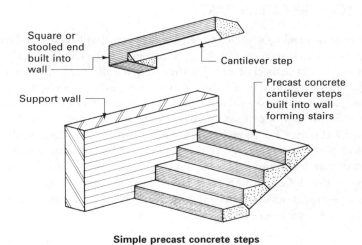

Square or
stooled end
built into
wall

Cantilever step

Support wall

Precast concrete
cantilever steps
built into wall
forming stairs

Simple precast concrete steps

Figure 7.6.1   Precast concrete stairs and steps.

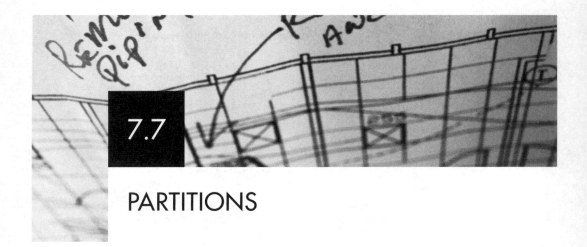

7.7

PARTITIONS

Internal walls that divide the interior of a building into areas of accommodation and circulation are called **partitions**, and these can be classified as loadbearing or non-loadbearing partitions.

■■■ LOADBEARING PARTITIONS

These are designed and constructed to receive superimposed loadings and transmit these loads to a foundation (see Fig. 4.1.2). Generally, loadbearing partitions are constructed of bricks or blocks bonded to the external walls (see Figs 3.3.1 and 3.3.2). Openings are formed in the same manner as for external walls; a lintel spans the opening, carrying the load above to the reveals on either side of the opening. Fixings can usually be made direct into block walls, whereas walls of bricks require to be drilled and plugged to receive nails or screws. Apart from receiving direct fixings, block partitions are lighter, cheaper, quicker to build and have better thermal insulation properties than brick walls, but their sound insulation values are slightly lower. For these reasons blockwork is usually specified for loadbearing partitions in domestic work. Loadbearing partitions, because of their method of construction, are considered to be permanently positioned.

■■■ NON-LOADBEARING PARTITIONS

These partitions, like loadbearing partitions, must be designed and constructed to carry their own weight and any fittings or fixings that may be attached to them, but they must not, in any circumstances, be used to carry or assist in the transmission of structural loadings. They must be designed to provide a suitable division between internal spaces and be able to resist impact loadings on their faces, also any vibrations set up by doors being closed or slammed.

Bricks or blocks can be used for the construction of non-loadbearing partitions. They may be built directly off a concrete floor with a thickened slab or steel reinforcement to support the wall's self-weight. At ceiling level stability is obtained by bracketing between or to the joists above. If the partition is built off a suspended timber floor a larger joist section or two joists side by side must be used to carry the load of the partition to the floor support walls. Openings are constructed as for loadbearing walls; alternatively a storey-height frame could be used.

Timber stud partitions with suitable facings are lighter than brick or block partitions but are less efficient as sound or thermal insulators. However, a high standard of sound and thermal insulation and fire resistance can be achieved by facing with plasterboard, filling the voids between studs and noggins with glass or mineral wool. They are easy to construct and provide a good fixing background, and because of their lightness are suitable for building off a suspended timber floor. The basic principle is to construct a simple framed grid of timber to which a dry lining such as plywood, plasterboard or hardboard can be attached. The lining material will determine the spacing of the uprights or studs to save undue wastage in cutting the boards to terminate on the centre line of a stud. To achieve a good finish it is advisable to use studs that have equal thicknesses, because the thin lining materials will follow any irregularity of the face of the studwork. Openings are formed by framing a head between two studs and fixing a lining or door frame into the opening in the stud partition; typical details are shown in Fig. 7.7.1. Where plasterboard is the specified facing/lining material, the extent of nailing will depend on the stud spacing, the facing thickness and its width. Table 7.7.1 provides guidance when using galvanised steel taper head nails at 150 mm spacing.

Proprietary partitions of plasterboard bonded on either side of a strong cellular core to form rigid panels are suitable as non-loadbearing partitions. These are fixed to wall and ceiling battens and supported on a timber sole plate. Timber blocks are inserted into the core to provide the fixing medium for door frames or linings and skirtings (see Fig. 7.7.2). This form of partition is available with facings suitable for direct decoration, plastercoat finish or with a plastic face as a self-finish.

Compressed strawboard panels provide another alternative method for non-loadbearing partitions. The storey-height panels are secured to a sole plate and a head plate by skew or tosh nailing through the leading edge. The 3 mm joint between consecutive panels is made with an adhesive supplied by the

Table 7.7.1 Guidance when using galvanised steel taper nails

Plasterboard thickness	Plasterboard width	Stud or batten spacing	Nails
9.5	900	450	2×30 long
	1200	400	2×30 long
12.5	600	600	2×40 long
	900	450	2×40 long
19.0	1200	600	2.6×40 long
	600	600	2.6×40 long

Note: All dimensions in millimetres.

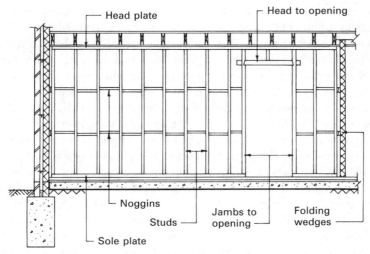

Typical arrangement

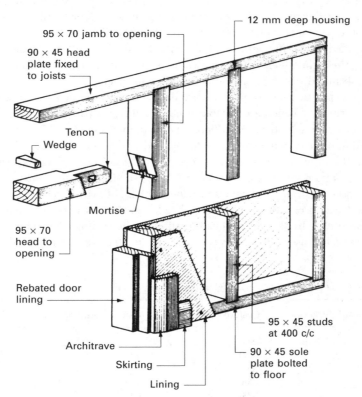

Figure 7.7.1   Typical timber stud partition.

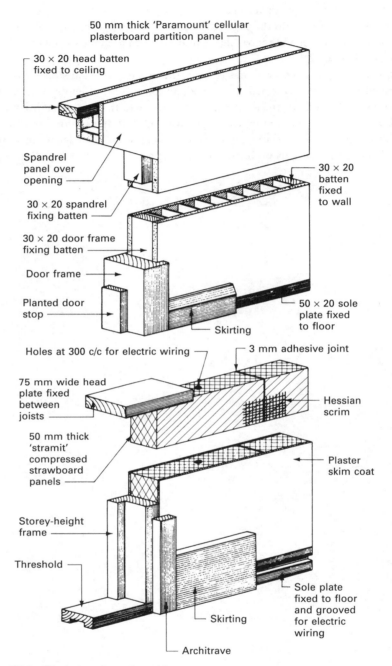

50 mm thick 'Paramount' cellular plasterboard partition panel

30 × 20 head batten fixed to ceiling

Spandrel panel over opening

30 × 20 spandrel fixing batten

30 × 20 door frame fixing batten

Door frame

Planted door stop

30 × 20 batten fixed to wall

50 × 20 sole plate fixed to floor

Skirting

Holes at 300 c/c for electric wiring

3 mm adhesive joint

75 mm wide head plate fixed between joists

Hessian scrim

50 mm thick 'stramit' compressed strawboard panels

Plaster skim coat

Storey-height frame

Threshold

Sole plate fixed to floor and grooved for electric wiring

Skirting

Architrave

Figure 7.7.2　Typical preformed partitions.

manufacturer. Openings are formed by using storey-height frames fixed with 100 mm screws direct into the edge of the panel. The joints are covered with a strip of hessian scrim, and the whole partition is given a skim coat of board plaster finish (see Fig. 7.7.2).

Partially prefabricated partitions, such as the plasterboard and strawboard described above, can be erected on site without undue mess. Being mainly a dry construction they reduce the drying time required with the traditional brick and block walls. They also have the advantage, like timber stud partitions, of being capable of removal or repositioning without causing serious damage to the structure or without causing serious problems to a contractor.

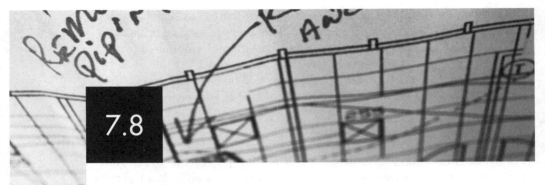

FINISHES: FLOOR, WALL AND CEILING

■■■ FLOOR FINISHES

The type of floor finish to be applied to a floor will depend upon a number of factors such as type of base, room usage, degree of comfort required, maintenance problems, cost, appearance, safety and individual preference.

Floor finishes can be considered under three main headings:

- *in-situ* floor finishes – those finishes that are mixed on site, laid in a fluid state, allowed to dry, and set to form a hard jointless surface;
- applied floor finishes – those finishes that are supplied in tile or sheet form and are laid onto a suitably prepared base;
- timber floor finishes – boards, sheets and blocks of timber laid on or attached to a suitable structural frame or base.

IN-SITU FLOOR FINISHES

Mastic asphalt – BS 6925 and BS 8204-5

This is a naturally occurring bituminous material obtained from asphalt lakes like those in Trinidad; it can also be derived from crude oil residues. Trinidad lake asphalt is used as a matrix or cement to a suitably graded mineral aggregate to form mastic asphalt as a material suitable for floor finishing. When laid, mastic asphalt is impervious to water and is ideal for situations such as sculleries, wash rooms and balconies. It also forms a very good surface on which to apply thin tile and sheet finishes (for example, PVC), and will at the same time fulfil the function of a damp-proof membrane or waterproof decking.

Mastic asphalt is a thermoplastic material and has to be melted before it can be applied to form a jointless floor finish. Hot mastic asphalt is applied by means of a float at a temperature of between 180 and 210 °C in a single 13 mm coat as a base for applied finishes or in a 16 mm single coat for a self-finish. Any sound and rigid subfloor is suitable, but a layer of ordinary black sheathing felt should be included between the subfloor and mastic asphalt to overcome the problems caused by differential movement. The finish obtained is smooth and hard, but the colour range is limited to dark colours such as red, brown and black. A matt surface can be produced by giving the top surface a dusting of sand or powdered stone.

Pitch mastic

Pitch mastic is a similar material to mastic asphalt but is produced from a mixture of calcareous and/or siliceous aggregates bonded with coal tar pitch. It is laid to a similar thickness and in a similar manner to mastic asphalt with a polished or matt finish. Pitch mastic floors are resistant to water but have a better resistance to oil and fats than mastic asphalt, and are therefore suitable for sculleries, washrooms and kitchens.

Granolithic

This is a mixture of Portland cement and granite chippings, which can be applied to a 'green' concrete subfloor or to a cured concrete subfloor. **Green concrete** is a term used to describe newly laid concrete that is not more than 3 hours old. A typical mix for granolithic is 1 part cement: 1 part sand: 2 parts granite chippings (5–10 mm free from dust) by volume. The finish obtained is very hard wearing, noisy and cold to touch; it is used mainly in situations where easy maintenance and durability are paramount, such as a common entrance hall to a block of flats.

If granolithic is being applied to a green concrete subfloor as a topping it is applied in a single layer approximately 20 mm thick in bay sizes not exceeding 28 m^2, and trowelled to a smooth surface. This method will result in a monolithic floor and finish construction.

The surface of mature concrete will need to be prepared by hacking the entire area and brushing well to remove all the laitance before laying a single layer of granolithic, which should be at least 40 mm thick. The finish should be laid, on a wet cement slurry coating or PVA bonding agent to improve the bond, in bay sizes not exceeding 14 m^2.

Magnesium oxychloride

This is a composition flooring that is sometimes used as a substitute for asphalt, as it has similar wearing and appearance properties. It is mixed on site using a solution of magnesium chloride with burned magnesite and fillers such as wood flour, sawdust, powdered chalk or limestone. It is essential that the ingredients are thoroughly mixed in a dry state before the correct amount of solution is added. The mixed material is laid in one or two layers, giving a total thickness of approximately 20 mm. The subfloor must be absolutely dry, otherwise the finish will lift and crack: therefore a damp-proof membrane in the subfloor is essential.

APPLIED FLOOR FINISHES

Many of the applied floor finishes are thin flexible materials, and should be laid on a subfloor with a smooth finish. This is often achieved by laying a cement/sand bed or screed with a steel float finish to the concrete subfloor. The usual screed mix is 1:3 cement/sand, and great care must be taken to ensure that a good bond is obtained with the subfloor. If the screed is laid on green concrete a thickness of 12 mm is usually specified, whereas screeds laid on mature concrete require a thickness of 40 mm. A mature concrete subfloor must be clean, free from dust, and dampened with water to reduce the suction before applying the bonding agent to receive the screed. To reduce the possibility of drying shrinkage cracks, screeds should not be laid in bays exceeding 15 m² in area. Screeds laid directly over rigid insulation should be at least 50 mm in thickness, reinforced with light steel mesh (chicken wire).

Flexible PVC tiles and sheet – BS EN 649 and BS 8203

Flexible PVC is a popular hardwearing floor finish produced by a mixture of polyvinyl chloride resin, pigments, mineral fillers and plasticisers to control flexibility. It is produced as 300 mm × 300 mm square tiles or in sheet form up to 2400 mm wide with a range of thicknesses from 1.5 to 4.5 mm. The floor tiles and sheet are fixed with an adhesive recommended by the manufacturer, and produce a surface suitable for most situations. PVC tiles, like all other small unit coverings, should be laid from the centre of the area towards the edges, so that if the area is not an exact tile module an even border of cut tiles is obtained.

Thermoplastic tiles

These are sometimes called **asphalt tiles** and are produced from coumarone-indene resins, fillers and pigments; the earlier use of asphalt binders limited the range of colours available. These tiles are hardwearing and moisture resistant, and are suitable for most situations, being produced generally as a 225 mm square tile either 3 or 4.5 mm thick. To make them pliable they are usually heated before being fixed with a bituminous adhesive. As they age they can become brittle and fracture, detaching from their base adhesive. The development of PVC tiles has largely displaced thermoplastic tiles as they are more resilient and offer a brighter colour range.

Rubber tiles and sheet – BS 1711 and BS 3187

Solid rubber tiles and sheet are produced from natural or synthetic rubber compounded with fillers to give colour and texture, the rubber content being not less than 35%. The covering is hardwearing, quiet and water resistant, and suitable for bathrooms and washrooms. Thicknesses range from 3 to 6.5 mm with square tile sizes ranging from 150 to 1200 mm; sheet widths range from 900 to 1800 mm. Fixing is by rubber-, epoxy- or neoprene-based adhesives, as recommended by the manufacturer, to a smooth surface.

Linoleum – BS EN 12104

Linoleum is produced in sheet or tile form from a mixture of drying oils, resins, fillers and pigments, which is pressed onto a hessian or bitumen-saturated felt paper backing. Good-quality linoleum has the pattern inlaid or continuing through the thickness, whereas the cheaper quality has only a printed surface pattern. Linoleum gives a quiet, resilient and hardwearing surface suitable for most domestic floors. Thicknesses vary from 2 to 6.5 mm for a standard sheet width of 1800 mm; tiles are usually 300 mm square with the same range of thicknesses. Fixing of linoleum tiles and sheet is by adhesive to any dry smooth surface, although the adhesive is sometimes omitted with the thicker sheets.

Carpet – BS 4223, BS 5808 and BS 5325

The chief materials used in the production of carpets are nylon, acrylics and wool, or mixtures of these materials. There is a vast range of styles, types, patterns, colours, qualities and sizes available for general domestic use in dry situations, as the resistance of carpets to dampness is generally poor. To obtain maximum life carpets should be laid over an underlay of felt or latex and secured by adhesives, nailing around the perimeter or being stretched over and attached to special edge-fixing strips. Carpet is supplied in narrow or wide rolls. Carpet squares (600 mm × 600 mm × 25 mm thick) are also available for covering floors without the use of adhesives; these rely on the interlocking of the edge fibres to form a continuous covering.

Cork tiles and carpet – BS EN 12104

Cork tiles are cut from baked blocks of granulated cork; the natural resins act as the binder. The tiles are generally 300 mm square with thicknesses of 5 mm upwards according to the wearing quality required, and are supplied in three natural shades. They are hardwearing, quiet and resilient but, unless treated with a surface sealant, they may collect dirt and grit. Fixing is by manufacturer's adhesive.

Cork carpet is a similar material, but it is made pliable by bonding the cork granules with linseed oil and resins onto a jute canvas backing. It is laid in the same manner as described above for linoleum, and should be treated with a surface sealant to resist dirt and grit penetration.

Quarry tiles – BS 6431

The term 'quarry' is derived from the Italian word *quadro* meaning square; it does not mean that the tiles are cut or won from an open excavation or quarry. They are made from ordinary or unrefined clays worked into a plastic form, pressed into shape, and hard burnt. Being hardwearing and with a good resistance to water they are suitable for kitchens and entrance halls, but they tend to be noisy and cold. Quarry tiles are produced as square tiles in sizes ranging from 100 mm × 100 mm × 20 mm to 225 mm × 225 mm × 32 mm thick.

Three methods of laying quarry tiles are recommended to allow for differential movement due to drying shrinkage or thermal movement of the screed. The first method is to bed the tiles in a 10–15 mm thick bed of cement mortar over a separating layer of sheet material such as polythene, which could perform the dual function of damp-proof membrane and separating layer. To avoid the use of a separating layer the 'thick bed' method can be used. With this method the concrete subfloor should be dampened to reduce suction and then covered with a semi-dry 1:4 cement/sand mix to a thickness of approximately 40 mm. The top surface of the compacted semi-dry 'thick bed' should be treated with a 1:1 cement/sand grout before tapping the tiles into the grout with a wood beater. Alternatively a dry cement may be trowelled into the 'thick bed' surface before bedding in the tiles. Special cement-based adhesives are also available for bedding tiles to a screed, and these should be used in accordance with the manufacturer's instructions to provide a thin bed fixing.

Clay tiles may expand, probably as a result of physical adsorption of water and chemical hydration, and for this reason an expansion joint of compressible material should be incorporated around the perimeter of the floor (see Fig. 7.8.1). In no circumstances should the length or width of a quarry tile floor exceed 7500 mm without an expansion joint. The joints between quarry tiles are usually grouted with a 1:1 cement/sand grout and then, after cleaning, the floor should be protected with sand or sawdust for 4 or 5 days. There is a wide variety of tread patterns available to provide a non-slip surface; also a wide range of fittings are available to form skirtings and edge coves (see Fig. 7.8.1).

Plain clay or ceramic floor tiles – BS 6431

These are similar to quarry tiles but are produced from refined natural clays, which are pressed after grinding and tempering into the desired shape before being fired at a high temperature. Plain clay floor tiles, being denser than quarry tiles, are made as smaller and thinner units ranging from 50 mm × 50 mm to 300 mm × 300 mm in thicknesses of 9.5–13 mm.

Laying, finishes and fittings available are all as described for quarry tiles.

TIMBER FLOOR FINISHES

Timber is a very popular floor finish with both designer and user because of its natural appearance, resilience and warmth. It is available as a board, strip, sheet or block finish, and if attached to joists, as in the case of a suspended timber floor, it also acts as the structural decking.

Timber boards

Softwood timber floorboards are joined together by tongued and grooved joints along their edges, and are fixed by nailing to the support joists or fillets attached to a solid floor. The boards are butt-jointed in their length, the joints being positioned centrally over the supports and staggered so that butt joints do not occur in the

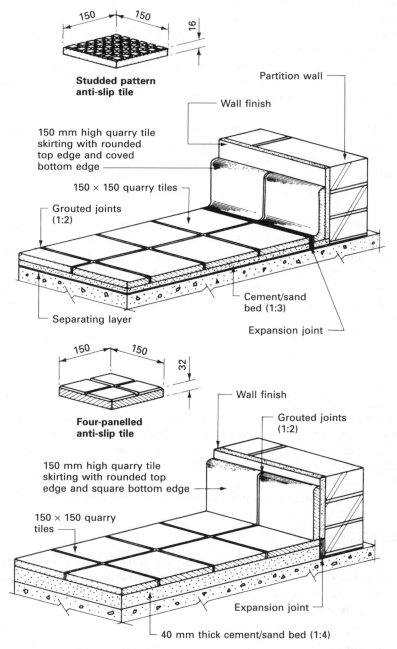

Figure 7.8.1 Typical quarry tile floors.

same position in consecutive lengths. The support spacing will be governed by the spanning properties of the board (which is controlled by its thickness), and supports placed at 400 mm centres are usual for boards 19 and 22 mm thick. The tongue is positioned slightly off centre, and the boards should be laid with the narrow shoulder on the underside to give maximum wear. It is essential that the boards are well cramped together before being fixed to form a tight joint, and that they are laid in a position where they will not be affected by dampness. Timber is a hygroscopic material and will therefore swell and shrink as its moisture content varies; ideally this should be maintained at around 12%.

Timber strip

These are narrow boards, being under 100 mm wide to reduce the amount of shrinkage and consequent opening of the joints. Timber strip can be supplied in softwood or hardwood, and is considered to be a superior floor finish to boards. Jointing and laying is as described for boards, except that hardwood strip is very often laid one strip at a time and secret-nailed (see Fig. 7.8.2).

Timber sheet floor finish – BS EN 312

Chipboard is manufactured from wood chips or shavings bonded together with thermosetting synthetic resins, and forms rigid sheets 18 and 22 mm thick, which are suitable as a floor finish. The sheets are fixed by nailing or screwing to support joists or fillets. If the sheet is used as an exposed finish a coating of sealer should be used. Alternatively chipboard can be used as a decking to which a thin tile or carpet finish can be applied.

Tongued and grooved boards of 600 mm width are also available as a floor-decking material that does not need to be jointed over a joist.

Wood blocks

These are small blocks of timber, usually of hardwood, which are designed to be laid in set patterns. Lengths range from 150 to 300 mm with widths up to 89 mm; the width is proportional to the length to enable the various patterns to be created. Block thicknesses range from 20 to 30 mm, and the final thickness after sanding and polishing is about 5–10 mm less.

The blocks are jointed along their edges with a tongued and grooved joint, and have a rebate, chamfer or dovetail along the bottom longitudinal edges to take up any surplus adhesive used for fixing. Two methods can be used for fixing wood blocks: the first uses hot bitumen and the second a cold latex bitumen emulsion. If hot bitumen is used, the upper surface of the subfloor is first primed with black varnish to improve adhesion, and then, before laying, the bottom face of the block is dipped into the hot bitumen. The cold adhesive does not require a priming coat to the subfloor. Blocks, like tiled floors, should be laid from the centre of the floor towards the perimeter, which is generally terminated with a margin border.

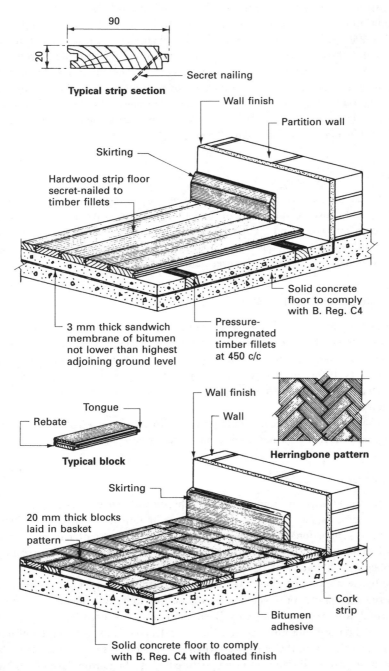

Figure 7.8.2   Typical timber floor finishes.

To allow for moisture movement a cork expansion strip should be placed around the entire edge of the block floor (see Fig. 7.8.2).

Parquet

This is a superior form of wood block flooring made from specially selected hardwoods chosen mainly for their decorative appearance. Parquet blocks are generally smaller and thinner than hardwood blocks and are usually fixed to a timber subfloor that is level and smooth. Fixing can be by adhesives or secret nailing; alternatively they can be supplied as a patterned panel fixed to a suitable backing sheet in panel sizes from 300 to 600 mm square.

Wood veneer and laminate

These finishes are achieved with a tongued and grooved interlocking composite strip floor finish system. Strips are generally 1200 mm long, in widths of 60–200 mm and thicknesses between 10 and 15 mm depending on the quality. Some systems are glued along the t & g and others simply snap-lock together. End jointing is staggered between adjacent strips to avoid continuity of joints.

Veneered floors have a thin hardwood facing bonded to an HDFB core with a paper underlayer. Laminate floors comprise a thin but dense aluminium oxide clear protective surface over a photographic image of wood, an HDFB core and a paper backing. If damaged, veneers can be sanded to some extent, but laminated artificial veneers are less easily repaired.

- **Underlay** Damp-proof plastic foil of 0.15 mm minimum thickness, with damp-proof adhesive-taped joints. Treated cork or foam alternatives are also available.
- **Sublayer** Structural concrete or screed must dry out and be level; suspended timber must be stable and level, i.e. floor boards screwed to joists and surface irregularities sanded.
- **Movement** An 8–10 mm expansion gap is required between strips and room periphery. With new build this gap can be accommodated under the skirting board. Where there is an existing skirting a moulding is used to cover the gap.

▣▣▣ WALL FINISHES

External brickwork with an exposed face of facing bricks is a self-finish and requires no further treatment. External walls of common bricks or blocks can be treated to give an acceptable appearance by the application of paint or an applied wall finish such as rendering, or can be clad with boards or tiles. Internal walls or partitions can be built with a fair face of natural materials such as bricks or stone, but generally it is cheaper to use a material such as blocks with an applied finish such as plaster, drylining or glazed tiles.

EXTERNAL RENDERING – BS 5262

This is a form of plastering using a mixture of cement and sand, or cement, lime and sand, applied to the face of a building to give extra protection against the penetration of moisture or to provide a desired texture. It can also be used in the dual capacity of providing protection and appearance.

The rendering must have the properties of durability, moisture resistance and an acceptable appearance. The factors to be taken into account in achieving the above requirements are mix design, bond to the backing material, texture of surface, degree of exposure of the building, and the standard of workmanship in applying the rendering.

Cement and sand mixes will produce a strong moisture-resistant rendering, but one that is subject to cracking because of high drying shrinkage. These mixes are used mainly on members that may be vulnerable to impact damage, such as columns. Cement, lime and sand mixes have a lower drying shrinkage but are more absorbent than cement and sand mixes; they will, however, dry out rapidly after periods of rain and are therefore the mix recommended for general use.

The two factors that govern the proportions to be used in a mix are:

- background to which the rendering is to be applied;
- degree of exposure of the building.

The two common volume mix ratios are:

- $1:\frac{1}{2}:4-4\frac{1}{2}$ cement:lime:sand, which is used for dense, strong backgrounds of moderate to severe exposure and for application to metal lathing or expanded metal backgrounds. Equivalent cement and sand with a plasticiser is 1:3–4.
- 1:1:5–6 cement:lime:sand, which is for general use. Equivalent cement and sand with a plasticiser is 1:5–6.

If the rendering is to be applied in a cold, damp situation the cement content of the mix ratio should be increased by 1. The final coat is usually a weaker mix than the undercoats: for example, when using an undercoat mix 1:1:6 (1:5–6 cement: sand with plasticiser), the final coat mix ratio should be 1:2:9 (1:7–8 cement:sand with plasticiser), to reduce the amount of shrinkage in the final coat. The number of coats required will depend upon the surface condition of the background and the degree of exposure. Generally a two-coat application is acceptable, except where the background is very irregular or the building is in a position of severe exposure, when a three-coat application would be specified. The thickness of any one coat should not exceed 15 mm, and each subsequent coat thickness is reduced by approximately 3 mm to give a final coat thickness of 6–10 mm. Wood float treatment produces a smooth flat surface, which may be treated with a felt-coated float or lamb's wool roller to raise the surface texture and provide better resistance to cracking.

Various textured surfaces can be obtained on renderings by surface treatments such as scraping the surface with combs, saw blades or similar tools to remove a surface skin of mortar. These operations are carried out some 3–4 hours after the initial application of the rendering and before the final set takes place.

Alternative treatments are as follows:

- **Roughcast** A wet plaster mix of 1 part cement: $\frac{1}{2}$ part lime: $1\frac{1}{2}$ parts shingle:3 parts sand, which is thrown onto a porous coat of rendering to give an even distribution.
- **Pebble or dry dash** Selected aggregate such as pea shingle is dashed or thrown onto a rendering background before it has set, and is tamped into the surface with a wood float to obtain a good bond.
- **Spattered finishes** These are finishes applied by a machine (which can be hand operated), guns or sprays using special mixes prepared by the machine manufacturers.

PLASTERING

Plastering, like brickwork, is one of the oldest-established crafts in this country, having been introduced by the Romans. The plaster used was a lime plaster, which generally has been superseded by gypsum plasters. The disadvantages of lime plastering are:

- drying shrinkage, which causes cracking;
- slow drying-out process, which can take several weeks, causing delays for the following trades;
- need to apply lime plaster in several coats, usually termed render, float and set, to reduce the amount of shrinkage.

Any plaster finish must smooth out irregularities in the backing wall, provide a continuous surface that is suitable for direct decoration, and be sufficiently hard to resist damage by impact upon its surface; gypsum plasters fulfil these requirements.

Gypsum is a crystalline combination of calcium sulphate and water. Deposits of suitable raw material are found in several parts of England, and after crushing and screening the gypsum is heated to dehydrate the material. The amount of water remaining at the end of this process defines its class under BS 1191: *Specification for gypsum building plasters*. If powdered gypsum is heated to about 170 °C it loses about three-quarters of its combined water and is called hemihydrate gypsum plaster, but is probably better known as **Plaster of Paris**. If a retarder is added to the hemihydrate plaster a new class of finishing plaster is formed to which the addition of expanded perlite and other additives will form a one-coat or universal plaster, a renovating-grade plaster or a spray plaster for application by a spray machine.

BS 1191 establishes four classes of plaster:

- Class A – Plaster of Paris;
- Class B – retarded hemihydrate gypsum plaster in two categories:
  (a) undercoat plasters;
  (b) finish coat plasters;
- Class C – anhydrous gypsum plaster (completely dehydrated gypsum);
- Class D – Keene's plaster (completely dehydrated gypsum, very slow set, high-quality finish).

Table 7.8.1   Categories of plaster

Gypsum-based plasters	BS ref.	Trade name
Class A:		
Plaster of Paris hemihydrate	BS 1191-1	
Class B:		
Retarded hemihydrate	BS 1191-1	
Undercoats:		
Browning	– section a1	Thistle browning
		Thistle slow set
		Thistle fibred
Metal lathing	– section a2	Thistle metal lathing
Finishes:		
Finish	– section b1	Thistle finish, Sirapite B
Board finish	– section b2	Thistle board finish
Class C:		
Anhydrous finish only	BS 1191-1	Sirapite
Class D:		
Keene's plaster	BS 1191-1	Standard Keene's
		Fine, polar
		white cement
Premixed lightweight	BS 1191-2	
Undercoats:		
Browning	– section a1	Carlite browning
		Carlite browning (HSB)
Metal lathing	– section a2	Carlite metal lathing
Bonding	– section a3	Carlite bonding
		Carlite welter-weight bonding
Finishes:		
Finish	– section b1	Carlite finish, Limelight finishing

Table 7.8.1 lists British Standard plaster classifications, but omits non-standard plasters for special applications. These include:

■ Thin-wall plasters – for skimming and filling. These may contain organic binders that could be incompatible with some backgrounds. Check with manufacturer before application.

■ Projection plasters – for spray machine application.

■ X-ray plasters – contain barium sulphate aggregate. This is a heavy metal with salts that provide a degree of exposure protection.

■ Acoustic plaster – contain porous gaps, which will absorb sound. Surface treatments that obscure should be avoided.

■ Resinous plaster – 'Proderite' formula S, a cement-based plaster especially formulated for squash courts as a substitute for Class D – Keene's plaster.

Gypsum plasters are not suitable for use in temperatures exceeding 43 °C, and should not be applied to frozen backgrounds. However, plasters can be applied under frosty conditions provided the surfaces are adequately protected from freezing after application.

Gypsum plasters are supplied in multi-walled paper sacks, which require careful handling. Damage to the bags will admit moisture and initiate commencement of the setting process. Plaster in an advanced state of set will have a short working time and may lack considerable strength. Characteristics are manifested in lack of bond and surface irregularities. Plaster should be stored in a dry location on pallets or other means of avoiding the ground. The dry, bagged plaster in store will be unaffected by low temperatures. Deterioration with age must also be avoided by regular checks on the manufacturer's date stamps, and by applying a system of strict rotation.

The plaster should be mixed in a clean plastic or rubber bucket using clean water only. Cleanliness is imperative, because any set plaster left in the mixing bucket from a previous mixing will shorten the setting time, which may reduce the strength of the plaster when set.

Premixed plasters incorporate lightweight aggregates such as expanded perlite and exfoliated vermiculite. Perlite is a glassy volcanic rock combining in its chemical composition a small percentage of water. When subjected to a high temperature the water turns into steam, thus expanding the natural perlite to many times its original volume. Vermiculite is a form of mica with many ultra-thin layers, between which are minute amounts of water, which expand or exfoliate the natural material to many times its original volume when subjected to high temperatures by the water turning into steam and forcing the layers of flakes apart. The density of a lightweight plaster is about one-third that of a comparable sanded plaster, and it has a thermal value of about three times that of sanded plasters, resulting in a reduction of heat loss, less condensation, and a reduction in the risk of pattern staining. It also has superior adhesion properties to all backgrounds including smooth dry clean concrete.

The choice of plaster mix, type and number of coats will depend upon the background or surface to which the plaster is to be applied. Roughness and suction properties are two of the major considerations. The roughness can affect the keying of the plaster to its background. Special keyed bricks are available for this purpose (see Fig. 3.2.3); alternatively the joints can be raked out to a depth of 15–20 mm as the wall is being built. *In-situ* concrete can be cast using sawn formwork giving a rough texture, and hence forming a suitable key. Generally all lightweight concrete blocks provide a suitable key for the direct application of plasters. Bonding agents in the form of resin emulsions are available for smooth surfaces; these must be applied in strict accordance with the manufacturer's instructions to achieve satisfactory results.

The suction properties of the background can affect the drying rate of the plaster by absorbing the moisture of the mix. Too much suction can result in the plaster failing to set properly, thus losing its adhesion to its background; too little suction can give rise to drying shrinkage cracks due to the retention of excess water in the plaster.

Undercoat plasters are applied by means of a wooden float or rule worked between dots or runs of plaster to give a true and level surface. The runs or rules and dots are of the same mix as the backing coat and are positioned over the background at suitable intervals to an accurate level so that the main application of plaster can be worked around the guide points. The upper surface of the undercoat plaster should be scored or scratched to provide a suitable key for the finishing coat. The thin finishing coat of plaster is applied to a depth of approximately 3 mm and finished

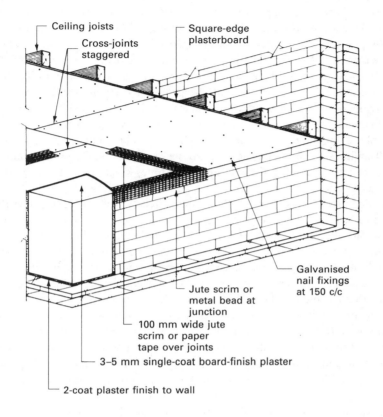

Ceiling joists

Cross-joints staggered

Square-edge plasterboard

Galvanised nail fixings at 150 c/c

Jute scrim or metal bead at junction

100 mm wide jute scrim or paper tape over joints

3–5 mm single-coat board-finish plaster

2-coat plaster finish to wall

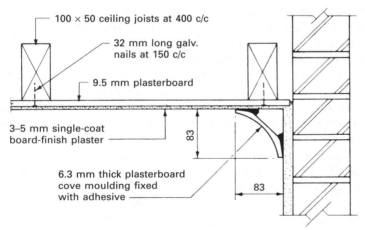

100 × 50 ceiling joists at 400 c/c

32 mm long galv. nails at 150 c/c

9.5 mm plasterboard

3–5 mm single-coat board-finish plaster

83

6.3 mm thick plasterboard cove moulding fixed with adhesive

83

Figure 7.8.3   Plasterboard ceilings.

with a steel float to give a smooth surface. In hot and dry conditions care must be taken to ensure that water is not allowed to evaporate from the mix. Water is very important in the setting process, and if applied plaster is allowed to dry too quickly, loss of strength, surface adhesion and surface irregularities will become apparent.

Most paints and wall coverings are compatible with plaster finishes, provided the plaster is thoroughly dried out prior to application. This normally takes several months: therefore, as an interim measure, permeable water-based emulsion paint could be used.

Guidance in application and standards of workmanship can be found in BS 5492: *Code of practice for internal plastering* and BS 8000-10: *Workmanship on building sites. Code of practice for plastering and rendering.*

DRYLINING TECHNIQUES

External walls or internal walls and partitions can be drylined with a variety of materials, which can be self-finished, ready for direct decoration, or have a surface suitable for a single final coat of board finish plaster. The main advantages of drylining techniques are: speed; reduction in the amount of water used in the construction of buildings, thus reducing the drying out period; and in some cases increased thermal insulation.

Suitable materials are hardboard, plywood, chipboard and plasterboard. Hardboard, plywood and chipboard are fixed to timber battens attached to the wall at centres to suit the spanning properties and module size of the board. Finishing can be a direct application of paint, varnish or wallpaper, but masking the fixings and joints may present problems. As an alternative the joints can be made a feature of the design by the use of edge chamfers or by using moulded cover fillets.

Plasterboard consists of an aerated gypsum core encased in and bonded to specially prepared bonded paper liners. The grey-coloured liner is intended for a skim coat of plaster, and the ivory-coloured liner is for direct decoration.

Gypsum plasterboards are manufactured in a variety of specifications to suit numerous applications. They are generally available in metric coordinated widths of 900 and 1200 mm with coordinated lengths from 1800 to 3000 mm in thicknesses of 9.5 and 12.5 mm. Boards can be obtained with a tapered edge for a seamless joint or a square edge for cover fillet treatment. Some variants are listed in Table 7.8.2.

Baseboard is produced with square edges as a suitable base for a single coat of board finish plaster. One standard width is manufactured with a thickness of 9.5 mm; lengths are 1200, 1219 and 1372 mm. Before the plaster coat is applied the joints should be covered with a jute scrim 100 mm wide (see Fig. 7.8.3).

Plasterboard lath is a narrow plasterboard with rounded edges, which removes the need for a jute scrim over the joints. The standard width is 406 mm, with similar lengths to baseboard and thicknesses of 9.5 and 12.5 mm.

The fixing of drylinings is usually by nails to timber battens suitably spaced, plumbed and levelled to overcome any irregularities in the background. Fixing battens are placed vertically between the horizontal battens fixed at the junctions of the ceiling and floor with the wall. It is advisable to treat all fixing battens with an insecticide and a fungicide to lessen the risk of beetle infestation and fungal attack.

Table 7.8.2   Common types of plasterboard

Type	Thickness (mm)	Width (mm)	Application
Plank	19	600	Sound insulation
Moisture resistant	9.5 and 12.5	1200	Bathroom/kitchen
Contoured	6	1200	Curved surfaces
Vapour check	9.5 and 12.5	900 and 1200	External wall/roof
Firecheck	12.5	900 and 1200	Fire cladding
Lath	9.5 and 12.5	400 and 600	Ceiling lining
Thermal	12.5*	1200	Thermal insulation
Glass reinforced	4–35	Various	Fire protection

* Overall thickness, including laminates of expanded polystyrene, extruded polystyrene or phenolic foam, can be between 25 and 50 mm.

The spacing of the battens will be governed by the spanning properties of the lining material: fixing battens at 450 mm centres are required for 9.5 mm thick plasterboards and at 600 mm centres for 12.5 mm thick plasterboards, so placed that they coincide with the board joints.

12.5 mm plasterboard can also be fixed to brick or block masonry with dabs of plaster. Dabs of board-finish plaster about 50–75 mm wide by 250 mm long are close spaced vertically at a horizontal spacing (max. 600 mm) to suit the board width. Intermediate dabs are similarly spaced with a continuous spread at ceiling and ground levels. The plasterboard is then placed in position, tapped and horizontally aligned with a spirit level until it is firmly in contact with the dabs. Double-headed (duplex) nails can be used to secure the boards temporarily while the dabs set, after which they are removed and joints are made good. It is recommended that tapered edge boards are used for this method of fixing.

GLAZED WALL TILES

Two processes are used in the manufacture of these hard glazed tiles. First, the body or 'biscuit' tile is made from materials such as china clay, ball clay, flint and limestone, which are mixed by careful processes into a fine powder before being heavily pressed into the required shape and size. The tile is then fired at a temperature of up to 1150 °C. Second, glazing is applied in the form of a mixed liquid consisting of fine particles of glaze and water. The coated tiles are then fired for a second time at a temperature of approximately 1050 °C, when the glazing coating fuses to the surface of the tile. Gloss, matt and eggshell finishes are available, together with a wide choice of colours, designs and patterns.

A range of fittings with rounded edges and mitres are produced to blend with the standard 150 mm × 150 mm square tiles, 5 or 6 mm thick; also a similar range of fittings are made for the 108 mm × 108 mm, 4 or 6 mm thick, square tiles. The appearance and easily cleaned surface of glazed tiles makes them suitable for the complete or partial tiling of bathrooms and the provision of splashbacks for sinks and basins.

Tiles are fixed with a suitable adhesive, which can be of a thin bed of mastic adhesive or a bed of a cement-based adhesive; the former requires a flat surface such as a mortar screed. Ceramic or glazed tiles are considered to be practically inert: therefore careful selection of the right adhesive to suit the backing and final condition is essential. Manufacturer's instructions or the recommendations of BS 5385 should always be carefully followed.

Glazed tiles can be cut easily using the same method as employed for glass. The tile is scored on its upper face with a glass cutter; this is followed by tapping the back of the tile behind the scored line over a rigid and sharp angle such as a flat straight edge.

■ ■ ■ CEILING FINISHES

Ceilings can be finished by any of the drylining techniques previously described for walls. The usual method is a plasterboard base with a skim coat of plaster. The plasterboards are secured to the underside of the floor or ceiling joists with galvanised plasterboard nails to reduce the risk of corrosion to the fixings. At the abutment of boards, a 3–5 mm gap is filled with jointing plaster and bridged with jute scrim or special jointing tape. The most vulnerable point in a ceiling to cracking is at the junction between the ceiling and wall; this junction should be strengthened with a jute scrim or jointing tape around the internal angle (see Fig. 7.8.3); alternatively the junction can be masked with a decorative plasterboard or polystyrene cove moulding.

The cove moulding is made in a similar manner to plasterboard and is intended for direct decoration. Plasterboard cove moulding is jointed at internal and external angles with a mitred joint and with a butt joint in the running length. Any clean, dry and rigid background is suitable for the attachment of plasterboard cove, which can be fixed by using a special water-mixed adhesive applied to the contact edges of the moulding, which is pressed into position; any surplus adhesive should be removed from the edges before it sets. A typical plasterboard cove detail is shown in Fig. 7.8.3.

Many forms of ceiling tiles are available for application to a joisted ceiling or solid ceiling with a sheet or solid background. Fixing to joists should be by concealed or secret nailing through the tongued and grooved joint. If the background is solid such as a concrete slab then dabs of a recommended adhesive are used to secure the tiles. Materials available include expanded polystyrene, mineral fibre, fibreboard and glass fibre with a rigid vinyl face.

Other forms of finish that may be applied to ceilings are sprayed plasters, which can be of a thick or thin coat variety. Spray plasters are usually of a proprietary mixture, applied by spraying apparatus directly on to the soffit, giving a coarse texture which can be trowelled smooth, patterned with combs or stippled with a stiff brush (Artexing). Application can be with a brush if preferred. Various patterned ceiling papers are produced to give a textured finish. These papers are applied directly to the soffit or over a stout lining paper. Some ceiling papers are designed to be a self-finish but others require one or more coats of emulsion paint.

7.9

INTERNAL FIXINGS AND SHELVES

■■■ INTERNAL FIXINGS

These consist of trims, in the form of skirtings, dado rails, frieze or picture rails, architraves and cornices, whereas fittings would include such things as cupboards and shelves.

SKIRTINGS

A skirting is a horizontal member fixed around the skirt or base of a wall, primarily to mask the junction between the wall finish and the floor. It can be an integral part of the floor finish, such as quarry tile or PVC skirtings, or it can be made from timber, metal or plastic. Timber is the most common material used, and is fixed by nails direct to the background, but if this is a dense material that will not accept nails, plugs or special fixing bricks can be built into the wall. External angles in skirtings are formed by mitres, but internal angles are usually scribed (see Fig. 7.9.1).

ARCHITRAVES

These are mouldings cut and fixed around door and window openings to mask the joint between the wall finish and the frame. Like skirtings the usual material is timber, but metal and plastic mouldings are available. Architraves are fixed with nails to the frame or lining, and to the wall in a similar manner to skirtings if the architrave section is large (see Fig. 7.9.1).

DADO RAILS

These are horizontal mouldings fixed in a position to prevent the walls from being damaged by the backs of chairs pushed against them. They are very seldom used

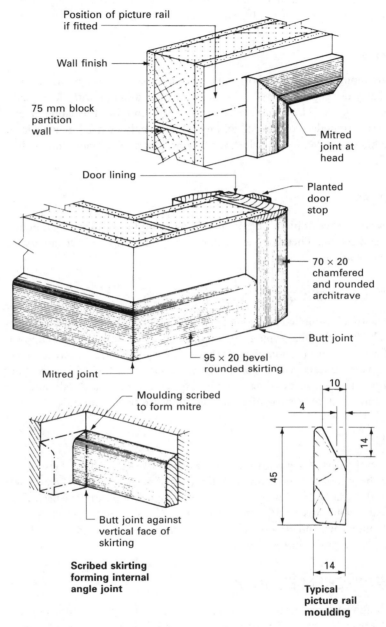

Position of picture rail if fitted

Wall finish

75 mm block partition wall

Mitred joint at head

Door lining

Planted door stop

70 × 20 chamfered and rounded architrave

Butt joint

95 × 20 bevel rounded skirting

Mitred joint

Moulding scribed to form mitre

Butt joint against vertical face of skirting

Scribed skirting forming internal angle joint

10

4

14

45

14

Typical picture rail moulding

Figure 7.9.1 Typical wood trim details.

today, as modern chair design renders them unnecessary. If used, they are fixed by nails directly to the wall or to plugs inserted in the wall.

PICTURE RAILS

These are moulded rails fixed horizontally around the walls of a room, from which pictures may be suspended, and are usually positioned in line with the top edges of the door architrave. They can be of timber or metal and, like the dado rail, are very seldom used in modern domestic buildings. They would be fixed by nails in the same manner as dado rails and skirtings; a typical picture rail moulding is shown in Fig. 7.9.1.

CORNICES

Cornices are timber or plaster ornate mouldings used to mask the junction between the wall and ceiling. These are very seldom used today, having been superseded by the cove mouldings.

CUPBOARD FITTINGS

These are usually supplied as a complete fitting and only require positioning on site; they can be free standing or plugged and screwed to the wall. Built-in cupboards can be formed by placing a cupboard front in the form of a frame across a recess and then hanging suitable doors to the frame. Another method of forming built-in cupboards is to use a recessed partition wall to serve as cupboard walls and room divider, and attach to this partition suitable cupboard fronts and doors.

SHELVES

Shelves can be part of a cupboard fitting or can be fixed to wall brackets or bearers that have been plugged and screwed to the wall. Timber is the usual material for shelving; this can be faced with a large variety of modern plastic finishes. Shelves are classed as solid or slat: the latter are made of 45 mm × 20 mm slats, spaced about 20 mm apart, and are used mainly in airing and drying cupboards. Typical shelf details are shown in Fig. 7.9.2.

■■■ KITCHEN FITTINGS

Kitchen cupboards and drawer units are manufactured in preassembled or flat pack form. Flat pack is more convenient for storage and delivery, but requires time for site assembly before fitting. Materials vary; they may be selected timber or its less expensive derivatives and composites of plywood, chipboard or medium-density fibreboard. Purpose-made plastic and metal brackets are used to assemble the components.

Disposition of fitments and general design layout is a specialist area of design for the application of standardised metric units shown in Fig. 7.9.3. Special

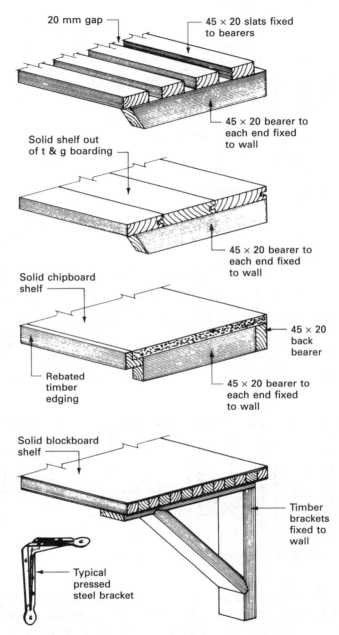

Figure 7.9.2 Shelves and supports.

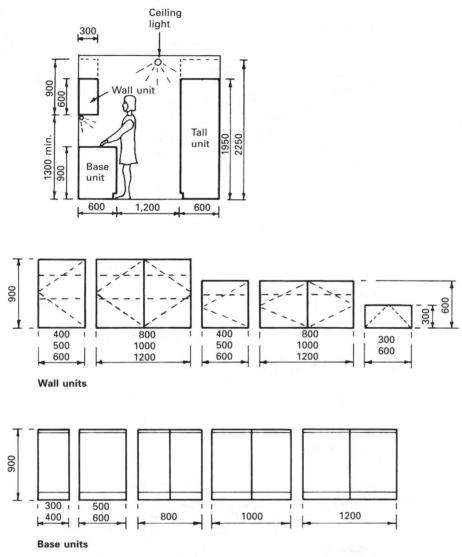

Figure 7.9.3   Standard kitchen units (dimensions in mm).

components such as corner pieces and purpose-made base units, e.g. wine racks, are not included. These vary in style and measurements between manufacturers, but retain overall dimensions to coordinate with the standard range.

Heights of floor-supported units are nominal, with adjustable legs fitted as standard. Surfaces or worktops are secured to the base units with small angle brackets. The worktops are available in laminates of solid timber (usually beech), although the less expensive pattern and/or grain effect plastic-faced particle board is most often used. Thickness varies between 28 and 40 mm, with the option of a rounded edge or wood trimming.

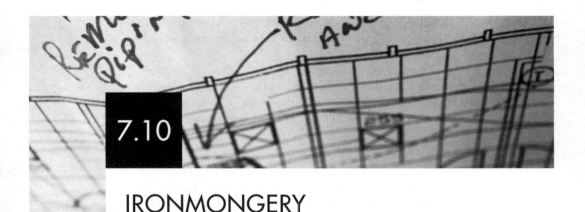

7.10

IRONMONGERY

Ironmongery is a general term that is applied to builder's hardware and includes such items as nails, screws, bolts, hinges, locks, catches, window and door fittings.

■■■ NAILS

A nail is a fixing device that relies on the grip around its shank and the shear strength of its cross-section to fulfil its function. It is therefore important to select the right type and size of nail for any particular situation. Nails are specified by their type, length and diameter range given in millimetres. The diameter range is comparable to the old British Standard Wire Gauge. The complete range of nails is given in BS 1202. Steel is the main material used in the manufacture of nails; other metals used are copper and aluminium alloy.

Nails in common use are:

- **Cut clasp nails**   These are made from black rolled steel and are used for general carcassing work.
- **Cut floor brads**   These are made from black rolled steel and are used mainly for fixing floorboards, because their rectangular section reduces the tendency of thin boards to split.
- **Round plain head**   Also known as round wire nails, these are made in a wide variety of lengths. They are used mainly for general carpentry work, but have a tendency to split thin members.
- **Oval brad head**   These are made with an oval cross-section to lessen the risk of splitting the timber members. They are used for the same purpose as round wire nails but have the advantage of being able to be driven below the surface of the timber. A similar nail with a smaller head is also produced and is called an oval lost head nail.

- **Clout nails**   Also called slate nails, these have a large flat head and are used for fixing tiles and slates. The felt nail is similar but has a larger head and is only available in lengths up to 38 mm.
- **Panel pins**   These are fine wire nails with a small head that can be driven below the surface. They are used mainly for fixing plywood and hardboard.
- **Plasterboard nails**   The holding power of these nails, with their countersunk head and jagged shank, makes them suitable for fixing ceiling and similar boards.

■■■ WOOD SCREWS

A wood screw is a fixing device used mainly in joinery, and relies upon its length and thread for holding power and resistance to direct extraction. For a screw to function properly it must be inserted by rotation and not be driven in with a hammer. It is usually necessary to drill pilot holes for the shank and/or core of the screw.

Wood screws are manufactured from cold-drawn wire, steel, brass, stainless steel, aluminium alloy, silica bronze and nickel–copper alloy. In addition to the many different materials a wide range of painted and plated finishes are available, such as an enamelled finish known as black japanned. Plated screws are used mainly to match the fitting that they are being used to fix, and include such finishes as galvanised steel, copper plated, nickel plated and bronze metal antique.

Screws are specified by their material, type, length and gauge. The screw gauge is the diameter of the shank, and is designated by a number, but unlike the gauge used for nails the larger the screw gauge number the greater the diameter of the shank. Various head designs are available for all types of wood screw, each having a specific purpose:

- **Countersunk head**   This is the most common form of screw head, and is used for a flush finish, the receiving component being countersunk to receive the screw head.
- **Raised head**   This is used mainly with good-quality fixtures; the rounded portion of the head with its mill slot remains above the surface of the fixture, ensuring that the driving tool does not come into contact with the surface, causing damage to the finish.
- **Round head**   The head, being fully exposed, makes these screws suitable for fittings of material that is too thin to be countersunk.
- **Recessed head**   Screws with a countersunk head and a recessed cruciform slot, giving a positive drive with a specially shaped screwdriver.
- **Coach screws**   These are made of mild steel with a square head for driving in with the aid of a spanner, and are used mainly for heavy carpentry work.

■■■ CAVITY FIXINGS

Various fixing devices are available for fixing components to thin materials of low structural strength, such as plasterboard and hardboard. Cavity fixings are designed to spread the load over a wide area of the board. Typical examples are as follows:

- **Steel spring toggles** Spring-actuated wings open out when the toggle fixing has been inserted through a hole in the board and spread out on the reverse side of the board. Spring toggles are specially suited to suspending fixtures from a ceiling.
- **Steel gravity toggles** When inserted horizontally into a hole in the board the long end of the toggle drops and is pulled against the reverse side of the board when the screw is tightened.
- **Rubber cavity fixings** A rubber or neoprene sleeve, in which a nut is embedded, is inserted horizontally through a hole in the board; the tightening of the screw causes the sleeve to compress and grip the reverse side of the board. This fixing device forms an airtight, waterproof and vibration-resistant fixing.

Typical examples of nails, screws and cavity fixings are shown in Fig. 7.10.1.

■ ■ ■ HINGES

Hinges are devices used to attach doors, windows and gates to a frame, lining or post so that they are able to pivot about one edge. It is of the utmost importance to specify and use the correct number and type of hinges in any particular situation to ensure correct operation of the door, window or gate. Hinges are classified by their function, length of flap, material used and sometimes by the method of manufacture. Materials used for hinges are steel, brass, cast iron, aluminium and nylon with metal pins. Typical examples of hinges in common use are as follows:

- **Steel butt hinge** These are the most common type in general use, and are made from steel strip, which is cut to form the flaps and is pressed around a steel pin.
- **Steel double-flap butt hinge** Similar to the steel butt hinge but made from two steel strips to give extra strength.
- **Rising butt hinge** This is used to make the door level rise as it is opened, to clear carpets and similar floor coverings. The door will also act as a gravity self-closing door when fitted with these butts, which are sometimes called skew butt hinges.
- **Parliament hinge** This is a form of butt hinge with a projecting knuckle and pin, enabling the door to swing through 180° to clear architraves and narrow reveals.
- **Tee hinge** Sometimes called a cross garnet, this type of hinge is used mainly for hanging matchboarded doors, where the weight is distributed over a large area.
- **Band and hook** A stronger type of tee hinge made from wrought steel and used for heavy doors and gates. A similar hinge is produced with a pin that projects from the top and bottom of the band, and is secured with two retaining cups screwed to the post; these are called reversible hinges.

Typical examples of common hinges are shown in Fig. 7.10.2.

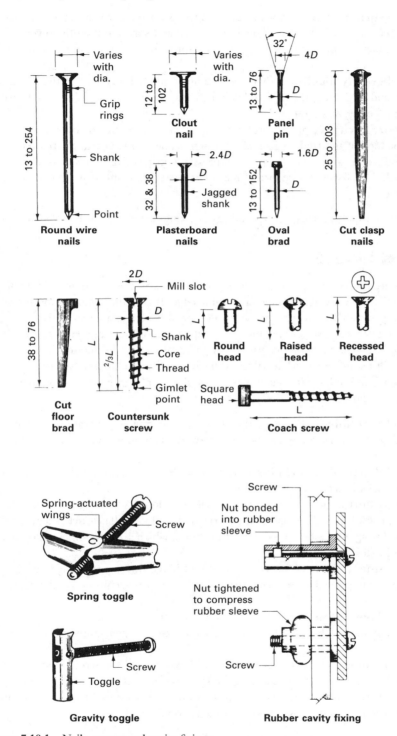

Figure 7.10.1   Nails, screws and cavity fixings.

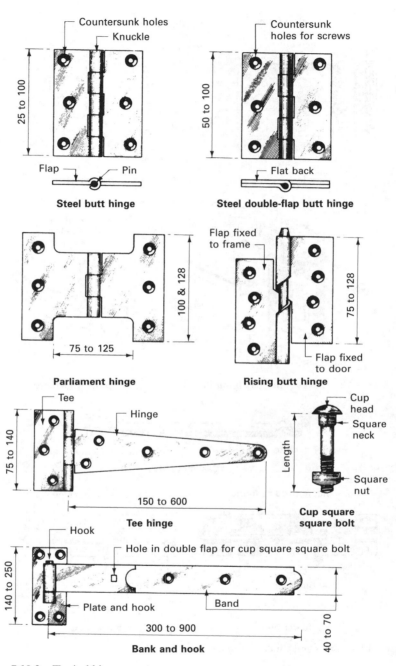

Figure 7.10.2 Typical hinges.

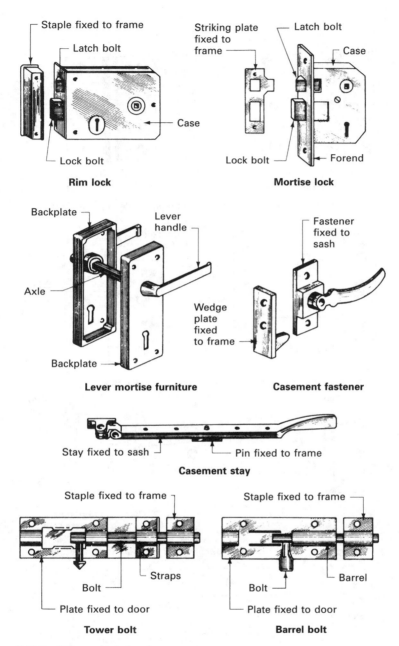

Figure 7.10.3   Door and window ironmongery.

▨ ■ ■ LOCKS AND LATCHES

Any device used to keep a door in the closed position can be classed as a lock or latch. A lock is activated by means of a key, whereas a latch is operated by a lever or bar. Latches used on lightweight cupboard doors are usually referred to as catches. Locks can be obtained with a latch bolt so that the door can be kept in the closed position without using a key; these are known as **deadlocks**.

Locks and latches can be fixed to the face of the door with a staple or kept fixed to the frame, when they are termed **rim locks or latches**. Alternatively they can be fixed within the body of the door, when they are called **mortise locks or latches**. When this form of lock or latch is used, the bolts are retained in mortises cut behind the striking plate fixed to the frame (see Fig. 7.10.3).

Cylindrical night latches are fitted to the stile of a door, and a connecting bar withdraws the latch when the key is turned. Most night latches have an internal device to stop the bolt from being activated from outside by the use of a key.

Door handles, cover plates and axles used to complete lock or latch fittings are collectively called **door furniture** and are supplied in a wide range of patterns and materials.

▨ ■ ■ DOOR BOLTS

Door bolts are security devices fixed to the inside faces of doors, and consist of a slide or bolt operated by hand to locate in a keep fixed to the frame. Two general patterns are produced: the **tower bolt**, which is the cheaper form; and the stronger, but dearer, **barrel bolt**. The bolt of a tower bolt is retained with staples or straps along its length, whereas in a barrel bolt it is completely enclosed along its length (see Fig. 7.10.3).

▨ ■ ■ CASEMENT WINDOW FURNITURE

Two fittings are required for opening sashes: the **fastener**, which is the security device; and the **stay**, which holds the sash in the opened position. Fasteners operate by the blade being secured in a mortise cut into the frame, or by the blade locating over a projecting wedge or pin fixed to the frame (see Fig. 7.10.3). Casement stays can be obtained to hold the sash open in a number of set positions by using a pin fixed to the frame and having a series of locating holes in the stay, or they can be fully adjustable by the stay sliding through a screw-down stop fixed to the frame (see Fig. 7.10.3).

Shootbolt multi-point espagnolette locking systems provide improved security for window casements. They can also be used for securing doors. They have two or three bolt functions (number depends on casement size) operating through one application of the casement fastener or lock mechanism. The bolts slide within a recess in the casement sash and lock into corresponding keeps secured to the casement frame. Figure 7.10.4 illustrates the principle.

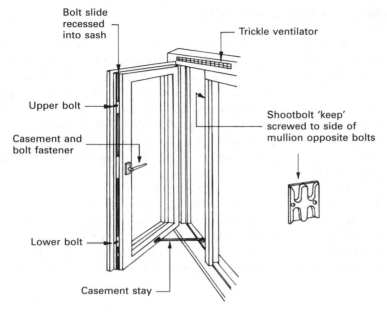

Bolt slide
recessed
into sash

Trickle ventilator

Upper bolt

Shootbolt 'keep'
screwed to side of
mullion opposite bolts

Casement and
bolt fastener

Lower bolt

Casement stay

Figure 7.10.4   Shootbolt multi–point espagnolette casement fastener.

■■■ LETTER PLATES

These are the hinged covers attached to the outside of a door, which cover the opening made to enable letters to be delivered through the door. The minimum opening size for letter plates recommended by the postal services is 200 mm × 45 mm; the bottom of the opening should be sited not lower than 750 mm from the bottom edge of the door and not higher than 1450 mm to the upper edge of the opening. A wide range of designs are available in steel, aluminium alloy and plastic, some of which have a postal knocker incorporated in the face design (see BS EN 13724).

BS EN 1906 and BS EN 12209 cover builder's hardware for housing, and give recommendations for materials, finishes and dimensions to a wide range of ironmongery items not covered by a specific standard. These include such fittings as finger plates, cabin hooks, gate latches, cupboard catches and drawer pulls.

PAINTING AND DECORATING

The application of coats of paint to the elements, components, trims and fittings of a building has two functions. The paint will impart colour and, at the same time, provide a protective coating that will increase the durability of the member to which it has been applied. The covering of wall and ceiling surfaces with paper or fabric is basically to provide colour, contrast and atmosphere. To achieve a good durable finish with any form of decoration the preparation of the surface and the correct application of the paint or paper are of the utmost importance.

■■■ PAINT

Paint is a mixture of a liquid or medium and a colouring or pigment. Mediums used in paint manufacture range from thin liquids to stiff jellies, and can be composed of linseed oil, drying oils, synthetic resins and water. The various combinations of these materials form the type or class of paint. The medium's function is to provide the means of spreading and binding the pigment over the surface to be painted. The pigment provides the body, colour and durability of the paint. White lead is a pigment that gives good durability and moisture resistance, but it is poisonous: therefore its use is confined mainly to priming and undercoating paints in situations where it is not exposed to human contact. Paints containing a lead pigment are required by law to state this fact on the can. The general pigment used for finishing paint is titanium dioxide, which gives good obliteration of the undercoating but is not poisonous.

OIL-BASED PAINTS

Priming paints

These are first-coat paints used to seal the surface, protect the surface against damp air, act as a barrier to prevent any chemical action between the surface and the finishing coats, and give a smooth surface for the subsequent coats. Priming paints are produced for application to wood, metal and plastered surfaces.

Undercoating paints

These are used to build up the protective coating and to provide the correct surface for the finishing coat(s). Undercoat paints contain a greater percentage of pigment than finishing paints, and as a result have a matt or flat finish. To obtain a good finishing colour it is essential to use an undercoat of the type and colour recommended by the manufacturer.

Finishing paints

A wide range of colours and finishes including matt, semi-matt, eggshell, satin, gloss and enamel are available. These paints usually contain a synthetic resin that enables them to be easily applied, be quick drying, and have good adhesive properties. Gloss paints have less pigment than the matt finishes and consequently less obliterating power.

POLYURETHANE PAINTS

These are quick-drying paints based on polyurethane resins giving a hard, heat-resisting surface. They can be used on timber surfaces as a primer and undercoat, but metal surfaces will require a base coat of metal primer; the matt finish with its higher pigment content is best for this 'one paint for all coats' treatment. Other finishes available are gloss and eggshell.

WATER-BASED PAINTS

Most of the water-based paints in general use come under a general classification of emulsion paints: they are quick drying and can be obtained in matt, eggshell, semi-gloss and gloss finishes. The water medium has additives such as polyvinyl acetate and alkyd resin to produce the various finishes. Except for application to ironwork, which must be primed with a metal primer, emulsion paints can be used for priming, undercoating and as a finishing application. Their general use is for large flat areas such as ceilings and walls.

VARNISHES AND STAINS

Varnishes form a clear, glossy or matt, tough film over a surface, and are a solution of resin and oil, their application being similar to that of oil-based paints. The type of resin used, together with the correct ratio of oil content, forms the various durabilities and finishes available.

Stains can be used to colour or tone the surface of timber before applying a clear coat of varnish; they are basically a dye in a spirit and are therefore quick drying.

PAINT SUPPLY

Paints are supplied in metal and plastic containers of 5 litres, 2.5 litres, 1 litre, 500 millilitres and 250 millilitres capacity. They are usually in one of the 100 colours recommended for building purposes in BS 4800. BS 381C is also a useful reference, giving 91 standard colours for special purposes such as determining colours used by HM services. The colour range in BS 381C can be used for purposes other than paint identification.

The British Standard colour range is derived from the Munsell scale of colours and includes black and white. The Munsell system is used internationally to provide an accurate definition of colours. It is based on hue (base colour), value (light or darkness) and chroma (intensity compared with greyness).

KNOTTING

Knots or streaks in softwood timber may exude resins, which may soften and discolour paint finishes; generally an effective barrier is provided by two coats of knotting applied before the priming coats of paint. Knotting is a uniform dispersion of shellac in a natural or synthetic resin in a suitable solvent. It should be noted that the higher the grade of timber specified, with its lower proportion of knots, the lower will be the risk of resin disfigurement of paint finishes.

■ ■ ■ WALLPAPERS

Most wallpapers consist of a base paper of 80% mechanical wood pulp and 20% unbleached sulphite pulp printed on one side with a colour and/or design. Papers with vinyl plastic coatings making them washable are also available, together with a wide range of fabrics suitably coated with vinyl to give various textures and finishes for the decoration of walls and ceilings.

The preparation of the surface to receive wallpaper is very important; ideally a smooth clean and consistent surface is required. Walls that are poor can be lined with a lightweight lining paper for traditional wallpaper and a heavier esparto lining paper for the heavier classes of wallpaper. Lining papers should be applied in the opposite direction to the direction of the final paper covering.

Wallpapers are attached to the surface by means of an adhesive such as starch/ flour paste or cellulose pastes, which are mixed with water to the manufacturer's recommendations. Heavy washable papers and woven fabrics should always be hung with an adhesive recommended by the paper manufacturer. New plaster and lined surfaces should be treated to provide the necessary 'slip' required for successful paper hanging. A glue size is usually suitable for a starch/flour paste, whereas a diluted cellulose paste should always be used when a cellulose paste is used as the adhesive.

Standard wallpapers are supplied in rolls with a printed width of 530 mm and a length of 10 m, giving an approximate coverage of 4.5 m^2; the actual coverage will be governed by the 'drop' of the pattern, which may be as much as 2 m. Decorative borders and motifs are also available for use in conjunction with standard wallpapers.

TYPES OF WALLPAPER

Underlayers

Lining paper is used to provide uniform suction on uneven surfaces and to reinforce cracked plaster before applying decorative wallpaper finishes. The quality varies from common pulp for normal use to a stronger brown paper for badly cracked surfaces. A pitch-impregnated lining paper is also available for damp walls, but it is not intended as a substitute for remedial treatment to a source of dampness. Cotton-backed paper is produced for use on boarded backgrounds that are likely to move. Lining papers are manufactured 11 m long by 560 or 760 mm in width.

Finishes

- **Flocks**   These are manufactured with a raised pile pattern produced by blowing coloured wool, cotton or nylon fibres onto an adhesive printed pattern.
- **Metallic finishes**   Paper coated with metal powders or a metal foil sheet material.
- **Moires**   A watered silk effect finish.
- **Ingrained**   An effect produced by incorporating sawdust or other fibres in the paper surface (wood-chip).
- **Textured effect**   An imitation texture achieved by embossing the surface.
- **Washable papers**   These include varnish- and plastic-coated paper, in addition to expanded PVC foam and textured PVC sheet.
- **Relief papers**   These are in two categories – high relief (Anaglypta) and low relief (Lincrusta). Anaglypta is a quality wallpaper moulded while wet to provide embossed patterns that give the paper strength and resilience. The depressions on the back of the paper can be filled with a paste/sawdust mix to resist impact damage. Lincrusta is a paper composed of whiting, wood flour, lithopone, wax, resin and linseed oil, which is superimposed hot on kraft paper. The result is a relatively heavy paper requiring a dense paste for adhesion to the wall.

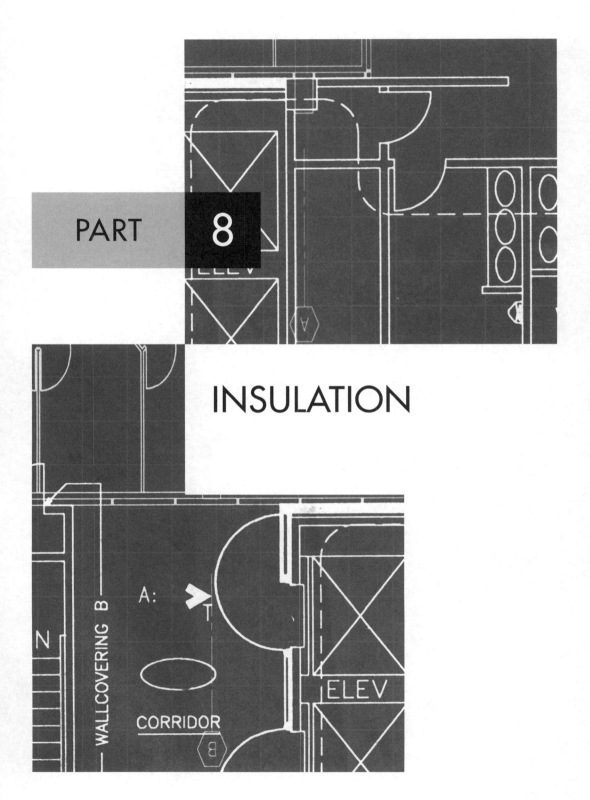

PART 8

INSULATION

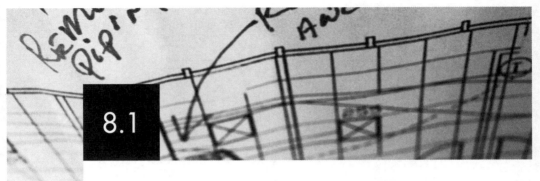

8.1

SOUND INSULATION

■■■ THE DEFINITION OF SOUND

Anything that can be heard is a sound, whether it is made by conversation, machinery, or walking on a hard surface. All sounds are produced by a vibrating object that moves rapidly to and fro, causing movement of the tiny particles of air surrounding the vibrating source. The displaced air particles collide with adjacent particles, setting them in motion and in unison with the vibrating object. Air particles move only to and fro, but the sound wave produced travels through the air until at some distance from the source the movement of the particles is so slight that the sound produced is inaudible.

For a sound to be transmitted over a distance a substance, called the **sound medium**, is required. It can be shown by experiments that sound cannot travel through a vacuum, but it can be transmitted through solids and liquids.

Sounds can differ in two important ways: by loudness and by pitch. The loudness of a sound depends on the distance through which the vibrating object moves to and fro as it vibrates – the greater the movement the louder the sound. The loudness with which a sound is heard depends upon how far away from the source the receiver or ear is. The unit of subjective loudness is a **phon**, and the objective unit is called a **decibel** (dB). Although the loudness of a sound will vary with the frequency of the note, for practical building purposes the phon and the decibel are considered to be equal over the range of 0 phons, the threshold of hearing, to 130 phons, the threshold of painful hearing.

The pitch of a sound depends on the rate at which the vibrating object oscillates. The number of vibrations in a second is called the **frequency**, and the higher the frequency the higher the pitch. The lowest-pitched note that the human ear can hear has a frequency of approximately 16 hertz (Hz, cycles per second) whereas the highest-pitched note that can be heard by the human ear has a frequency of approximately 20 000 Hz.

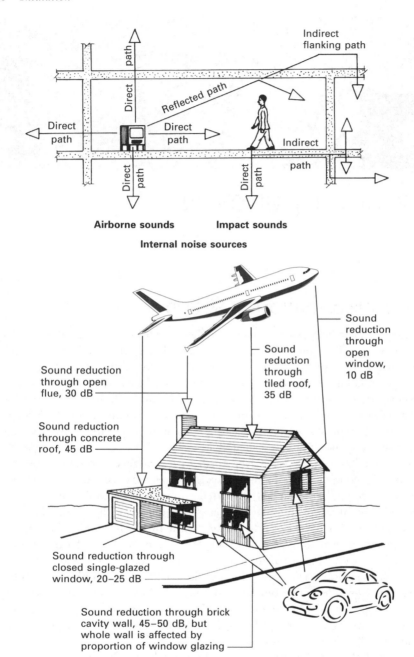

Figure 8.1.1   Sources of noise within and around buildings.

When a sound is produced within a building three reactions can occur:

1. The pressure or sound waves can come into contact with the walls, floor and ceiling and be reflected back into the building.
2. Some of the sound can be absorbed by these surfaces and/or the furnishes. Note that sound absorption normally benefits only the occupants of the room in which the sound is generated, as its function is to reduce the amount of reflected sound.
3. The sound waves upon reaching the walls, floor and ceiling can set these members vibrating in unison and thus transmit the sound to adjacent rooms.

Note also that sounds can enter a building from external sources such as traffic and low-flying aircraft (see Fig. 8.1.1).

Sounds may be defined as either **impact sounds**, caused by direct contact with the structure such as footsteps or hammering on walls, which will set that part of the structure vibrating, or **airborne sounds**, such as conversation or radio, which sets the structure vibrating only when the sound waves emitted from the source reach the structural enclosure.

A noise can be defined as any undesired sound, and may have any one of the following effects:

- annoyance;
- disturbance of sleep;
- interference with the ability to hold a normal conversation;
- damage to hearing.

It is difficult to measure annoyance, as it is a subjective attitude and will depend upon the mental and physical well-being of the listener, together with the experience of being subjected to such types of noise. Damage to hearing can be caused by a sudden noise, such as a loud explosion, or by gradual damage resulting from continual noise over a period of years.

The solution to noise or sound problems can only therefore be reasonable to cater for the average person and conditions. The approach to solving a noise problem can be threefold:

- Reduce the noise emitted at the source, by such devices as attenuators, or by mounting machinery on resilient pads;
- Provide a reasonable degree of sound insulation to reduce the amount of sound transmitted;
- Isolate the source and the receiver.

■ ■ ■ SOUND INSULATION

The most effective barrier to the passage of sound is a material of high mass. With modern materials and methods this form of construction is both impracticable and uneconomic. Unfortunately modern living with its methods of transportation and entertainment generates a considerable volume of noise, and therefore some degree of sound insulation in most buildings is not only desirable but also mandatory under building law.

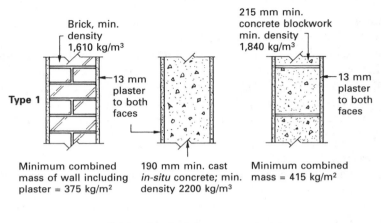

Type 1

Brick, min. density 1,610 kg/m³

13 mm plaster to both faces

Minimum combined mass of wall including plaster = 375 kg/m²

190 mm min. cast *in-situ* concrete; min. density 2200 kg/m³

215 mm min. concrete blockwork min. density 1,840 kg/m³

13 mm plaster to both faces

Minimum combined mass = 415 kg/m²

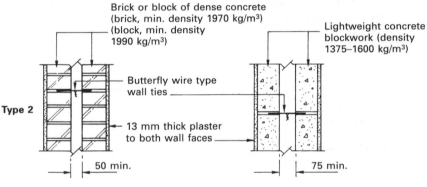

Type 2

Brick or block of dense concrete (brick, min. density 1970 kg/m³) (block, min. density 1990 kg/m³)

Butterfly wire type wall ties

13 mm thick plaster to both wall faces

50 min.

Minimum combined mass including plaster = 415 kg/m²

Lightweight concrete blockwork (density 1375–1600 kg/m³)

75 min.

Minimum combined mass including plaster = 300 kg/m²

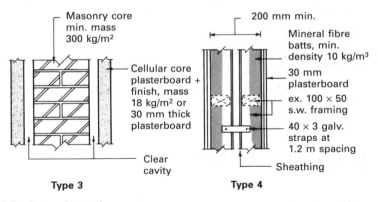

Masonry core min. mass 300 kg/m²

Cellular core plasterboard + finish, mass 18 kg/m² or 30 mm thick plasterboard

Clear cavity

Type 3

200 mm min.

Mineral fibre batts, min. density 10 kg/m³

30 mm plasterboard

ex. 100 × 50 s.w. framing

40 × 3 galv. straps at 1.2 m spacing

Sheathing

Type 4

Figure 8.1.2 Separating walls.

■■■ BUILDING REGULATIONS

Parts E1 to E3 of the Building Regulations consider sound insulation to all types of dwelling. This includes hostels, hotels, student accommodation, nurses' homes and homes for the elderly, but excludes hospitals and prisons. Part E4 is specifically to control acoustic conditions in new schools, i.e. between rooms, from external noise and to create appropriate sound reverberation characteristics for clarity of speech.

The Approved Documents supporting these regulations are as follows:

■ E1 *Protection against sound from adjoining dwellings or buildings etc.*;
■ E2 *Protection against sound within a dwelling etc.*;
■ E3 *Reverberation in the common internal parts of buildings containing dwellings etc.*;
■ E4 *Acoustic conditions in schools.*

PROTECTION AGAINST SOUND FROM ADJOINING DWELLINGS OR BUILDINGS (FIGS 8.1.2 AND 8.1.3)

This will be satisfied with separating elements of structure (walls, floors and stairs between dwellings and residential buildings) having the following minimum standards of sound insulation:

■ dwellings purpose built:
  ■ airborne sound 45 dB, impact sound (floors/stairs) 62 dB;
■ dwellings created by change of use:
  ■ airborne sound 43 dB, impact sound (floors/stairs) 64 dB;
■ residential purpose built:
  ■ airborne sound (walls) 43 dB, impact sound (walls) n/a;
  ■ airborne sound (floors/stairs) 45 dB, impact sound (floors/stairs) 62 dB;
■ residential created by change of use:
  ■ airborne sound (walls) 43 dB, impact sound (walls) n/a;
  ■ airborne sound (floors/stairs) 43 dB, impact sound (floors/stairs) 64 dB.

PROTECTION AGAINST SOUND WITHIN A DWELLING

This applies specifically to: walls (it should also apply to floors) between a WC compartment and habitable rooms; walls between bedrooms, and between bedrooms and other rooms; and floors between bedrooms and other bedrooms and other rooms. Performance is satisfied if the minimum airborne sound insulation between these walls, floors and stairs is 40 dB. This is verified by types of construction assembled, tested and measured in laboratory-controlled conditions. Examples include:

■ walls:
  ■ timber (75 mm min.) or metal stud (45 mm min.) framing with two layers of plasterboard (min. mass 9 kg/m^2 per sheet) each side;
  ■ concrete blockwork (min. mass 120 kg/m^2) with plaster or plasterboard finish both sides;

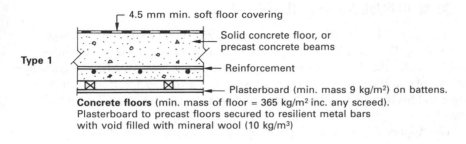

Type 1

4.5 mm min. soft floor covering

Solid concrete floor, or precast concrete beams

Reinforcement

Plasterboard (min. mass 9 kg/m²) on battens.

Concrete floors (min. mass of floor = 365 kg/m² inc. any screed). Plasterboard to precast floors secured to resilient metal bars with void filled with mineral wool (10 kg/m³)

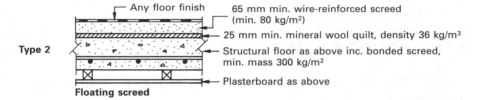

Type 2

Any floor finish

65 mm min. wire-reinforced screed (min. 80 kg/m²)

25 mm min. mineral wool quilt, density 36 kg/m³

Structural floor as above inc. bonded screed, min. mass 300 kg/m²

Plasterboard as above

Floating screed

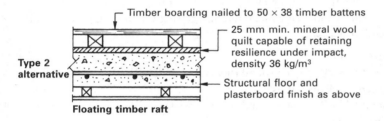

Type 2 alternative

Timber boarding nailed to 50 × 38 timber battens

25 mm min. mineral wool quilt capable of retaining resilience under impact, density 36 kg/m³

Structural floor and plasterboard finish as above

Floating timber raft

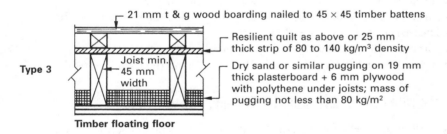

Type 3

21 mm t & g wood boarding nailed to 45 × 45 timber battens

Resilient quilt as above or 25 mm thick strip of 80 to 140 kg/m³ density

Dry sand or similar pugging on 19 mm thick plasterboard + 6 mm plywood with polythene under joists; mass of pugging not less than 80 kg/m²

Joist min. 45 mm width

Timber floating floor

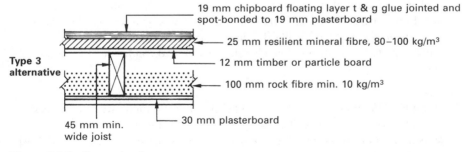

Type 3 alternative

19 mm chipboard floating layer t & g glue jointed and spot-bonded to 19 mm plasterboard

25 mm resilient mineral fibre, 80–100 kg/m³

12 mm timber or particle board

100 mm rock fibre min. 10 kg/m³

30 mm plasterboard

45 mm min. wide joist

Figure 8.1.3   Separating floors.

- floors:
    - timber joists with wood board (min. mass 15 kg/m^2) upper surface, plasterboard (min. mass 9 kg/m^2) ceiling and 100 mm mineral wool (min. density 10 kg/m^3) in void;
    - 40 mm min. screed over precast concrete beams and infill blocks (min. mass exc. screed 220 kg/m^2) with timber battens and plasterboard (min. mass 9 kg/m^2) soffit.

PROTECTION AGAINST NOISE FROM EXTERNAL SOURCES

This requirement is generally satisfied where external elements are constructed in accordance with structural and thermal insulation requirements (Building Regulations A and L respectively). For example, brick and block cavity masonry or brick-clad timber-framed walls, tiled or slated roof containing insulation and 12.5 mm plasterboard ceiling, double-glazed windows with trickle vents (see additional section headed EXTERNAL NOISE on this page).

REVERBERATION IN THE COMMON INTERNAL PARTS OF BUILDINGS CONTAINING DWELLINGS

This is satisfied by applying absorptive materials, e.g. carpet, as a surface finish. Different applications, i.e. entrance halls, corridors, etc., will have suitable coefficients of sound absorption depending on the building purpose. These are listed in the Approved Document to the Building Regulations and correspond to a range of octave frequencies. Worked examples are also shown.

ACOUSTIC CONDITIONS IN SCHOOLS

Standards of construction should be adequate to satisfy guidelines for sound insulation, reverberation time and internal noise as determined in the publications:

Building Bulletin 87 – *Guidelines for Environmental Design in Schools*, and Building Bulletin 93 – *Acoustic Design of Schools*.

Both are produced by the DfES and published by the Stationery Office.

[*Note*: Where the local building control authority or other control body is required to determine conformity to standards, random sample testing of some houses on an estate development is deemed adequate. Site equipment is a simple hand-held meter, which can be used to indicate the sound level from an established noise source on one side of a separating element and that occurring on the other. Variations in air pressure are converted to a metered scale in volts, which are depicted in units of decibels as the **root mean square (RMS)** of the sound. The RMS of sound is a type of uniform average.]

◼◼◼ EXTERNAL NOISE

The main barrier to external noise is provided by the shell or envelope of the building, the three main factors being:

- the mass of the enclosing structure;
- the continuity of the structure;
- isolation by double-leaf construction when using lightweight materials.

Generally the main problem for insulation against external noise is the windows, particularly if these can be opened for ventilation purposes. Windows cannot provide the dual function of insulation against noise and ventilation, because the admission of air will also admit noise. Any type of window when opened will give a sound reduction of about 10 dB as opposed to the 45–50 dB reduction of the traditional cavity wall. A closed window containing single glazing will give a reduction of about 20 dB, or approximately half that of the surrounding wall. It is obvious that the window-to-wall ratio will affect the overall sound reduction of the enclosing structure.

Double glazing can greatly improve the sound insulation properties of windows provided the following points are observed:

- Sound insulation increases with the distance between the glazed units; for a reduction of 40 dB the airspace should be 150–200 mm wide.
- If the double windows are capable of being opened they should be weather-stripped.
- Sound insulation increases with glass thickness, particularly if the windows are fixed; this may mean the use of special ventilators having specific performances for ventilation and acoustics.

Double glazing, designed specifically to improve the thermal properties of a window, has no real value for sound insulation except by increasing the mass of resistance.

Roofs of traditional construction and of reinforced concrete generally give an acceptable level of sound insulation, but the inclusion of rooflights can affect the resistance provided by the general roof structure. Lightweight roofing such as profiled steel or fibre cement will provide only a 15–20 dB reduction, but is generally acceptable on industrial buildings where noise is generated internally by the manufacturing process. Furthermore, the mandatory inclusion of thermal insulation in the roof slope will benefit sound reduction by its inherent absorbing properties. The inclusion of rooflights in this type of roof generally has no adverse effects in an uninsulated roof, as the sound insulation values of the rooflight materials are similar to those of the coverings. However, this is not so if the main elements of the roof are insulated.

Modern buildings can be designed to give reasonable sound insulation and consequent comfort to the occupiers, but improvements and upgrading to existing properties can present practical problems. In addition to the Building Regulations, Approved Documents E1 to E5, the following provide a useful source of information on reducing noise in existing buildings:

- BS 8233: *Sound insulation and noise reduction for buildings. Code of practice*;
- BRE Report 238: *Sound control for homes*;
- BRE Digest 293: *Improving the sound insulation of separating walls and floors* (see also Digests 333 and 334);
- BRE Information Paper 6/88: *Methods for improving the sound insulation between converted flats*.

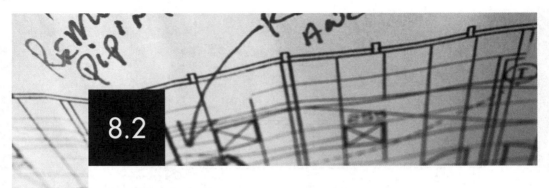

8.2

THERMAL INSULATION

Thermal insulation may be defined as a barrier to the natural flow of heat from an area of high temperature to an area of low temperature. In buildings this flow is generally from the interior to the exterior. Heat is a form of energy consisting of the ceaseless movement of tiny particles of matter called molecules; if these particles are moving fast they collide frequently with one another and the substance becomes hot. Temperature is the measure of hotness and should not be confused with heat.

The transfer of heat can occur in three ways:

- **Conduction**   Vibrating molecules come into contact with adjoining molecules and set them vibrating faster: hence they become hotter. This process is carried on throughout the substance without appreciable displacement of the particles.
- **Convection**   Transmission of heat within a fluid (gas or liquid) caused by the movement of particles. The fluid becomes less dense when heated and rises, thus setting up a current or circulation.
- **Radiation**   The transmission of thermal energy by electromagnetic waves from one body to another. The rate at which heat is emitted and absorbed will depend on the respective surface temperatures.

In a building all three methods of heat transfer can take place, because the heat will be conducted through the fabric of the building and dissipated on the external surface by convection and/or radiation.

The traditional thick and solid building materials used in the past had a natural resistance to the passage of heat in large quantities, whereas the lighter and thinner materials used today generally have a low resistance to the transfer of heat. Therefore to maintain a comfortable and healthy internal temperature the external fabric of a building must be constructed of a combination of materials that will provide an adequate barrier to the transfer of heat.

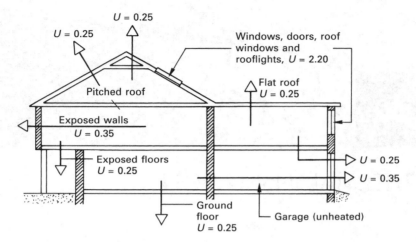

Worst acceptable area weighted elemental *U* values for dwellings

Note: Area-weighted average allows for individual areas of components.
e.g. 1, glass: frame area in window frames.
e.g. 2, openings in walls such as meter cupboards.

Worst acceptable individual elemental *U* values:
Wall, 0.45 (e.g. meter cupboard enclosure)
Floor, 0.45
Roof, 0.35
Windows, etc., 3.30

Objective insulation values:

Element	U value (W/m² K)
Pitched roof (insulation between rafters)	0.16
Pitched roof (insulation between joists)	0.13
Flat roof	0.16
External wall	0.27
Floor (ground & exposed)	0.22
Windows, doors, roof windows and rooflight*	1.80 (area weighted ave.)**

*   Should not exceed 25% of the dwelling total floor area.
** Alternatively, a window energy rating of not less than −30 may be acceptable.
     Ref. British Fenestration Rating Council. website: www.bfrc.org.

Figure 8.2.1   Insulation to dwellings.

Thermal insulation of buildings will give the following advantages:

■ reduction in the rate of heat loss;
■ lower capital costs for heating equipment;
■ lower fuel costs;
■ reduction in the risk of pattern staining;
■ reduction of condensation and draughts, thus improving the comfort of the occupants;
■ less maintenance and replacement costs of heating equipment;
■ conservation of diminishing fossil fuel reserves;
■ reduction of atmospheric pollution, greenhouse effect and ozone depletion.

■ ■ ■ BUILDING REGULATIONS

The Building Regulations Part L the states that reasonable provision shall be made for the conservation of fuel and power in dwellings. The requirements of this regulation can be partly satisfied by limiting the areas of windows, doors and rooflights and by not exceeding the worst acceptable U values for elements, which are given in Fig. 8.2.1. U values are expressed in $W/m^2 K$, which is the rate of heat transfer in watts (joules/sec.) through 1 m^2 of the structure for one unit of temperature difference between the air on the two sides of the structure.

To calculate a U value the complete constructional detail must be known, together with the following thermal properties of the materials and voids involved:

■ **Thermal conductivity** This is represented by the symbol λ (lambda). It is the measure of a material's ability to transmit heat, and is expressed as the energy flow in watts per square metre of surface area for a temperature gradient of one Kelvin per metre thickness, i.e. W/m K.
■ **Thermal resistance** This is symbolized by the letter R, as representative of a material's thermal resistance achieved by dividing its thickness in metres by its thermal conductivity, i.e. $R = m/\lambda$ or $m^2 K/W$.
■ **Surface or standard resistances** These are values for surface and air space (cavity) resistances. They vary with direction of energy flow, building elevation, surface emissivity and degree of exposure. They are expressed in the same units as thermal resistance. In calculations they are represented by:
  ■ R_{si} = surface resistance inside (typical value 0.12 $m^2 K/W$)
  ■ R_{so} = surface resistance outside (typical value 0.06 $m^2 K/W$)
  ■ R_a = surface resistance of the air space (typical value 0.18 $m^2 K/W$)

[*Note*: Values for λ and R can be obtained by reference to tables published by the Chartered Institution of Building Services Engineers or from listings in product manufacturer's catalogues.] See also Figs. 8.2.2 to 8.2.5.

The method for calculating the U value of any combination of materials will depend on the construction. For example, a traditional rendered and plastered one-brick wall is consistent: therefore the U value is expressed as the reciprocal of

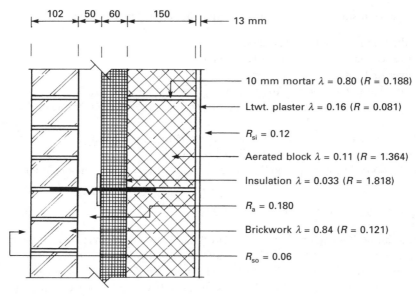

Figure 8.2.2   Calculation of *U* value through an external insulated cavity wall.

the summation of resistances, i.e. $U = 1/\Sigma R$. ΣR comprises internal surface resistance (R_{si}), thermal resistance of materials [R_1 = plaster, R_2 = brickwork (inc. mortar) and R_3 = render] and external surface resistance (R_{so}). If an element of structure is inconsistent, e.g. lightweight concrete blockwork with dense mortar joints, or timber framing with insulation infilling, cold bridging will occur through the denser parts. Inconsistencies can be incorporated into the calculation using the Combined Method from BS EN ISO 6946: *Building components and building elements. Thermal resistance and thermal transmittance. Calculation method.*

Cavity wall *U* value calculation (ref. Fig. 8.2.2):

1. Butterfly pattern twisted wire wall ties can be used where the gap between masonry leaves does not exceed 75 mm. With these ties no allowance for thermal conductivity or bridging is considered necessary. In Fig. 8.2.2 where the gap exceeds 75 mm, vertical twist pattern ties are generally specified and an allowance of 0.02 W/m^2 K should be added to the *U* value calculation.
2. As the thermal resistances of brick and mortar are similar, i.e. < 0.1 m^2 K/W, the outer leaf is considered consistent and unbridged.
3. Fractional area of bridged layer:

   Standard block nominal area = $450 \times 225 = 101\ 250$ mm^2
   Standard block format area   = $440 \times 215 = \underline{94\ 600}$ mm^2
   Difference or mortar area   = $= \ \ \ 6\ 650$ mm^2

   Block format = 94 600 /101 250 = 0.934 or 93.4%
   Mortar area = 6650/101 250 = 0.066 or 6.6%

4. The example shown has only one bridged layer, the mortar joints in the inner leaf of brickwork.

Material	Thermal resistance (R)
Internal surface	0.120
Plaster	0.081
Blockwork – mortar (6.6%)	0.188
Blockwork – blocks (93.4%)	1.364
Insulation	1.818
Cavity	0.180
Brickwork (inc. mortar)	0.121
External surface	0.060
	3.932 m² K/W

5. Upper resistance limit:
Heat energy loss has two paths in the wall, through the blocks and through the mortar.

   (a) Resistance through blocks, $3.932 - 0.188 = 3.744$ m² K/W [R_1]. Fractional area, 93.4% or 0.934 [F_1]
   (b) Resistance through mortar, $3.932 - 1.364 = 2.568$ m² K/W [R_2]. Fractional area, 6.6% or 0.066 [F_2]

$$\text{Upper limit of resistance} = \frac{1}{\dfrac{F_1}{R_1} + \dfrac{F_2}{R_2}} = \frac{1}{\dfrac{0.934}{3.744} + \dfrac{0.066}{2.568}}$$

$$= 3.634 \text{ m}^2 \text{ K/W}$$

6. Lower limit of resistance is a summation of resistance of all parts of the wall:

Internal surface	= 0.120
Plaster	= 0.081
Blockwork $1/[(0.934/1.364) + (0.066/0.188)]$	= 0.965
Insulation	= 1.818
Cavity	= 0.180
Brickwork	= 0.121
External surface	= 0.060
	3.345 m² K/W

7. The total resistance of the wall is the average of upper and lower resistances:

$(3.634 + 3.345)/2 = 3.489$ m² K/W
U value $= 1/R = 1/3.489 = 0.286$ W/m² K

To this value, add 0.020 (wall tie allowance) $= 0.306$ W/m² K

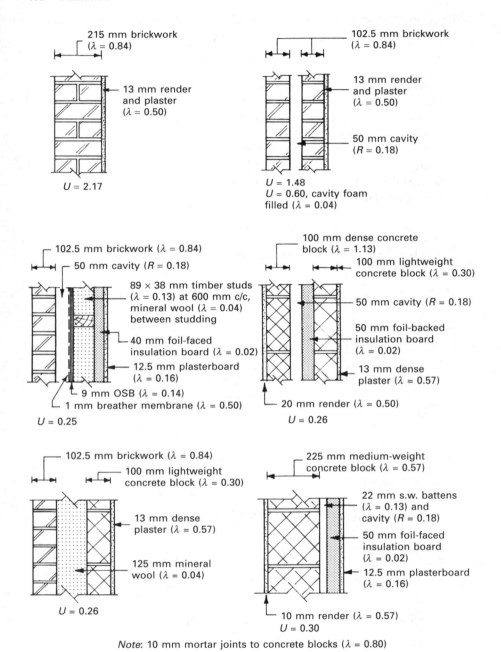

Figure 8.2.3   Wall construction, typical U values (W/m²K).

The thermal insulation of roofs can be effected between and/or over the rafters beneath the covering or at ceiling level. Generally rafter level insulation will use more materials but can be applied as a combined roofing felt and insulation to economise in application costs. The roof void will be warm, and on sheltered sites it should not be necessary to protect the water cistern and pipework against frost. Applying the insulation at ceiling level is simpler, but as the roof void is unheated the plumbing housed within this space will require insulation from frost damage. Cold roofs will need ventilation to prevent condensation in accordance with Building Regulation F2 and BS 5250: *Code of practice for control of condensation in buildings.*

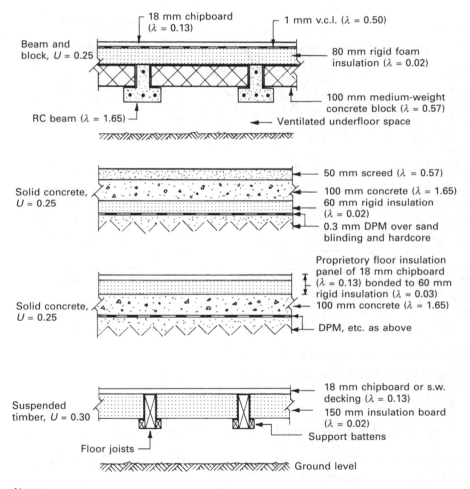

Notes:
1. *U* values will vary significantly for different floor area and perimeter ratios. See data/tables in insulation manufacturers' publications.
2. Rigid insulation boards may be produced from extruded polystyrene, polyisocyanurate, urethane or phenolic materials (conductivity λ range, 0.02 to 0.03).

Figure 8.2.4   Floor construction, typical *U* values (W/m^{2}K).

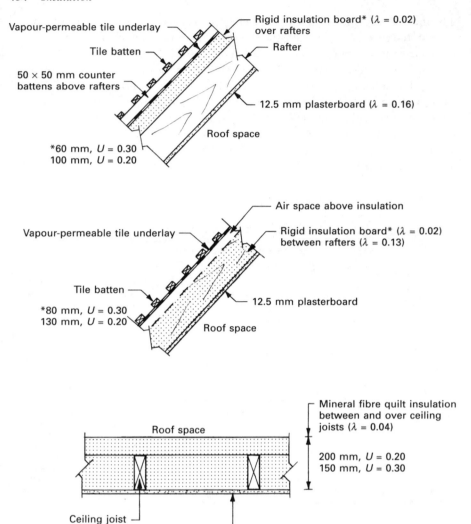

Vapour-permeable tile underlay

Rigid insulation board* (λ = 0.02) over rafters

Tile batten

Rafter

50 × 50 mm counter battens above rafters

12.5 mm plasterboard (λ = 0.16)

Roof space

*60 mm, U = 0.30
100 mm, U = 0.20

Air space above insulation

Vapour-permeable tile underlay

Rigid insulation board* (λ = 0.02) between rafters (λ = 0.13)

Tile batten

12.5 mm plasterboard

*80 mm, U = 0.30
130 mm, U = 0.20

Roof space

Mineral fibre quilt insulation between and over ceiling joists (λ = 0.04)

Roof space

200 mm, U = 0.20
150 mm, U = 0.30

Ceiling joist

12.5 mm vapour-check plasterboard ceiling

Figure 8.2.5   Roof construction, typical U values (W/m^2 K).

■■■ ENERGY RATING OF DWELLINGS

Part L of the Building Regulations requires new build homes and refurbishment projects such as house conversions to flats to be assessed in terms of energy efficiency and performance of the whole building. Calculation of U values of the elements of construction is an important contributory factor, but the overall objective is to satisfy energy performance targets for building type and size relative to atmospheric contamination by carbon dioxide (CO_2) expressed in kg per m^2 of floor area.

DCER, TCER AND SAP

The measure of CO_2 is known as the **dwelling carbon emissions rate** (DCER). All new homes are built to an acceptable DCER measured against a **target carbon emissions rate** (TCER). Methods and formulae for calculation are published in the government's Standard Assessment Procedure for Energy Rating of Dwellings, known as SAP. SAP incorporates a number of considerations relating to energy consumption. In addition to U values and type of insulation, these include the effect of incidental solar gains, ventilation characteristics, boiler type, hot water and heating systems, energy controls and accessories, amount and orientation of glazing, fuel prices and chosen fuel.

Part L also places emphasis on:

- quality of construction and workmanship;
- external lighting;
- interior fixed lighting;
- heating and hot water equipment;
- solar overheating;
- thermal bridges.

Quality of construction and workmanship

This is essential to minimise energy loss due to air leakage through the structure. Areas most vulnerable are the construction interface between wall and floor, wall and roof, wall and window, etc. For guidance, Robust Standard Details (RSDs) are published to supplement the Building Regulation Approved Documents. RSDs include criteria such as density and thickness of materials and application of sealants at junctions and openings (see Chapter 8.4). Where RSDs are specified, a sample of finished dwellings is subject to air pressure test. Where alternative construction to RSDs is used, pre-completion air pressure testing is applied to a greater proportion of houses. For all new dwellings the worst acceptable air permeability is $10 \text{ m}^3/\text{h}/\text{m}^2$ envelope area at 50 pascals (Pa) pressure.

External lighting

The maximum rating of each light fitting/luminaire is 150 watts, with an efficacy greater than 40 lumens per circuit watt. Control should be with an automatic light detector (photoelectric cell) and time switch to extinguish the light when not required. The recommended type of lamp is compact fluorescent and not the standard tungsten filament pattern.

Interior fixed lighting

In most-used locations, such as the living room, light fittings/luminaires should also be fitted with lamps of minimum efficacy, 40 lumens per circuit watt. See Table 8.2.1.

Table 8.2.1   Guide to disposition of efficient lights.

No. of rooms in a dwelling	Minimum no. of efficient lights
1–3	1
4–6	2
7–9	3
10–12	4

Garage and outbuilding not included.
Hall, stairs and landing regarded as one room.

Heating and hot water equipment

All new installations must have high-efficiency boilers, graded on the SEDBUK scale at A or B energy rating. (SEDBUK is the acronym for Seasonal Efficiency of a Domestic Boiler in the United Kingdom.) All manufacturers' boilers are listed on a database developed by the government's Energy Efficiency Best Practice programme. This provides an impartial comparison of different manufacturers' models (see the website, www.boilers.org.uk). Boilers rated at levels A or B are over 86% efficient, i.e. not more than 14% fuel energy potential is lost in converting the fuel to heat energy in water. These high-efficiency boilers are otherwise known as **condensing boilers** and operate on the principle of a double heating effect. After the initial heat exchange from burning the fuel, the flue gases are circulated by fan around the outside of the heat exchanger for a secondary effect before discharging to atmosphere.

Other considerations include solar heating and insulation of pipes and ducting.

Solar overheating

Solar overheating in summer must be avoided in the interests of comfort, otherwise high-energy-consuming air conditioning may be an option, which contradicts the principles of conserving energy. Control can be achieved by limiting window areas, installing awnings and shading systems, and limiting glazing exposed to the south.

Thermal bridges

Thermal bridges must be avoided. See Chapter 8.3. RSDs can be consulted.

House builders and designers now have far more to consider when specifying the construction of new dwellings. No longer can energy efficiency be determined by insulation alone. There are now many other factors to consider, creating many benefits in terms of greater flexibility of design. CO_2 reduction is the key, and this may be achieved by including renewable energy technologies into the construction as a trade off against perhaps limited glazing area or extensive insulation, for example. Photovoltaic solar panels and heat pumps rate highly in SAP terms, but even if these save energy in the long term, there is an inevitable initial increase in capital costs of construction. Designing new houses to be energy efficient is a complex and specialised exercise.

■ ■ ■ INSULATING MATERIALS

When selecting or specifying thermal insulation materials the following must be taken into consideration:

- Thermal resistance of the material.
- Need for a vapour control layer, because insulating materials that become damp or wet, generally from condensation, rapidly lose their insulation properties: therefore, if condensation is likely to occur, a suitable vapour control layer should be included in the detail. Vapour control layers should always be located on the warm side of the construction.
- Availability of material chosen.
- Ease of fixing or including the material in the general construction.
- Appearance if visible.
- Cost in relation to the end result and ultimate savings on fuel and/or heating installation.
- Fire risk: all wall and ceiling surfaces must comply with the requirements of Building Regulation B2: *Internal fire spread (linings)*.

Insulating materials are made from a wide variety of materials and are available in a number of forms, as follows.

INSULATING CONCRETE

Basically a concrete of low density containing a large number of voids. This can be achieved by using lightweight aggregates such as clinker, foamed slag, expanded clay, sintered pulverised fuel ash, exfoliated vermiculite and expanded perlite, or alternatively an aerated concrete made by the introduction of air or gas into the mix. No fines concrete made by using lightweight or gravel aggregates between 20 and 10 mm size and omitting the fine aggregate is suitable for loadbearing walls. Generally lightweight insulating concrete is used in the form of an *in-situ* screed to a structural roof or as lightweight concrete blocks for walls.

LOOSE FILLS

Materials that can be easily poured from a bag and levelled off between the joists with a shaped template. Materials include exfoliated vermiculite, fine glass fibrewool, mineral wool and cork granules. The depth required to satisfy current legislation is usually well above the top level of ceiling joists: therefore this type of insulation is now better used as a vertical fill between stud framing to dormer windows and walls in loft conversions. Most loose fills are rot and vermin proof as well as being classed as non-combustible.

BOARDS

Used mainly as drylinings to walls and ceilings, either for self-finish or for direct decoration. Types include metallised polyester-lined plasterboard, woodwool slabs, expanded polystyrene boards, thermal-backed (expanded polystyrene/extruded polystyrene) plasterboard and fibreboards. Insulating fibreboards should be conditioned on site before fixing to prevent buckling and distortion after fixing. A suitable method is to expose the boards on all sides so that the air has a free passage around the sheets for at least 24 hours before fixing. During this conditioning period the boards must not be allowed to become wet or damp.

QUILTS

Made from glass fibre or mineral wool bonded or stitched between outer paper coverings for easy handling. The quilts are supplied in rolls from 6.000 to 13.000 m long and cut to suit standard joist spacings. They are laid between and over the ceiling joists and can be obtained in two thicknesses, 100 mm for general use and 150 mm for use in roof spaces. They can be placed in two layers, the lower layer between the joists with another superimposed at right angles to the joists.

INSULATING PLASTERS

Factory-produced premixed plasters that have lightweight perlite and vermiculite expanded minerals as aggregates, and require only the addition of clean water before application. They are only one-third the weight of sanded plasters, have three times the thermal insulation value and are highly resistant to fire. However, they can only be considered as a supplement to other insulation products as they have insufficient thickness to provide full insulation.

FOAMED CAVITY FILL

A method of improving the thermal insulation properties of an external cavity wall by filling the cavity wall with urea–formaldehyde resin foamed on site. The foam is formed using special apparatus by combining urea–formaldehyde resin, a hardener, a foaming agent and warm water. Careful control with the mixing and application is of paramount importance if a successful result is to be achieved; specialist contractors are normally employed. The foam can be introduced into the cavity by means of 25 mm boreholes spaced 1.000 m apart in all directions, or by direct introduction into the open end of the cavity. The foam is a white cellular material containing approximately 99% by volume of air with open cells. The foam is considered to be impermeable, and therefore unless fissures or cracks have occurred during the application it will not constitute a bridge of the cavity in the practical sense. Non-combustible water-repellent glass or rock wool fibres are alternative cavity fill materials. Manufacturer's approved installers use compressors to blow the fibres into new or existing cavity walls where both inner and outer leaves are constructed of masonry. The application of cavity fill must comply with the requirements of Building Regulation D1.

The most effective method of improving thermal comfort conditions within a building is to ensure that the inside surface is at a reasonably high temperature, and this is best achieved by fixing insulating materials in this position.

Thermal insulation for buildings other than dwellings is also contained in the Building Regulations under Part L2. This is considered in *Advanced Construction Technology*.

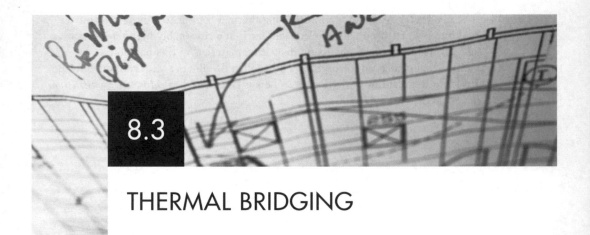

8.3

THERMAL BRIDGING

Consistency of construction in the external elements is desirable to achieve a uniform heat transfer. It is, nevertheless, impractical to build in uniform layers, and the routine is interrupted with 'bridges' between air voids and across solids such as timber in framed walls. This results in variations in thermal transmittance values (see proportional area calculation – previous chapter), which should be avoided as much as possible. Examples include metal or dense concrete lintels, and gaps in the continuity of cavity insulation. Where this does occur, the thermal or cold bridge will incur a cooler surface temperature than the remainder of the wall, greater heat loss at this point and a possible dew-point temperature manifesting in water droplets or condensation. The moisture will attract dust and dirt to contrast with adjacent clean areas. This is known as **pattern staining**.

The Approved Document to Part L of the Building Regulations makes specific reference to thermal bridging, indicating the areas most exposed. Further reference to the Building Regulations support document, *Thermal insulation: avoiding risks* (2nd edition), provides numerous examples of acceptable forms of construction at vulnerable places.

Apart from examples of bad practice, such as using bricks to make up coursing in lightweight insulating concrete block inner leaf walls, the most common areas for thermal bridging are around door and window openings, at junctions between ground floor and wall, and the junction between wall and roof. These are shown in Fig. 8.3.1.

Attention to detail at construction interfaces is the key to preventing thermal bridging. The details shown in Fig. 8.3.2 incorporate the necessary features to promote good practice.

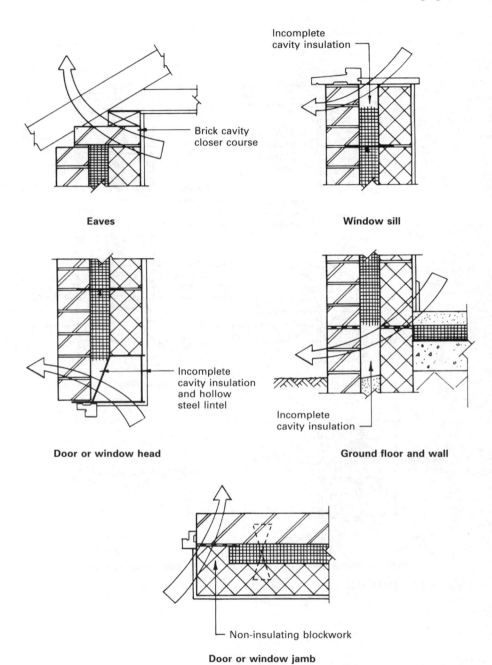

Eaves

Window sill

Incomplete
cavity insulation

Brick cavity
closer course

Door or window head

Ground floor and wall

Incomplete
cavity insulation
and hollow
steel lintel

Incomplete
cavity insulation

Door or window jamb

Non-insulating blockwork

Figure 8.3.1   Thermal bridging – areas for concern.

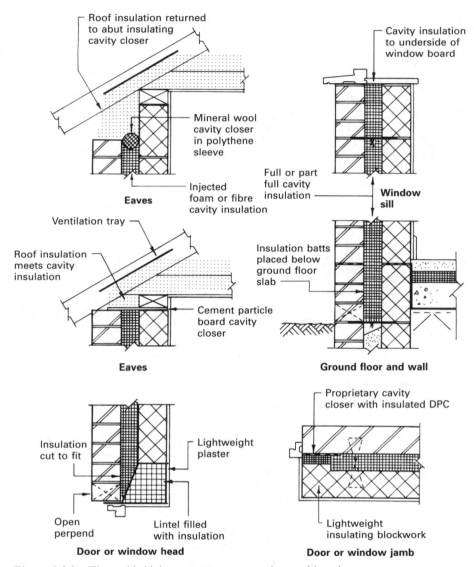

Figure 8.3.2   Thermal bridging – treatment at openings and junctions.

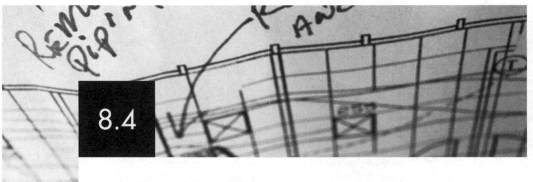

8.4

DRAUGHT-PROOFING AND AIR PERMEABILITY

Draught-proofing is required to prevent the infiltration of cold external air through leakage in the building envelope. It is also to limit heat losses through breaks in the continuity of construction.

The areas most exposed occur at the junction of different components in the same element, e.g. door and window abutments with walls, access hatches into the roof space or eaves cupboards. The fit between window sashes and frames, and between doors and frames, is also very vulnerable to air leakage. The closeness of fit should be the subject of quality control at the point of manufacture, but this alone is insufficient to satisfy Part L of the Building Regulations. All opening units should have purpose-made seals or draught excluders.

Particular attention must be provided at the points where service pipes and ducts penetrate the structure. Gaps must be made good with appropriate filling and a flexible sealant applied between pipe or duct and adjacent filling. Drylinings, too, must receive special attention. It is not sufficient to secure these to background walls with only plaster dabs. Continuous plaster seals are required at abutments with walls, floors, ceilings and openings for services, windows and doors. Some areas and methods of treatment are shown in Fig. 8.4.1.

Other areas vulnerable to air permeability occur at junctions between different elements of construction, particularly where materials with differing movement characteristics meet, e.g. timber and masonry. A specific area for attention in where floor joists build into the inner leaf of external cavity walls. See Fig. 8.4.2.

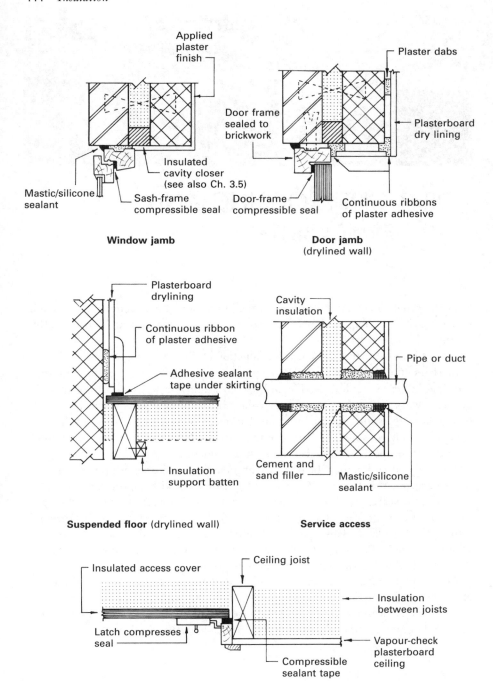

Figure 8.4.1   Prevention of air infiltration.

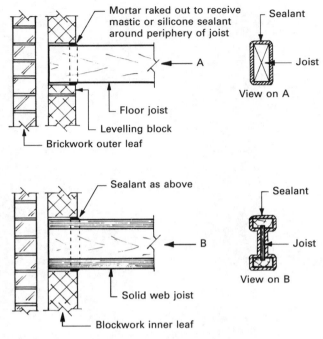

Figure 8.4.2 Air permeability seal at junction of floor joists and external wall.

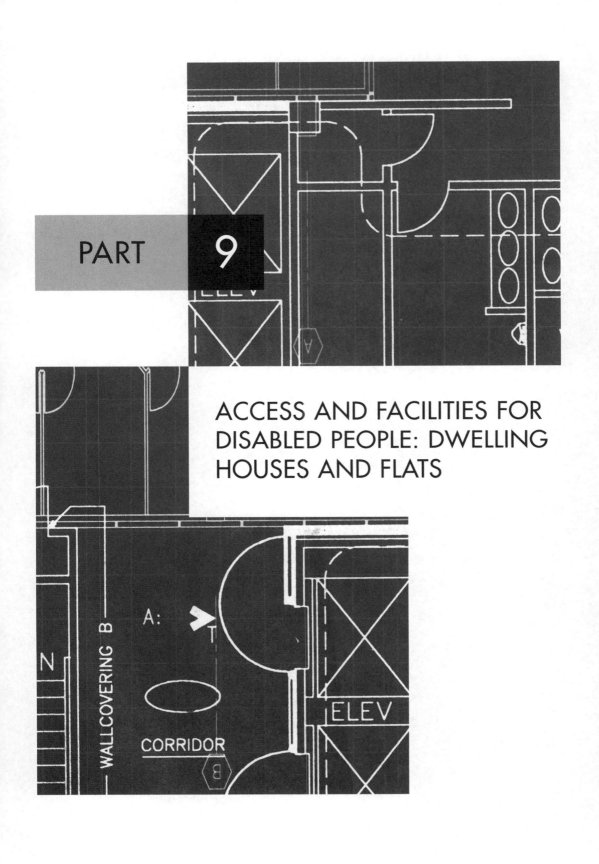

PART 9

ACCESS AND FACILITIES FOR
DISABLED PEOPLE: DWELLING
HOUSES AND FLATS

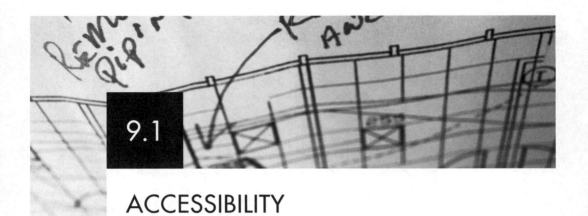

9.1

ACCESSIBILITY

All new, refurbished and adapted dwellings must have reasonable provisions to allow accessibility for disabled persons. The objective is to provide wheelchair users with an easily negotiable approach from the point of alighting a vehicle to the dwelling entrance.

◼◼◼ STEPPED APPROACHES

Steps are to be avoided. However, if the site has a gradient exceeding 1 in 15, steps designed to suit ambulant disabled persons may be acceptable. Figure 9.1.1 shows external step profiles that satisfy the Building Regulations.

Additional requirements for steps:

- unobstructed width, min. 900 mm;
- flight rise between landings, max. 1800 mm;
- landing length, min. 900 mm;
- tapered treads to be avoided, but if used dimension for tread measured 270 mm from the inside of tread;
- for three or more risers, a handrail should be provided between 850 and 1000 mm above the step pitch line (an imaginary line linking the tread nosings);
- handrail to extend 300 mm beyond top and bottom nosings. Figure 9.1.2 shows acceptable design limitations for handrails.

◼◼◼ LEVEL OR RAMPED APPROACH

A level approach from a parked vehicle should satisfy the following:

- gradient not greater than 1 in 20;
- surface firm, even and not containing any loose shingle or other unstable material;
- unobstructed width not less than 900 mm.

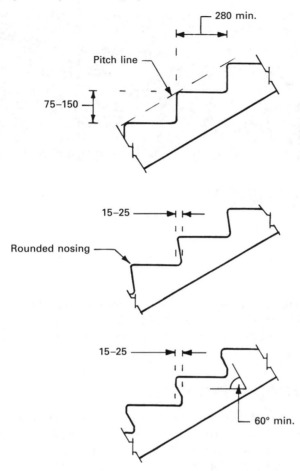

Figure 9.1.1   Acceptable external step profiles.

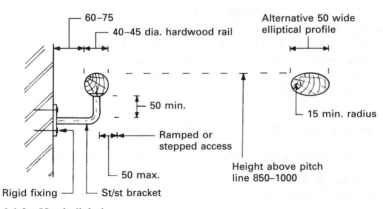

Figure 9.1.2   Handrail design.

A ramped approach from a parked vehicle to dwelling access is defined as having a measured slope greater than 1 in 20, but not greater than 1 in 15 overall. A ramp should have:

- a firm and even surface, free of loose shingle or other unstable material;
- unobstructed width of not less than 900 mm;
- maximum flight length of 10 m for gradients not greater than 1 in 15;
- maximum flight length of 5 m where a gradient within the overall plot is not steeper than 1 in 12;
- horizontal landings at top, bottom and if necessary at intermediate points at least 1200 mm long, not including the passage of any door opening onto the landing.

[*Note*: Crossfalls to level and ramped approaches should not exceed 1 in 40.]

▨▬■ DWELLING ACCESS

Stepped thresholds should be avoided. For ease of wheelchair manoeuvring at entrances, the maximum recommended vertical projection is 15 mm. However, if a step is unavoidable owing to specific site circumstances, a rise at the threshold not exceeding 150 mm may be considered. Figure 3.5.9 shows a means for facilitating wheelchair entry with a drainage slot to convey rainwater from the door face.

▨▬■ DWELLING ENTRANCE DOOR

The minimum acceptable width of door opening for wheelchairs is 775 mm. As shown in Fig. 9.1.3, this is measured from the face of the door stop, latch side of the frame, to the face of the door opened at 90° to the frame.

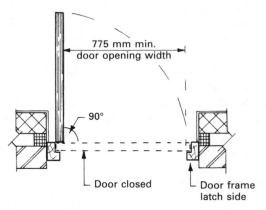

Figure 9.1.3   Entrance door.

9.2

CIRCULATION SPACE: PRINCIPAL STOREY

The principal storey within a dwelling is usually the entrance storey. At this level the objective is to provide access into habitable rooms and a WC compartment.

■ ■ ■ CORRIDORS

An adequate means of passage is required within the entrance storey to permit unobstructed circulation by a wheelchair user. Allowance for normal projections such as shelves and radiators should be considered when designing for minimum acceptable corridor width. The passageway widths given in Table 9.2.1 can be reduced to 750 mm for a distance of no more than 2 m, where a necessary permanent obstruction such as a radiator cannot be avoided. This relaxation is not acceptable where it would impede a wheelchair user from turning into or out of a room, as shown in Fig. 9.2.1.

Rooms and facilities off a room must also be unobstructed and wheelchair accessible. Internal door openings should have sufficient width to accommodate the manoeuvring of a wheelchair from corridors and passageways. Table 9.2.1 indicates the spatial requirements for internal doors relative to corridor width and direction of approach.

Table 9.2.1   Passageway and door width dimensions

Internal doorway width (mm) See Fig. 9.2.1	Unobstructed passageway width (mm) minimum
≥ 750	900 (direct approach)
750	1200 (oblique approach)
775	1050 (oblique approach)
800	900 (oblique approach)

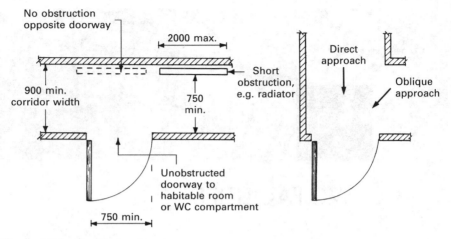

Figure 9.2.1   Corridors and passageways.

◼◼◼ STEPS

Where the slope of a site is considerable, it may be necessary to incorporate a stepped change of level into the entrance storey. A slight slope is preferred, but where steps are unavoidable a stair should have a clear width of at least 900 mm. If the stair has three or more rises, a continuous handrail is required both sides of the flight. Step rise and going should satisfy the private stair provisions in Approved Document K of the Building Regulations. See Chapter 7.4.

9.3

WC FACILITIES

In addition to a WC and bathroom that may be located elsewhere in a dwelling, a WC is required at the entrance storey containing a habitable room. If there are no habitable rooms in the entrance storey, a WC can be still provided here as well as within the principal storey. The WC does not have to be independent; it may be part of a bathroom suite. Access from habitable rooms to the WC compartment should avoid negotiation of stairs. To maximise space for a wheelchair user, an outward-opening door will provide clear access to the WC. The minimum widths of door openings are defined in Table 9.2.1. A basin of the hand wash type should be of minimal dimensions so as not to impede access to the WC. Clear space requirements in the vicinity of a WC are indicated in Fig. 9.3.1.

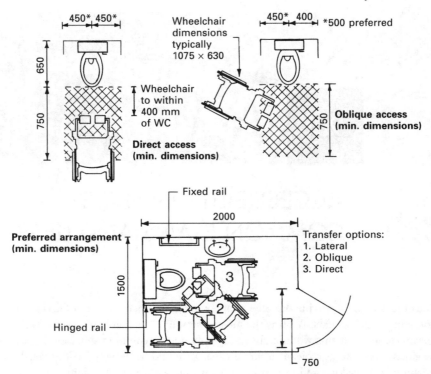

Figure 9.3.1 WC access for a wheelchair user.

References:

BS 8300: Design of buildings and their approaches to meet the needs of disabled
   people. Code of practice.
Building Regulations Part M and Approved Document: Access to and use of
   buildings.
Disability Discrimination Act.

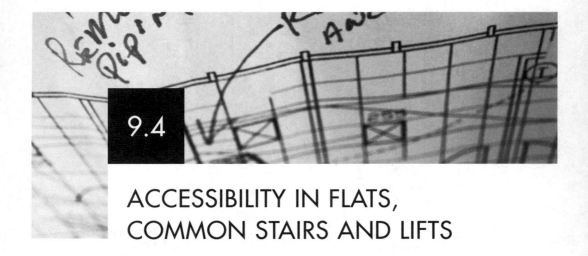

9.4

ACCESSIBILITY IN FLATS, COMMON STAIRS AND LIFTS

The objective is to provide disabled people with a reasonable means of access to any storey in a block of flats. A lift is the most desirable and suitable means for vertical transportation, but this will not always be provided. A common stairway is essential for general use and as a means of fire escape. It should be designed with regard for the ambulant disabled and in particular for people with limited vision.

■■■ STAIRS

The following guidelines should be incorporated into the design:

- Landings provided at the top and bottom of a stair flight. Minimum length not less than stair width and without permanent obstruction.
- Handrail on both sides of the flight and landings, where there are two or more rises. See Fig. 9.4.1.
- Any door that opens onto a landing must leave at least 400 mm clear space across the width of the landing.
- Uniform rise of step, maximum 170 mm.
- Risers full, i.e. no gaps under tread.
- Uniform going of tread, minimum 250 mm.
- Tapered treads to be avoided, but if used, going measured 270 mm from inside of tread.
- Tread nosing profiles as shown in Fig. 9.4.1.
- Nosings made prominent with contrasting colour and brightness.

■■■ LIFTS

Dimensions in and around lifts should be sufficient for the convenience of an unaccompanied wheelchair user. Control facilities should also be designed to be

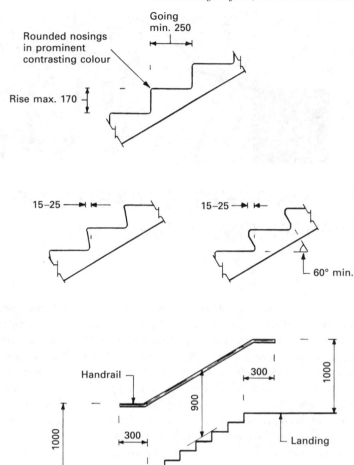

Figure 9.4.1 Step profiles for common stairs.

accessible and usable from a wheelchair and by people with a visual impairment. The following provides some design guidance:

- Lift door opening and closing time adequate for a wheelchair to access without contacting closing doors. Door fully open for at least 3 seconds.
- Minimum car load capacity of 400 kg.
- Landing space in front of lift doors, minimum 1500 mm × 1500 mm.
- Door opening width, minimum 800 mm.
- Lift car dimensions, minimum 900 mm wide × 1250 mm deep.
- Controls at landings and in car, minimum 900 mm, maximum 1200 mm above floor levels. Within car, minimum 400 mm from the front wall.
- Controls prominent/projecting to provide a tactile facility, i.e. floor numbers profiled.
- Visual indication that a lift is responding to a call.
- Over three storeys, visual and audible indication of floor level reached.

9.5

SWITCHES, SOCKETS AND GENERAL CONTROLS

This requirement covers a range of control devices normally installed within a dwelling, i.e. light switches, power sockets, telephone sockets, entry telephones, heating thermostats and radiator valves. It should also include door furniture and ironmongery. The objective is to provide these means of control at a level readily accessible for people with reach limitations. In addition to the registered disabled, this category will also include the elderly and infirm. Design of controls should facilitate ease of operation and prominent visibility, e.g. large rocker-type electrical switches and contrasting colours respectively. Interior design and layout of premises should ensure that control facilities are not impeded by obstructions.

A suitable wall mounting height for controls is recommended as between 450 and 1200 mm above finished floor level. Ceiling string pull switch pendants should also terminate within this range. Door handles align with switch controls as shown in Fig. 9.5.1.

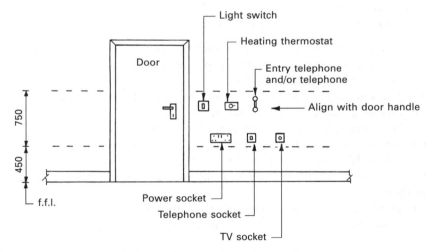

Figure 9.5.1 Recommended accessibility for domestic controls.

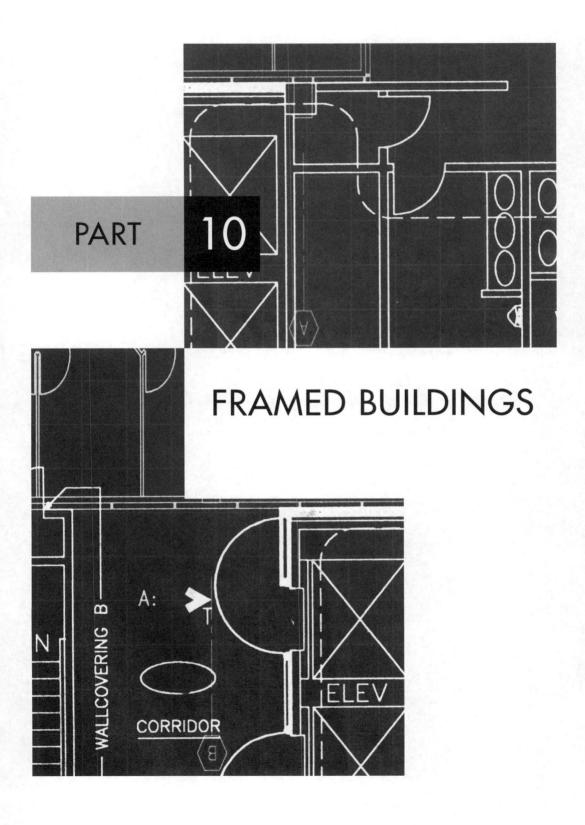

PART 10

FRAMED BUILDINGS

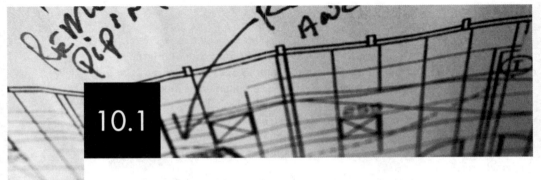

SIMPLE FRAMED BUILDINGS

The purpose of any framed building is to transfer the loads of the structure plus any imposed loads through the members of the frame to a suitable foundation. This form of construction can be clad externally with lightweight non-loadbearing walls to provide the necessary protection from the elements and to give the required degree of comfort in terms of sound and thermal insulation. Framed buildings are particularly suitable for medium- and high-rise structures and for industrialised low-rise buildings such as single-storey factory buildings.

Frames can be considered under three headings:

- plane frames;
- space frames;
- skeleton frames.

■ ■ ■ PLANE FRAMES

These are fabricated in a flat plane and are usually called **trusses** or **girders** according to their elevation shape. They are designed as a series of connected rigid triangles, which gives a lightweight structural member using the minimum amount of material; main uses are in roof construction and long-span beams of light loading.

■ ■ ■ SPACE FRAMES

These are similar in conception to a plane frame but are designed to span in two directions as opposed to the one-direction spanning of the plane frame. A variation of the space frame is the **space deck**, which consists of a series of linked pyramid frames forming a lightweight roof structure. For details of these forms of framing, refer to *Advanced Construction Technology*.

▩▩■ SKELETON FRAMES

Basically, these are a series of rectangular frames placed at right angles to one another so that the loads are transmitted from member to member until they are transferred through the foundations to the subsoil. Skeleton frames can be economically constructed of concrete or steel or a combination of the two. Timber skeleton frames, although possible, are generally considered to be uneconomic in this form. The choice of material for a framed structure can be the result of a number of factors such as site conditions, economics, availability of labour and materials, time factor, statutory regulations, capital costs, maintenance costs and personal preference.

FUNCTIONS OF SKELETON FRAME MEMBERS

- **Main beams**   Span between columns and transfer the live and imposed loads placed upon them to the columns.
- **Secondary beams**   Span between and transfer their loadings to the main beams. Primary function is to reduce the spans of the floors or roof being supported by the frame.
- **Tie beams**   Internal beams spanning between columns at right angles to the direction of the main beams and having the same function as a main beam.
- **Edge beams**   As tie beams but spanning between external columns.
- **Columns**   Vertical members that carry the loads transferred by the beams to the foundations.
- **Foundation**   The base(s) to which the columns are connected and which serve to transfer the loadings to a suitable loadbearing subsoil.
- **Floors**   May or may not be an integral part of the frame; they provide the platform on which equipment can be placed and on which people can circulate. Besides transmitting these live loads to the supporting beams they may also be required to provide a specific fire resistance, together with a degree of sound and thermal insulation.
- **Roof**   Similar to floors but its main function is to provide a weather-resistant covering to the uppermost floor.
- **Walls**   The envelope of the structure, which provides the resistance to the elements, entry of daylight, natural ventilation, fire resistance, thermal insulation and sound insulation.

The three major materials used in the construction of skeleton frames – reinforced concrete, precast concrete and structural steelwork – are considered in detail in the following chapters.

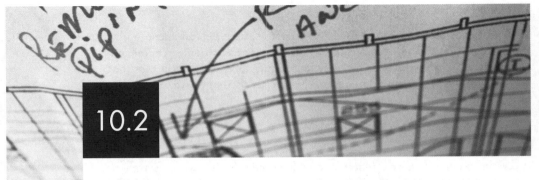

REINFORCED CONCRETE FRAMES

■■■ CONCRETE

Plain concrete is a mixture of cement, fine aggregate, coarse aggregate and water. Concrete sets to a rock-like mass because of a chemical reaction that takes place between the cement and water, resulting in a paste or matrix that binds the other constituents together. Concrete gradually increases its strength during the curing or hardening period to obtain its working strength in about 28 days if ordinary Portland cement is used. The strength achieved will depend on a number of factors, as follows.

Type of cement used

In all cases the cement should conform to the appropriate British Standard.

Type and size of aggregates

In general, aggregates should comply with BS EN 12620 for quality and grading. BS 812 and BS EN 932-5 provide tests for aggregates and define nominal maximum sizes into categories of 40, 20, 14 and 10 mm. 5 mm is also included, although used mainly for precast concrete products.

Water

This should be clean, free from harmful matter, and should comply with the requirements of BS EN 1008.

Use of admixtures

Individually, admixtures can be used as accelerators, for air entrainment to give a weight reduction and added protection against water penetration, chemical and fungal attack. The instructions given by the manufacturer or engineer must be carefully followed, as many admixtures if incorrectly used can have serious adverse effects on the hardened concrete.

Water/cement ratio

A certain amount of water is required to hydrate the cement, and any extra water is needed only to produce workability. The workability of fresh concrete should be such that it can be handled, placed and compacted so that it will surround any reinforcement and completely fill the formwork. Workability can be specified in terms of a slump test, compacting factor test or VB consistometer test, and should not vary beyond the values for the maximum free water/cement ratio. See BS EN 206-1: *Concrete. Specification, performance, production and conformity*.

Grade strength

Concrete should be specified as designated, designed or prescribed mix, with the grade required in accordance with BS EN 206-1 (see also Chapter 2.4 of this book). Grades are numerically related to characteristic strength or cube test strength taken at 28 days. This is for concrete with any type of cement except high–alumina cement, which hardens rapidly and has structural limitations – see Chapter 2.4. The grades and general application include:

- Grades 7 and 10 for plain unreinforced concrete. C10 is the minimum grade normally acceptable, provided the cement content is at least 175 kg/m^3 for designated mixes and 210 kg/m^3 for other concrete specifications.
- Grades 15 and 20 for reinforced concrete with lightweight aggregates.
- Grade 25 for reinforced concrete with dense aggregates.
- Grades 30 to 50 for prestressed concrete.

The characteristic strength of grade 7 concrete is 7.0 N/mm^2, and for grade 10 concrete it is 10.0 N/mm^2; similar values can be deduced for the other grades listed above.

Plain concrete, in common with other brittle materials, has a greater crushing or compressive strength than tensile strength. The actual ratio varies, but plain concrete is generally considered to be 10 times stronger in compression than in tension. If a plain concrete member is loaded so that tension is induced, it will fail in tension when the compressive strength of the concrete has only reached one-tenth of its ultimate value. If this weakness in tension can be reinforced in such a manner that the tensile resistance is raised to a similar value as its compressive strength the member will be able to support a load 10 times that of plain concrete; alternatively, for any given load a smaller section can be used if concrete is reinforced.

▨ ▨ ▨ REINFORCEMENT

Any material specified for use as a reinforcement to concrete must fulfil certain requirements if an economic structural member is to be constructed. These basic requirements are as follows:

- It must have tensile strength.
- It must be capable of achieving this tensile strength without undue strain.
- It must be of a material that can be easily bent to any required shape.
- Its surface must be capable of developing an adequate bond between the concrete and the reinforcement to ensure that the required design tensile strength is obtained.
- A similar coefficient of thermal expansion is required to prevent unwanted stresses being developed within the member owing to temperature changes.
- It must be available at a reasonable cost, which must be acceptable to the overall design concept.

The material that meets all the above requirements is steel in the form of bars, and is supplied in two basic types: mild steel and high-yield steel. Hot-rolled steel bars are covered by BS 4449, which specifies a characteristic strength of 250 N/mm^2 for mild steel and 460 N/mm^2 for high-yield steel. The surface of mild steel provides adequate bond, but the bond of high-yield bars, being more critical with the higher stresses developed, is generally increased by rolling longitudinal or transverse ribs onto the surface of the bar. Grade 460A can be identified by having at least two rows of parallel transverse ribs; grade 460B is similar but has at least one row of ribs contrary to the others. As an alternative to hot-rolled steel bars, cold-worked steel bars can be used. When bars are cold-worked they become harder and stiffer, and develop a higher tensile strength.

The range of nominal sizes available for both round and deformed bars is 6, 8, 10, 12, 16, 20, 25, 32 and 40 mm with length specified by the purchaser. For pricing purposes the 16 and 20 mm bars are taken as basic, with the diameters on either side becoming more expensive as the size increases or decreases. A good design will limit the range of diameters used together with the type of steel chosen to achieve an economic structure and to ease the site processes of handling, storage, buying and general confusion that can arise when the contractor is faced with a wide variety of similar materials.

The bending of reinforcement can be carried out on site by using a bending machine, which shapes the cold bars by pulling them round a mandrel. Small diameters can also be bent round simple jigs such as a board with dowels fixed to give the required profile; large diameters may need a power-assisted bending machine. Bars can also be supplied ready bent and labelled so that only the fabrication processes take place on site.

The bent reinforcement should be fabricated into cages for columns and beams and into mats for slabs and walls. Where the bars cross or intersect one another they should be tied with soft iron wire, fixed with special wire clips or tack-welded to maintain their relative positions. Structural members that require only small areas of reinforcement can be reinforced with steel fabric, which can be supplied in sheets or rolls.

Steel fabric for the reinforcement of concrete is covered by BS 4483, which gives four basic preferred types: 'A' for square mesh (floor slabs); 'B' for rectangular structural mesh (also for slabs); 'C' for long mesh (roads and pavements); and 'D', a small-diameter wrapping mesh (binding concrete fireproofing). The fabric is factory made by welding or interweaving wires complying with the requirements of BS 4482 to form sheets with a length of 4.800 m and a width of 2.400 m, or rolls 48.000 and 72.000 m long with a common width of 2.400 m. Each type has a letter prefix, which is followed by a reference number, which is the total cross-sectional area of main bars in mm^2 per metre width. Typical examples of reinforcing bars and fabric are shown in Fig. 10.2.1.

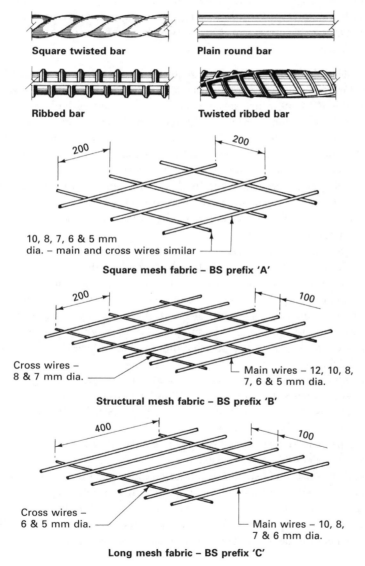

Square twisted bar

Plain round bar

Ribbed bar

Twisted ribbed bar

200   200

10, 8, 7, 6 & 5 mm
dia. – main and cross wires similar

Square mesh fabric – BS prefix 'A'

200   100

Cross wires –
8 & 7 mm dia.

Main wires – 12, 10, 8,
7, 6 & 5 mm dia.

Structural mesh fabric – BS prefix 'B'

400   100

Cross wires –
6 & 5 mm dia.

Main wires – 10, 8,
7 & 6 mm dia.

Long mesh fabric – BS prefix 'C'

Figure 10.2.1 Typical reinforcing bars and welded fabric.

Before placing reinforcement into the formwork it should be brushed free of all loose rust and mill scale and be free of grease or oil, as the presence of any of these on the surface could reduce the bond and hence the strength of the reinforced concrete. Reinforcement must have a minimum cover of concrete to give the steel protection from corrosion due to contact with moisture and to give the structural member a certain degree of fire resistance. Nominal cover to reinforcement should always be equal to the size of the bar being used or, where groups of bars are used, at least the size of the largest diameter. BS 8110, Tables 3.3 and 3.4, set out the recommended nominal amount of cover, in relationship to the exposure/durability and periods of fire-resistance criteria, respectively, for various concrete grades. The Building Regulations, Approved Document B, establishes the minimum fire resistance for various purpose groups of buildings. Note that the dimensions of structural members are also of importance, so that failure of the concrete due to the high temperatures encountered during a fire is avoided before the reinforcement reaches its critical temperature. Table A3 of Approved Document B and Table 3.4 and Fig. 3.2 in BS 8110-1 give suitable dimensions for various fire-resistance periods and methods of construction of reinforced concrete members.

To maintain the right amount of concrete cover during construction small blocks of concrete may be placed between the reinforcement and the formwork; alternatively, plastic clips or spacer rings can be used. Where top reinforcement has to be retained in position (such as in a slab), cradles or chairs made from reinforcing bar may have to be used. All forms of spacers must be of a material that will not lead to corrosion of the reinforcement or cause spalling of the hardened concrete.

DESIGN

The design of reinforced concrete is the prerogative of the structural engineer, and is a subject for special study, but the technologist should have an understanding of the principles involved. The designer, by assessing the possible dead and live loads on a structural member, can calculate the reactions and effects that such loadings will have on the member. This will indicate areas of tensile stress and critical locations for reinforcement. Calculations are based on the recommendations and design tables in BS 8110-1 to 3: *Structural use of concrete*, together with established formulae for determination of bending moments, shear forces and the area of steel required. Typical examples of these forces for simple situations are shown in Fig. 10.2.2. Some applications of simple reinforced concrete design are included at the end of this chapter.

REINFORCEMENT SCHEDULES AND DETAILS

Once the engineer has determined the reinforcement required, detail drawings can be prepared to give the contractor the information required to construct the structure. The drawings should give the following information:

- sufficient cross-reference to identify the member in relationship to the whole structure;
- all the necessary dimensions for design and fabrication of formwork;

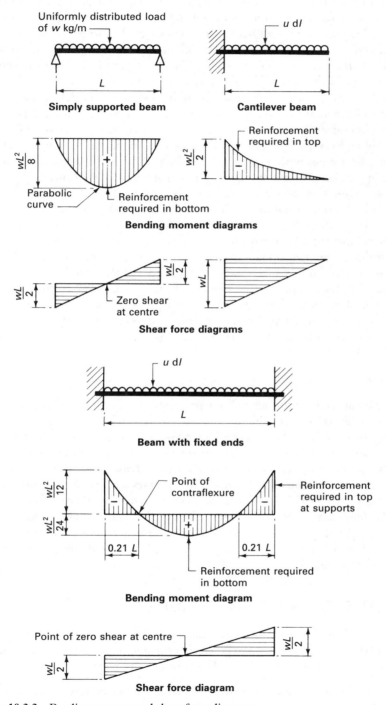

Figure 10.2.2   Bending moment and shear force diagrams.

- details of the reinforcement;
- minimum cover of concrete over reinforcement;
- concrete grade required if not already covered in the specification.

Reinforced concrete details should be prepared so that there is a distinct definition between the lines representing the outline of the member and those representing the reinforcement. Bars of a common diameter and shape are normally grouped together with the same reference number when included in the same member. To simplify the reading of reinforced concrete details it is common practice to show only one bar of each group in full, together with the last bar position (see Fig. 10.2.3).

The bars are normally bent and scheduled in accordance with the recommendations of BS 8666 and BS EN ISO 4066. These give details of the common bending shapes, the method of setting out the bending dimensions, the method of calculating the total length of the bar required, and a shape code for use with data-processing routines. A preferred form of bar schedule is also given, which has been designed to give easy cross-reference to the detail drawing.

Reinforcement on detail drawings is annotated by a coding system to simplify preparation and reading of the details. For example:

9 R 1201–300, which can be translated as:

9 = total number of bars in the group
12 = diameter in mm
01 = bar mark number
300 = spacing centre to centre
R = mild steel round bar (250 N/mm^2)

The code letter R could be replaced by code letter T, which is used for high-yield round bars and high-yield square twisted bars; other types of bars are coded S for stainless steel and W for wire reinforcement (4–12 mm dia.). A typical reinforced concrete beam detail and schedule is shown in Fig. 10.2.3. All other reinforced concrete details shown in this volume are intended to show patterns of reinforcement rather than detailing practice, and are therefore shown in full without reference to bar diameters and types.

HOOKS, BENDS AND LAPS

To prevent bond failure, bars should be extended beyond the section where there is no stress in the bar. The length of bar required will depend upon such factors as grade of concrete, whether the bar is in tension or compression, and whether the bar is deformed or plain. Hooks and bends can be used to reduce this anchorage length at the ends of bars, and should be formed in accordance with the recommendations of BS EN ISO 4066 (see Fig. 10.2.4).

Where a transfer of stress is required at the end of a bar the bars may be welded or lapped.

BS 8110 recommends that laps and joints should only be made by the methods specified in the contract and at the positions shown on the drawings and as agreed by the engineer.

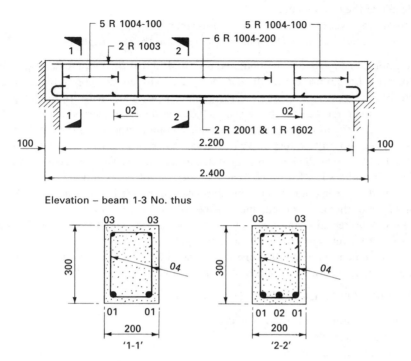

Elevation – beam 1-3 No. thus

'1-1'   '2-2'

Note: Cover to main bars 25 mm

Member	Bar mark	Type & size	No. of mbrs	No. in each	Total no.	Length of each bar[†]	Shape. All dimensions* are in accordance with BS 4466
Beam 1	1	R20	3	2	6	2.660	⌐ 2.300 ⌐
	2	R16	3	1	3	1.400	Straight
	3	R10	3	2	6	2.300	Straight
	4	R10	3	16	48	1.000	250 ☐ 150

[†] Specified to nearest 25 mm   * Specified to nearest 5 mm

Figure 10.2.3   Typical reinforced concrete beam details and schedule.

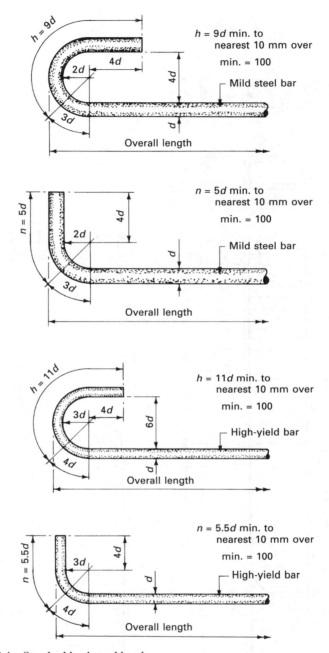

Figure 10.2.4 Standard hooks and bends.

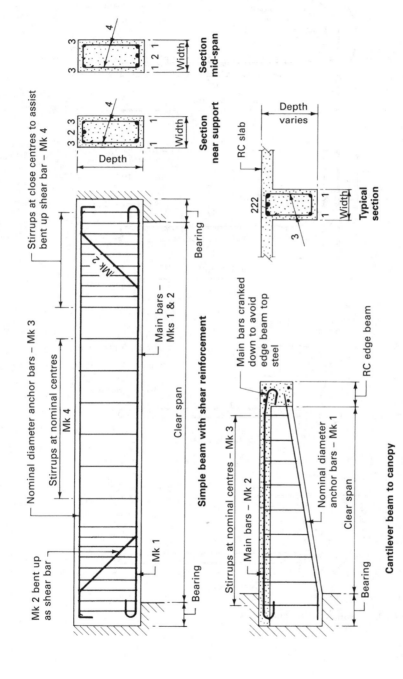

Figure 10.2.5 Simple reinforced concrete beams.

REINFORCED CONCRETE BEAMS

Beams can vary in their complexity of design and reinforcement, from the very simple beam formed over an isolated opening, such as those shown in Figs 10.2.3 and 10.2.5, to the more common form encountered in frames, where the beams transfer their loadings to the columns (see Fig. 10.2.6).

When tension is induced in a beam the fibres will lengthen until the ultimate tensile strength is reached, when cracking and subsequent failure will occur. With a uniformly distributed load the position and value of tensile stress can easily be calculated by the structural engineer, but the problem becomes more complex when heavy point loads are encountered.

The correct design of a reinforced concrete beam will ensure that it has sufficient strength to resist both the compression and tensile forces encountered in the outer fibres, but it can still fail in the 'web' connecting the compression and tension areas. This form of failure is called **shear failure** and is in fact diagonal tension. Concrete has a limited amount of resistance to shear failure, and if this is exceeded reinforcement must be added to provide extra resistance. Shear occurs at or near the supports as a diagonal failure line at an angle of approximately 45° to the horizontal and sloping downwards towards the support. A useful fact to remember is that zero shear occurs at the point of maximum bending (see Fig. 10.2.2).

Reinforcement to resist shearing force may be either stirrups or inclined bars, or both. The total shearing resistance is the sum of the shearing resistances of the inclined bars and the stirrups, calculated separately if both are provided. Inclined or bent-up bars should be at 45° to the horizontal and positioned to cut the anticipated shear failure plane at right angles. These may be separate bars, or main bars from the bottom of the beam that are no longer required to resist tension, and that can be bent up and carried over or onto the support to provide the shear resistance (see Figs 10.2.5 and 10.2.6). Stirrups or binders are provided in beams, even where not required for shear resistance, to minimise shrinkage cracking and to form a cage for easy handling. The nominal spacing for stirrups must be such that the spacing dimension used is not greater than the lever arm of the section, which is the depth of the beam from the centre of the compression area to the centre of the tension area or 0.75 times the effective depth of the beam, which is measured from the top of the beam to the centre of the tension reinforcement. If stirrups are spaced at a greater distance than the lever arm it would be possible for a shearing plane to occur between consecutive stirrups, but if the centres of the stirrups are reduced locally about the position at which shear is likely to occur, several stirrups may cut the shear plane, and therefore the total area of steel crossing the shear plane is increased to offer the tensile resistance to the shearing force (see Figs 10.2.5 and 10.2.6).

REINFORCED CONCRETE COLUMNS

A column is a vertical member carrying the beam and floor loadings to the foundation, and is a compression member. As concrete is strong in compression it may be concluded that, provided the compressive strength of the concrete is not

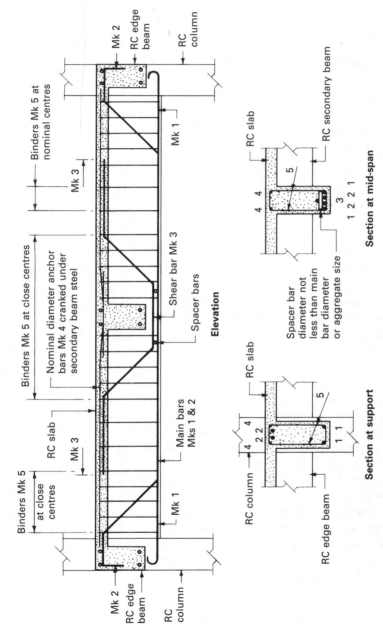

Figure 10.2.6 Reinforced concrete beam with heavy reinforcement.

exceeded, no reinforcement will be required. For this condition to be true the following conditions must exist:

- Loading must be axial.
- Column must be short, which can be defined as a column where the ratio of its effective height to its thickness does not exceed 12 (effective height may be calculated by formula and tables in Section 3.8.1.6 of BS 8110-1).
- Cross-section of the column must be large.

These conditions rarely occur in framed buildings: consequently bending is induced and the need for reinforcement to provide tensile strength is apparent. Bending in columns may be induced by one or more of the following conditions:

- load coupled with the slenderness of the column – a column is considered to be slender if the ratio of effective height to thickness exceeds 12;
- reaction to beams upon the columns – as the beam deflects it tends to pull the column towards itself, thus inducing tension in the far face;
- the reaction of the frame to wind loadings, both positive and negative.

The minimum number of main bars in a column should not be less than four for rectangular columns and six for circular columns, with a total cross-sectional area of not less than 0.8% of the cross-sectional area of the column and a minimum diameter of 12 mm. To prevent the slender main bars from buckling, and hence causing spalling of the concrete, links or binders are used as a restraint. These should be at least 6 mm diameter and not less than one quarter of the largest main bar diameter. The pitch or spacing is no greater than the least of:

- the least lateral column dimension;
- 12 times the smallest longitudinal reinforcement;
- 300 mm.

All bars in compression should be tied by a link passing around the bar in such a way that it tends to move the bar towards the centre of the column; typical arrangements are shown in Fig. 10.2.7.

Where the junction between beams and columns occurs there could be a clash of steel, as bars from the beam may well be in the same plane as bars in the columns. To avoid this situation one group of bars must be offset or cranked into another plane. It is generally considered that the best practical solution is to crank the column bars to avoid the beam steel: typical examples of this situation, together with a method using straight bars, are shown in Fig. 10.2.8. A similar situation can occur where beams of similar depth intersect: see cantilever beam example in Fig. 10.2.5.

REINFORCED CONCRETE SLABS

A reinforced concrete slab will behave in exactly the same manner as a reinforced concrete beam, and it is therefore designed in the same manner. The designer will analyse the loadings, bending moments, shear forces and reinforcement requirements on a slab strip 1.000 m wide. In practice the reinforcement will be fabricated to form a continuous mat. For light loadings a mat of welded fabric could be used.

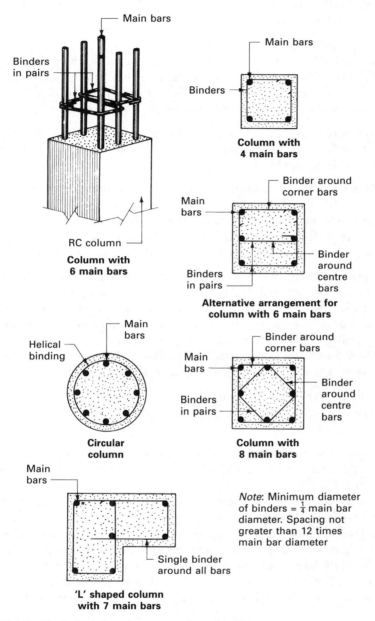

Figure 10.2.7   Typical reinforced concrete column binding arrangements.

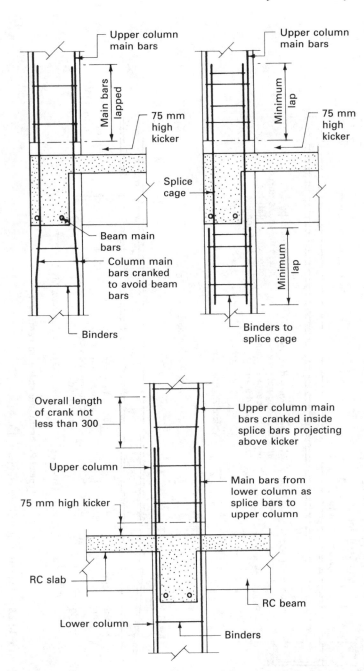

Figure 10.2.8   Reinforced concrete column and beam junctions.

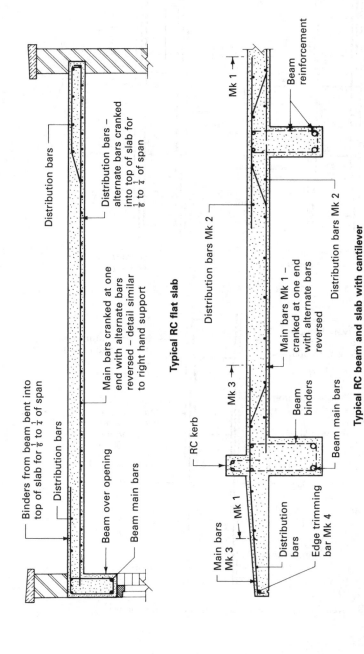

Typical RC flat slab

Distribution bars

Distribution bars – alternate bars cranked into top of slab for $\frac{1}{6}$ to $\frac{1}{4}$ of span

Main bars cranked at one end with alternate bars reversed – detail similar to right hand support

Binders from beam bent into top of slab for $\frac{1}{6}$ to $\frac{1}{4}$ of span

Distribution bars

Beam over opening

Beam main bars

Typical RC beam and slab with cantilever

Mk 1

Beam reinforcement

Distribution bars Mk 2

Distribution bars Mk 2

Main bars Mk 1 – cranked at one end with alternate bars reversed

Mk 3

Beam binders

Beam main bars

RC kerb

Main bars Mk 3

Mk 1

Distribution bars

Edge trimming bar Mk 4

Figure 10.2.9   Typical reinforced concrete slab details.

There are three basic forms of reinforced concrete slabs:

- flat slab floors or roofs;
- beam and slab floors or roofs;
- ribbed floors or roofs – see Chapter 4.4.

Flat slabs

These are basically slabs contained between two plain surfaces, and can be either simple or complex. The design of the complex form is based upon the slab acting as a plate in which the slab is divided into middle and column strips, the reinforcement being concentrated in the latter strips.

Simple flat slabs can be thick and heavy, but have the advantage of giving clear ceiling heights as there are no internal beams. They are generally economic up to spans of approximately 9.000 m, and can be designed to span one way – that is, across the shortest span – or to span in two directions. These simple slabs are generally designed to be simply supported: that is, there is no theoretical restraint at the edges, and therefore tension is not induced and reinforcement is not required. However, it is common practice to provide some top reinforcement at the supports as anti-crack steel should there, in practice, be a small degree of restraint. Generally this steel is 50% of the main steel requirement and extends into the slab for 0.2 m of the span. An economic method is to crank up 50% of the main steel or every alternate bar over the support, as the bending moment will have reduced to such a degree at this point that it is no longer required in the bottom of the slab. If there is an edge beam, the top steel can also be provided by extending the beam binders into the slab (see Fig. 10.2.9).

Beam and slab

By adopting this method of design, large spans are possible and the reinforcement is generally uncomplicated. A negative moment will occur over the internal supports, necessitating top reinforcement; as with the flat slabs, this can be provided by cranked bars (see Fig. 10.2.9). Each bar is in fact cranked, but alternate bars are reversed, thus simplifying bending and identification of the bars. Alternatively a separate mat of reinforcement supported on chairs can be used over the supports.

■ ■ ■ BASIC DESIGN PRINCIPLES

SLABS AND BEAMS

The simplest of design situations occurs where a slab or beam spans two opposing supports. The concrete will require reinforcement, mainly in the bottom to resist bending and tensile forces. A continuous slab or beam has several supports, incurring additional tensile forces in the top of the structure above the interim supports. Figure 10.2.10 illustrates the effect.

Simple slab and beam

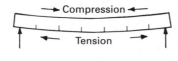

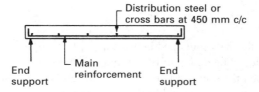

Continuous slab and beam

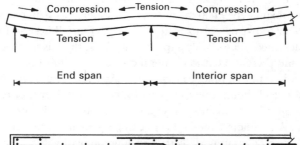

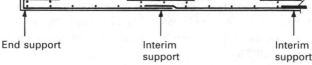

Figure 10.2.10   Slab and beam reinforcement.

Concrete compressive strength is usually specified in the region of 25–30 N/mm², with mild steel bar reinforcement occupying about 1% of the gross area. Table 10.2.1 provides a reference for estimating section sizes for modest imposed floor loading, typical of normal office conditions at 500 kg/m² (approx. 5 kN/m²). For comparison, domestic floor imposed loadings are taken as 150 kg/m² (approx. 1.5 kN/m²).

Table 10.2.1   Guidance for depth and span between opposing supports

Depth (mm)	Self-weight (kg/m²)	Imposed load (kg/m²)	Total load (kg/m²)	Total load (kN/m²)	Span (mm)	Steel bars (mm dia. @ 112 mm c/c)
100	240	500	740	7.26	2400	10
125	300	500	800	7.85	3000	12
150	360	500	860	8.44	3600	16

Note: Continuous support permits an increase in span of 20% for interior spans and 10% for end spans.

■ Greater loads: depth of concrete is increased in proportion to the square root of the loads. For example, total load of 2000 kg/m^2 over a 3.6 m span =

$$\sqrt{2000/860} \times 150 = 230 \text{ mm}$$

■ Greater spans: depth of concrete is increased relative to the proportional increase in span. For example, a 4 m span will have a depth =

$$125 + (125/3) = 167 \text{ mm}$$
$$\text{or} \quad 150 \times (4/3.6) = 167 \text{ mm}$$

COLUMNS

A column or pillar constructed from plain concrete without any reinforcement could in theory withstand substantial compressive loads. Where the loading is absolutely central, then the safe load on a column can be calculated. For example:

Concrete compressive strength = 25 N/mm^2
Safe working stress of concrete (safety factor of 5) = 5 N/mm^2
Column dimensions of say 300 mm × 300 mm square has a safe load of:

$$300 \times 300 \times 5 = 450\ 000 \text{ N or } 450 \text{ kN}$$
$$450/9.81 = 45.87 \text{ tonnes}$$

With reinforced columns the calculation must acknowledge the contribution that steel makes to withstanding the compressive load. This contribution is known by the modular ratio. It is usually between 15 and 18 depending on the quality of materials. For example, using the figure of 18 means that each square unit of steel carries as much load as 18 square units of concrete.

Using the preceding example of a 300 mm square column, shown in Fig. 10.2.11, the subsequent calculation shows the column to have greater load potential due to the reinforcement.

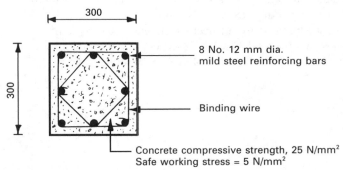

300

300

8 No. 12 mm dia.
mild steel reinforcing bars

Binding wire

Concrete compressive strength, 25 N/mm^2
Safe working stress = 5 N/mm^2

Min. recommended reinforcement, 0.8%
column area = 720 mm^2
Max. recommended reinforcement, 8%
column area = 7200 mm^2

Figure 10.2.11 Reinforced concrete column.

Calculation

8 reinforcing bars at 12 mm diameter.

$$8 \times \pi r^2 = 8 \times \pi \times 6 \times 6 = 905 \text{ mm}^2 \text{ (ignoring binders)}$$

Each 1 mm^2 of steel carries as much load as 18 mm^2 of concrete, but in so doing it displaces 1 mm^2 of concrete.

Therefore the net gain is 17 mm^2, so the steel area is multiplied by 17 to give $17 \times 905 = 15\,385$ mm^2 of concrete equivalent to 905 mm^2 of steel.

Adding 15 385 mm^2 to 90 000 mm^2 (300 mm $\times$ 300 mm column) of concrete, the equivalent area of steel and concrete combined is 105 385 mm^2.

Multiplying by the safe working stress of 5 N/mm^2, the safe load is

$$5 \times 105\,385 = 526\,925 \text{ N or } 526.93 \text{ kN}$$
$$526.93/9.81 = 53.71 \text{ tonnes}$$

Buckling, bending and eccentric loads

The preceding calculation still assumes a central load on the column. In practice it is unlikely that bending stresses and eccentric loads will be absent. Therefore, some allowance for buckling or bending will normally be applied. The effective length or height of a column will have some influence: see Fig. 10.2.12. The effective length is an adjustment to the actual measured length of a column, the amount determined by the means of fixing and securing at both ends. Where the effective length is less than 15 times the least lateral dimension of a column, the safe working stress of concrete is modified by a factor of 0.8, e.g. 5 N/mm^2 $\times$ 0.8 = 4 N/mm^2. Where the effective column length is greater than 15, a further correction for buckling and bending is required. See Table 10.2.2.

For example, continuing to use the same 300 mm $\times$ 300 mm square reinforced concrete column as before, with an effective length of 6 m:

Effective length/least lateral dimension: 6000/300 = 20
From table, buckling correction factor = 0.833 (by interpolation)
Concrete safe working stress = 5 N/mm^2
Equivalent combined area of steel + concrete = 105 385 mm^2
Therefore, $5 \times 0.8 \times 0.833 \times 105\,385 = 351\,143$ N or 351.14 kN
$$351.14/9.81 = 35.79 \text{ tonnes}$$

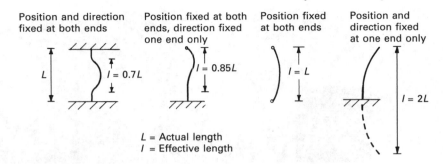

Figure 10.2.12 Effective length of columns.

Table 10.2.2 Correction factors for buckling and bending

Effective length/Least lateral dimension	Buckling correction
15	1.0
18	0.9
21	0.8
24	0.7
27	0.6
30	0.5
33	0.4
36	0.3
39	0.2
42	0.1
45	0

10.3

FORMWORK

Formwork for *in-situ* concrete work may be described as a mould or box into which wet concrete can be poured and compacted so that it will flow and finally set to the inner profile of the box or mould. It is important to remember that the inner profile must be opposite to that required for the finished concrete: so if, for example, a chamfer is required on the edge, a triangle fillet must be inserted into the formwork.

To be successful in its function, formwork must fulfil the following requirements:

- It should be strong enough to support the load of wet concrete, which is generally considered to be approximately 2400 kg/m^3.
- It must not be able to deflect under load, which would include the loading of wet concrete, self-weight, and any superimposed loads such as operatives and barrow runs over the formwork.
- It must be accurately set out; as concrete is a fluid when placed, it will take up the shape of the formwork, which must therefore be of the correct shape, size and in the right position.
- It must have grout-tight joints. Grout leakage can cause honeycombing of the surface, or produce fins, which have to be removed. The making-good of defective concrete surfaces is both time consuming and costly. Grout leakage can be prevented by using sheet materials and sealing the joints with flexible foamed polyurethane strip or by using a special self-adhesive tape.
- Form sizes should be designed so that they are the maximum size that can easily be handled by hand or by a mechanical lifting device.
- Material must be chosen so that it can be easily fixed using double-headed nails, round wire nails or wood screws. The common method is to use nails, and these should be at least two-and-a-half times the thickness of the timber being nailed, in length.
- The design of the formwork units should be such that they can easily be assembled and dismantled without any members being trapped.

▨▩■ MATERIALS

The requirements for formwork enumerated above make timber the most suitable material for general formwork. It can be of board form, either wrot or unwrot depending on whether a smooth or rough surface is required.

Softwood boards used to form panels for beam and column sides should be joined together by cross-members over their backs at centres not exceeding 24 times the board's thickness.

The moisture content of the timber should be between 15% and 20% so that the moisture movement of the formwork is reduced to a minimum. If the timber is dry it will absorb moisture from the wet concrete, which could weaken the resultant concrete member. It will also cause the formwork to swell and bulge, which could give an unwanted profile to the finished concrete. If timber with a high moisture content is used it will shrink and cup, which could result in open joints and a leakage of grout.

Plywood is extensively used to construct formwork units as it is strong, light, and supplied in sheets of 1.200 m wide with standard lengths of 2.400, 2.700 and 3.000 m. The quality selected should be an exterior grade and the thickness related to the anticipated pressures so that the minimum number of strengthening cleats on the back are required.

Chipboard can also be used as a formwork material but, because of its lower strength, will require more supports and stiffeners. The number of uses that can be obtained from chipboard forms is generally less than plywood, softwood boarding or steel.

Steel forms are generally based upon a manufacturer's patent system, and within the constraints of that system are an excellent material. Steel is not as adaptable as timber but if treated with care will give 30 or 40 uses, which is approximately double that of similar timber forms.

MOULD OILS AND EMULSIONS

Two defects that can occur on the surface of finished concrete are:

- **Blowholes**   These are small holes, of less than 15 mm in diameter, caused by air being trapped between the formwork and the concrete face.
- **Uneven colour**   This is caused by irregular absorption of water from the wet concrete by the formwork material. A mixture of old and new material very often accentuates this particular defect.

Mould oils can be applied to the inside surface of the formwork to alleviate these defects. To achieve a uniform colour an impervious material or lining is recommended, but this will increase the risk of blowholes. Mould oils are designed to overcome this problem when using steel forms or linings by encouraging the trapped air to slide up the face of the formwork. A neat oil will encourage blowholes but will discourage uneven colour, whereas a mould oil incorporating an emulsifying agent will discourage blowholes and reduce uneven colouring. Great care must be taken when applying some mould oils, because over-oiling may cause

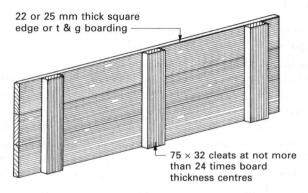

22 or 25 mm thick square
edge or t & g boarding

75 × 32 cleats at not more
than 24 times board
thickness centres

Typical boarded formwork panel

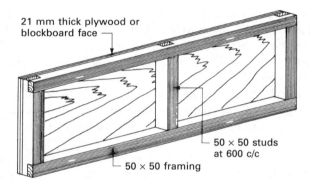

21 mm thick plywood or
blockboard face

50 × 50 studs
at 600 c/c

50 × 50 framing

Typical framed formwork panel

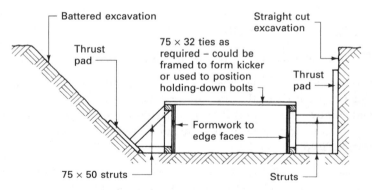

Battered excavation

Straight cut
excavation

Thrust
pad

75 × 32 ties as
required – could be
framed to form kicker
or used to position
holding-down bolts

Thrust
pad

Formwork to
edge faces

75 × 50 struts

Struts

Typical foundation formwork

Figure 10.3.1   Formwork to foundations.

some retardation of the setting of the cement. Emulsions are either drops of water in oil or, conversely, drops of oil in water and are easy to apply, but should not be used in conjunction with steel forms as they encourage rusting. Note that, generally, mould oils and emulsions also act as release agents, and therefore it is essential that the oil or emulsion is applied only to the formwork and not to the reinforcement, as this may cause a reduction of bond.

FORMWORK LININGS

To obtain smooth, patterned or textured surfaces the inside of a form can be lined with various materials, such as oil-tempered hardboard, moulded rubber, moulded PVC or glass-fibre-reinforced polyester; the last-named is also available as a complete form mould. When using any form of lining the manufacturer's instructions regarding sealing, fixing and the use of mould oils must be strictly followed to achieve a satisfactory result.

TYPES OF FORMWORK

FOUNDATION FORMWORK

Foundations to a framed building consist generally of a series of isolated bases or pads, although if these pads are close together it may be more practicable to merge them together to form a strip. If the subsoil is firm and hard it may be possible to excavate the trench or pit for the foundations to the size and depth required and cast the concrete against the excavated faces. Where this method is not practicable formwork will be required. Side and end panels will be required, and these should be firmly strutted against the excavation faces to resist the horizontal pressures of the wet concrete and to retain the formwork in the correct position. Ties will be required across the top of the form as a top restraint, and these can be utilised to form the kicker for a reinforced concrete column or as a template for casting in the holding-down bolts for precast concrete or structural steel columns (see Fig. 10.3.1).

COLUMN FORMWORK

A column form or box consists of a vertical mould that has to resist considerable horizontal pressures in the early stages of casting. The column box should be located against a 75 mm high plinth or kicker that has been cast monolithic with the base or floor. The kicker not only accurately positions the formwork but also prevents loss of grout from the bottom edge of the form. The panels forming the column sides can be strengthened by using horizontal cleats or vertical studs, which are sometimes called **soldiers**. The form can be constructed to the full storey-height of the column, with cut-outs at the top to receive the incoming beam forms. The thickness of the sides does not generally provide sufficient bearing for the beam boxes, and therefore the cut-outs have a margin piece fixed around the opening to provide extra bearing (see Fig. 10.3.2). It is general practice, however,

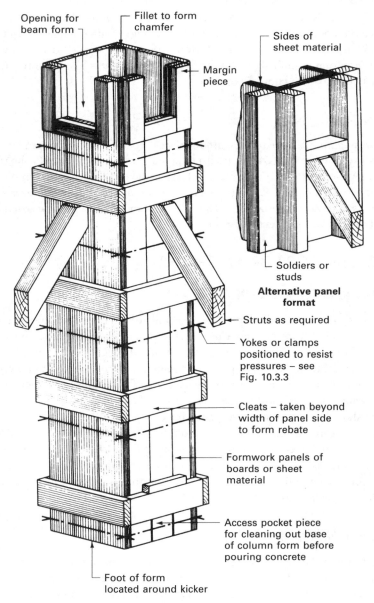

Opening for
beam form

Fillet to form
chamfer

Sides of
sheet material

Margin
piece

Soldiers or
studs

**Alternative panel
format**

Struts as required

Yokes or clamps
positioned to resist
pressures – see
Fig. 10.3.3

Cleats – taken beyond
width of panel side
to form rebate

Formwork panels of
boards or sheet
material

Access pocket piece
for cleaning out base
of column form before
pouring concrete

Foot of form
located around kicker

Figure 10.3.2   Column formwork principles.

to cast the columns up to the underside of the lowest beam soffit, and to complete the top of the column at the same time as the beam, using make-up pieces to complete the column and receive the beam intersections. The main advantage of casting full-height columns is the lateral restraint provided by the beam forms; the disadvantage is the complexity of the formwork involved.

Column forms are held together with collars of timber or metal called **yokes** in the case of timber and **clamps** when made of metal. Timber yokes are purpose made, whereas steel column clamps are adjustable within the limits of the blades (see Fig. 10.3.3).

The spacing of the yokes and clamps should vary with the anticipated pressures, the greatest pressure occurring at the base of the column box. The actual pressure will vary according to:

- rate of placing;
- type of mix being used – generally the richer the mix the greater the pressure,
- method of placing – if vibrators are used, pressures can increase up to 50% over hand placing and compacting;
- air temperature – the lower the temperature the slower is the hydration process and consequently higher pressures are encountered.

Some preliminary raking strutting is required to plumb and align the column forms in all situations. Free-standing columns will need permanent strutting until the concrete has hardened but with tied columns the need for permanent strutting must be considered for each individual case.

Shaped columns will need special yoke arrangements unless they are being formed using a patent system. Typical examples of shaped column forms are shown in Fig. 10.3.4.

BEAM FORMWORK

A beam form consists of a three-sided box, which is supported by cross-members called **headtrees**, which are propped to the underside of the soffit board. In the case of framed buildings support to the beam box is also provided by the column form. The soffit board should be thicker than the beam sides, because this member will carry the dead load until the beam has gained sufficient strength to be self-supporting. Soffit boards should be fixed inside the beam sides so that the latter can be removed at an early date; this will enable a flow of air to pass around the new concrete and speed up the hardening process, and also release the formwork for reuse at the earliest possible time. Generally the beam form is also used to support the slab formwork, and the two structural members are then cast together. The main advantage of this method is that only one concrete operation is involved, although the complexity of the formwork is increased. If the beams and slabs are carried out as separate operations there is the possibility of a shear plane developing between the beam and floor slab; it would be advisable to consult the engineer before adopting this method of construction. Typical examples of beam forms are shown in Fig. 10.3.5.

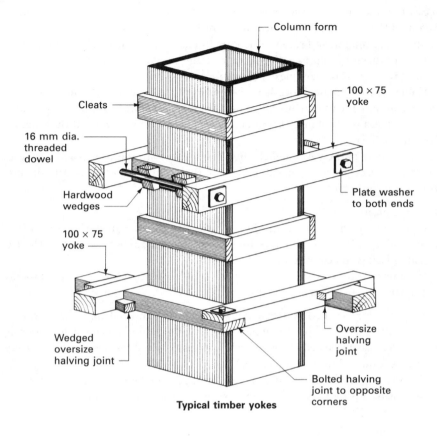

Column form

Cleats

16 mm dia.
threaded
dowel

100 × 75
yoke

Hardwood
wedges

Plate washer
to both ends

100 × 75
yoke

Wedged
oversize
halving joint

Oversize
halving
joint

Bolted halving
joint to opposite
corners

Typical timber yokes

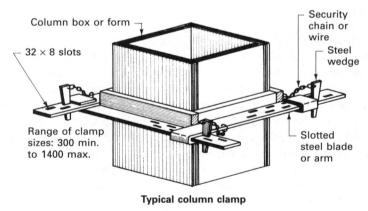

Column box or form

32 × 8 slots

Security
chain or
wire

Steel
wedge

Range of clamp
sizes: 300 min.
to 1400 max.

Slotted
steel blade
or arm

Typical column clamp

Figure 10.3.3   Typical column yokes and clamps.

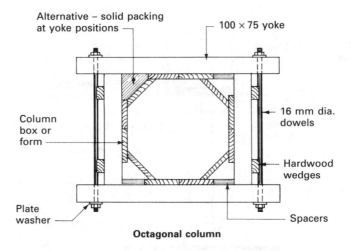

Alternative – solid packing at yoke positions

100 × 75 yoke

Column box or form

16 mm dia. dowels

Hardwood wedges

Plate washer

Spacers

Octagonal column

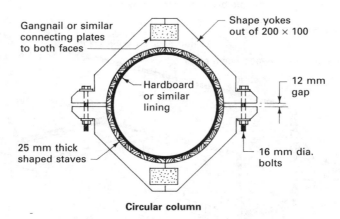

Gangnail or similar connecting plates to both faces

Shape yokes out of 200 × 100

Hardboard or similar lining

12 mm gap

25 mm thick shaped staves

16 mm dia. bolts

Circular column

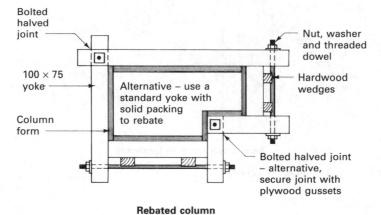

Bolted halved joint

Nut, washer and threaded dowel

100 × 75 yoke

Alternative – use a standard yoke with solid packing to rebate

Hardwood wedges

Column form

Bolted halved joint – alternative, secure joint with plywood gussets

Rebated column

Figure 10.3.4 Shaped column forms and yokes.

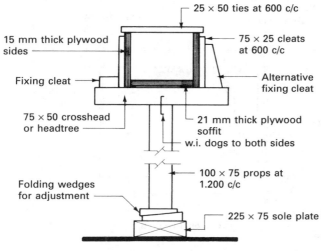

25 × 50 ties at 600 c/c

15 mm thick plywood sides

75 × 25 cleats at 600 c/c

Fixing cleat

Alternative fixing cleat

75 × 50 crosshead or headtree

21 mm thick plywood soffit

w.i. dogs to both sides

Folding wedges for adjustment

100 × 75 props at 1.200 c/c

225 × 75 sole plate

Simple beam or lintel formwork

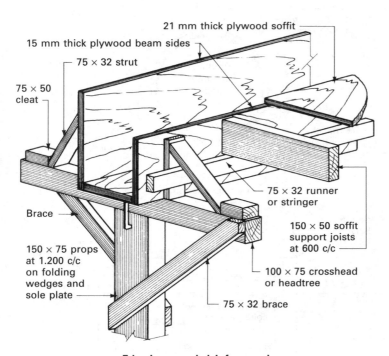

21 mm thick plywood soffit

15 mm thick plywood beam sides

75 × 32 strut

75 × 50 cleat

75 × 32 runner or stringer

Brace

150 × 50 soffit support joists at 600 c/c

150 × 75 props at 1.200 c/c on folding wedges and sole plate

100 × 75 crosshead or headtree

75 × 32 brace

Edge beam and slab formwork

Figure 10.3.5   Typical beam formwork.

Structural steelwork has to be protected against corrosion and fire; one method is to encase the steel section with concrete. The steel frame is erected before casting the concrete encasement and, in the case of beams, it is possible to suspend the form box from the steel section by using a metal device called a **hanger fixing** or alternatively using a steel column clamp or timber yoke (see Fig. 10.3.6). The hanger fixings are left embedded in the concrete encasing, but the bolts and plate washers are recoverable for reuse. If only a haunch is cast around the bottom flange then the projecting hanger fixing wires can be cut off level with the concrete haunch, or the floor units can be slotted to receive them.

SLAB FORMWORK

Floor or roof slab formwork is sometimes called **shuttering**, and consists of panels of a size that can be easily handled. The panels can be framed or joisted and supported by the beam forms with any intermediate propping that is required (see Fig. 10.3.7). Adjustment for levelling purposes can be carried out by using small folding wedges between the joists or framing and the beam box.

PERMANENT FORMWORK

The term **permanent formwork** applies to formwork or shuttering that is left in place. This can occur where removal is too difficult or uneconomic. It can also apply specifically to walls and beams, where an attached facing material combines as outer cladding and formwork. See Chapter 9 in *Advanced Construction Technology*.

Application to floor slabs is with profiled galvanised steel sheeting. Metal profiles are a self-supporting shuttering system, functioning initially as troughed floor slab decking to receive an *in-situ* concrete topping. The additional function is for the profiled and indented steel surface to bond with the wet concrete and combine as a composite floor slab when the concrete has matured. The structural contribution of the steel decking to the tensile strength of the slab is supplementary to standard mesh placed within the *in-situ* concrete.

In addition to a saving on formwork-striking costs, the floor section eliminates concrete where it would otherwise be redundant. This represents an economical benefit in material cost and handling time, and a significant reduction in dead loading on support columns and foundations. Fig. 10.3.8 shows various steel profiles and Fig. 10.3.9 a typical construction detail.

For lateral restraint and fixing to support beams of steel or concrete, profiles are secured with shot-fired pins. Alternatively, they may be drilled and bolted to steel beams, or drilled, plugged and screwed to concrete. Shear connection between steel support beams and *in-situ* concrete is provided by pairs of through-deck welded studs at beam support positions. Side and edge trims are produced to contain concrete around the slab periphery.

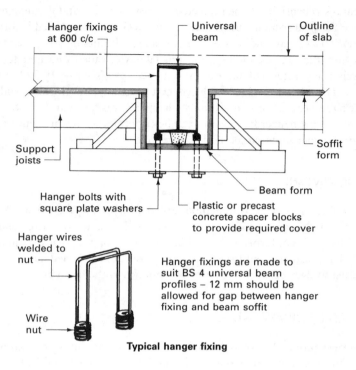

Typical hanger fixing

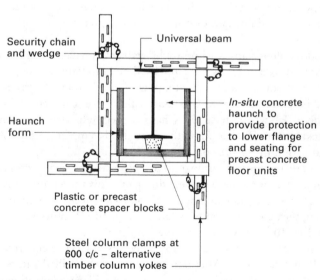

Figure 10.3.6   Suspended formwork.

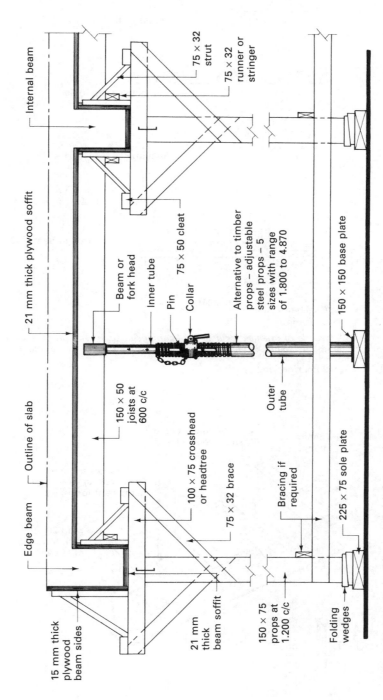

15 mm thick plywood beam sides

21 mm thick plywood soffit

Outline of slab

Edge beam

Internal beam

75 × 32 strut

75 × 32 runner or stringer

75 × 50 cleat

Beam or fork head

Inner tube

Pin

Collar

Alternative to timber props – adjustable steel props – 5 sizes with range of 1.800 to 4.870

150 × 150 base plate

150 × 50 joists at 600 c/c

100 × 75 crosshead or headtree

75 × 32 brace

21 mm thick beam soffit

150 × 75 props at 1.200 c/c

Outer tube

Bracing if required

225 × 75 sole plate

Folding wedges

Figure 10.3.7 Typical beam and slab formwork.

Standard metal profiles

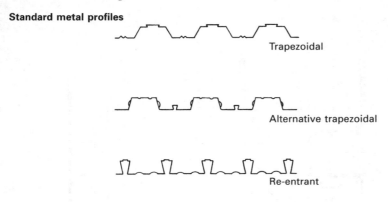

Trapezoidal

Alternative trapezoidal

Re-entrant

Floor section

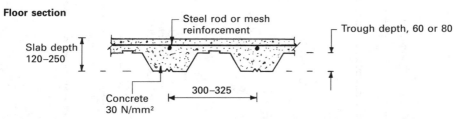

Steel rod or mesh reinforcement

Trough depth, 60 or 80

Slab depth 120–250

Concrete 30 N/mm²

300–325

Note: See BS 5950–4: *Code of practice for design of composite slabs with profiled steel sheeting* for slab depth and reinforcement requirements relative to span and loading.

Figure 10.3.8   Metal floor decking profiles.

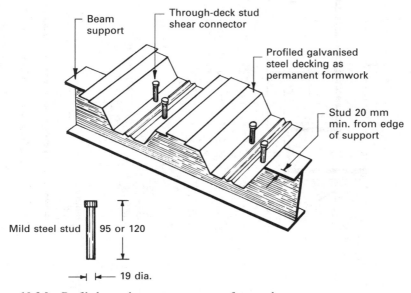

Beam support

Through-deck stud shear connector

Profiled galvanised steel decking as permanent formwork

Stud 20 mm min. from edge of support

Mild steel stud   95 or 120

19 dia.

Figure 10.3.9   Profiled trough pattern permanent formwork.

 SITEWORK

When the formwork has been fabricated and assembled, the interior of the forms should be cleared of all rubbish, dirt and grease before the application of any mould oil or releasing agent. All joints and holes should be checked to ensure that they are grout tight.

The distance from the mixer to the formwork should be kept as short as possible to maintain the workability of the mix and to avoid double handling as far as practicable. Care must be taken when placing and compacting the concrete to ensure that the reinforcement is not displaced. The depth of concrete that can be placed in one lift will depend upon the mix and section size. If vibrators are used as the means of compaction this should be continuous during the placing of each batch of concrete until the air expulsion has ceased, and care must be taken, as over-vibrating concrete can cause segregation of the mix.

The striking or removal of formwork should only take place upon instruction from the engineer or agent. The appropriate time at which it is safe to remove formwork can be assessed by tests on cubes taken from a similar batch mixed at the time the concrete was poured and cured under similar conditions. The characteristic cube strength should be 10 N/mm^2 or twice the stress to which the structure will then be submitted, whichever is the greater before striking the formwork. If test cubes are not available, the following table from BS 8110 can be used as a guide where ordinary Portland cement is used.

Location	Surface or air temperature	
	16 °C	7 °C
Vertical formwork	12 hours	18 hours
Slab soffits (props left under)	4 days	6 days
Removal of props	10 days	15 days
Beam soffits (props left under)	10 days	15 days
Removal of props	14 days	21 days

In very cold weather the above minimum periods should be doubled, and when using rapid-hardening Portland cement the above minimum periods can generally be halved.

Formwork must be removed slowly, as the sudden removal of the wedges is equivalent to a shock load being placed upon the partly hardened concrete. Materials and/or plant should not be placed on the partly hardened concrete without the engineer's permission. When the formwork has been removed it should be carefully cleaned to remove any concrete adhering to the face before being reused. If the forms are not required for immediate reuse they should be carefully stored and stacked to avoid twisting.

The method of curing the concrete will depend upon climatic conditions, type of cement used, and the average temperature during the curing period. The objective is to allow the concrete to cure and obtain its strength without undue distortion or cracking. It may be necessary to insulate the concrete by covering with polythene sheeting or an absorbent material that is kept damp to control the surface temperature and prevent the evaporation of water from the surface. Under normal conditions, using ordinary Portland cement and with an average air temperature of over 10 °C, this period would be 2 days, rising to 4 days during hot weather and days with prolonged drying winds.

10.4

PRECAST CONCRETE FRAMES

The overall concept of a precast concrete frame is the same as for any other framing material. Single- or multi-storey frames can be produced on the skeleton or box frame principle. Single- and two-storey buildings can also be produced as portal frames. Most precast concrete frames are produced as part of a 'system' building, and therefore it is only possible to generalise in an overall study of this method of framing.

Advantages
- Mixing, placing and curing of the concrete carried out under factory-controlled conditions, which results in uniform and accurate units. The casting, being an 'off-site' activity, will release site space that would have been needed for the storage of cement and aggregates, mixing position, timber store and fabrication area for formwork and the storage, bending and fabrication of the reinforcement.
- Repetitive standard units reduce costs: it must be appreciated that the moulds used in precast concrete factories are precision made, resulting in high capital costs. These costs must be apportioned over the number of units to be cast.
- Frames can be assembled on site in cold weather, which helps with the planning, programming and progressing of the building operations. This is important to the contractor, because delays can result in the monetary penalty clauses, for late completion of the contract, being invoked.
- In general the frames can be assembled by semi-skilled labour. With the high turnover rate of labour within the building industry, operatives can be recruited and quickly trained to carry out these activities.

Disadvantages
- System building is less flexible in its design concept than purpose-made structures. Note that there is a wide variety of choice of systems available to the designer, so most design briefs can be fulfilled without too much modification to the original concept.

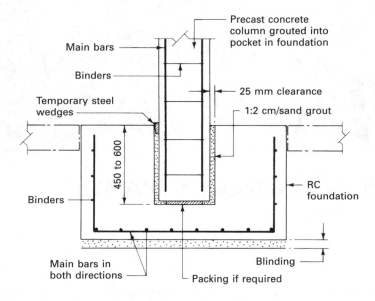

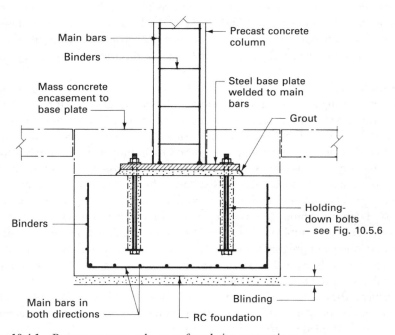

Figure 10.4.1   Precast concrete column to foundation connections.

- Mechanical lifting plant will be needed to position the units; this can add to the overall contracting costs, as generally larger plant is required for precast concrete structures than for *in-situ* concrete structures.
- Programming may be restricted by controls on delivery and unloading times laid down by the police. Restrictions on deliveries is a point that must be established at the tender period so that the tender programme can be formulated with a degree of accuracy and the cost of any special requirements can be included in the unit rates for pricing.
- Structural connections between the precast concrete units can present both design and contractual problems. The major points to be considered are protection against weather, fire and corrosion, appearance and the method of construction. The last named should be issued as an instruction to site, setting out in detail the sequence, temporary supports required and full details of the joint.

■ ■ ■ METHODS OF CONNECTIONS

FOUNDATION CONNECTIONS

Precast columns are connected to their foundations by one of two methods, depending mainly upon the magnitude of the load. For light and medium loads the foot of the column can be placed in a pocket left in the foundation. The column can be plumbed and positioned by fixing a collar around its perimeter and temporarily supporting the column from this collar by using raking adjustable props. Wedges can be used to give added rigidity while the column is being grouted into the pocket (see Fig. 10.4.1). The alternative method is to cast or weld on a base plate to the foot of the column and use holding-down bolts to secure the column to its foundation in the same manner as described in detail for structural steelwork (see Fig. 10.4.1).

COLUMN CONNECTIONS

The main principle involved in making column connections is to ensure continuity, and this can be achieved by a variety of methods. In simple connections a direct bearing and grouted dowel joint can be used, the dowel being positioned in the upper or lower column. Where continuity of reinforcement is required, the reinforcement from both upper and lower columns is left exposed and either lapped or welded together before completing the connection with *in-situ* concrete. A more complex method is to use a stud and plate connection, in which one set of threaded bars are connected through a steel plate welded to a set of bars projecting from the lower column; again the connection is completed with *in-situ* concrete. Typical column connections are shown in Fig. 10.4.2. Column connections should be made at floor levels but above the beam connections, a common dimension being 600 mm above structural floor level. The columns can be of single- or multi-storey height, the latter having provisions for beam connections at the intermediate floor levels.

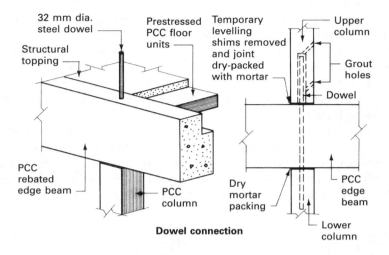

Dowel connection

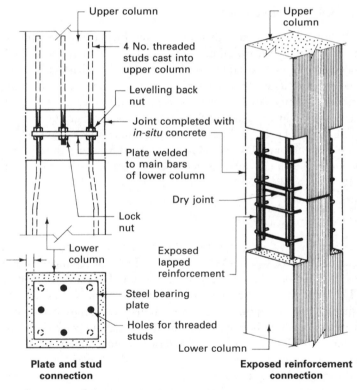

Plate and stud connection

Exposed reinforcement connection

Figure 10.4.2   Precast concrete column connections.

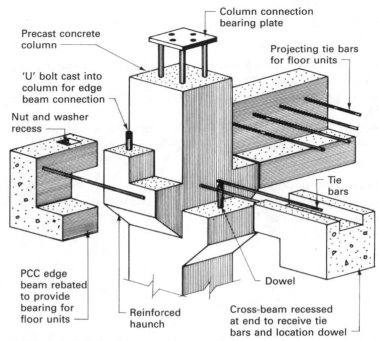

Column connection
bearing plate

Precast concrete
column

Projecting tie bars
for floor units

'U' bolt cast into
column for edge
beam connection

Nut and washer
recess

Tie
bars

PCC edge
beam rebated
to provide
bearing for
floor units

Reinforced
haunch

Dowel

Cross-beam recessed
at end to receive tie
bars and location dowel

Note: Beam recess filled with cement grout to complete connection

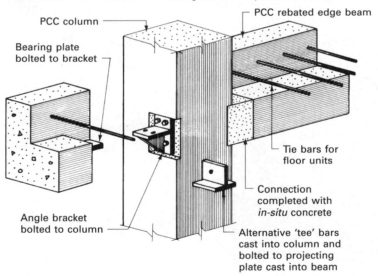

PCC column

PCC rebated edge beam

Bearing plate
bolted to bracket

Tie bars for
floor units

Connection
completed with
in-situ concrete

Angle bracket
bolted to column

Alternative 'tee' bars
cast into column and
bolted to projecting
plate cast into beam

Figure 10.4.3   Typical precast concrete beam connections.

BEAM CONNECTIONS

As with columns, the main emphasis is on continuity within the joint. Three basic methods are used:

- A projecting concrete haunch is cast onto the column with a locating dowel or stud bolt to fix the beam.
- A projecting metal corbel is fixed to the column, and the beam is bolted to the corbel.
- Column and beam reinforcement, generally in the form of hooks, are left exposed. The two members are hooked together and covered with *in-situ* concrete to complete the joint.

With most beam-to-column connections lateral restraint is provided by leaving projecting reinforcement from the beam sides to bond with the floor slab or precast concrete floor units (see Fig. 10.4.3).

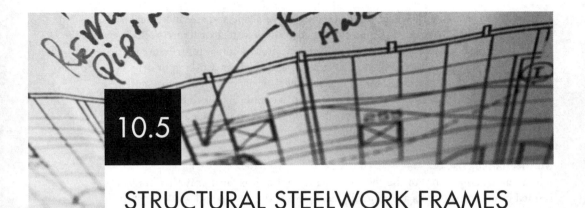

STRUCTURAL STEELWORK FRAMES

Structural steel, as a means of constructing a framed building, has been used since the beginning of the twentieth century, and was the major structural material used until the advent of the Second World War, which led to a shortage of the raw material. This shortage led to an increase in the use of *in-situ* and precast concrete frames. Today both systems are used, and this means that a comparison must be made before any particular framing medium is chosen. The main factors to be considered in making this choice are:

- site costs;
- construction costs;
- maintenance costs.

Site costs
A building owner will want to obtain a financial return on his or her capital investment as soon as possible: therefore speed of construction is of paramount importance. The use of a steel or precast concrete frame will permit the maximum amount of prefabrication off site, during which time the general contractor can be constructing the foundations in preparation for the erection of the frame. To obtain the maximum utilisation of a site the structure needs to be designed so that the maximum amount of rentable floor area is achieved. Generally, prefabricated section sizes are smaller than comparable *in-situ* concrete members, mainly because of the greater control over manufacture obtainable under factory conditions, and thus these will occupy less floor area.

Construction costs
The main factors are design considerations, availability of labour, availability of materials and site conditions. Concrete is an adaptable material, which allows the designer to be more creative than when working within the rigid confines of

standard steel sections. However, as the complexity of shape and size increases so too does the cost of formwork, and for the erection of a steel structure skilled labour is required, whereas activities involved with precast concrete structures can be carried out by the more readily available semi-skilled labour working under the direction of a competent person. The availability of materials fluctuates, and only a study of current market trends can give an accurate answer to this problem. Site conditions regarding storage space, fabrication areas and manoeuvrability around and over the site can well influence the framing method chosen.

Maintenance costs

These can be considered in the short or long term, but it is fair to say that in most framed buildings the costs are generally negligible if the design and workmanship are sound. Steelwork, because of its corrosion properties, will need some form of protective treatment, but as most steel structures have to be given a degree of fire resistance the fire protective method may well perform the dual function.

■■■ STRUCTURAL STEEL FRAMES

The design, fabrication, supply and erection of a structural steel frame are normally placed in the hands of a specialist subcontractor. The main contractor's task is to provide the foundation bases in the correct positions and to the correct levels with the necessary holding-down fixing bolts. The designer will calculate the loadings, stresses and reactions in the same manner as for reinforced concrete and then select a standard steel member whose section properties meet the design requirements.

Standard hot-rolled sections are given in BS 4 and BS EN 10067 and in the *Handbook of Structural Steelwork* published jointly by the British Constructional Steelwork Association and the Steel Construction Institute, which provide the following section types:

- **Universal beams**   These are a range of sections supplied with tapered or parallel flanges, and are designated by their serial size × mass in kilograms per metre run. To facilitate the rolling operation of universal beam sections the inner profile is a constant dimension for any given serial size. The serial size is therefore only an approximate width and breadth, and is given in millimetres.
- **Joists**   These are a range of small-size beams that have tapered flanges and are useful for lintels and small frames around openings. In the case of joists the serial size is the overall nominal dimension.
- **Universal columns**   These members are rolled with parallel flanges, and are designated in the same manner as universal beams. It is possible to design a column section to act as a beam and conversely a beam section to act as a column.
- **Channels**   These are rolled with tapered flanges and designated by their nominal overall dimension × mass per metre run. They can be used for trimming and bracing members or as a substitute for joist sections.

- **Angles** These are light framing and bracing sections with parallel flanges. The flange or leg lengths can be equal or unequal, and the sections are designated by the nominal overall leg lengths × nominal thickness of the flange.
- **T bars** These are used for the same purposes as angles, and are available as rolled sections with a short or long stalk; alternatively, they can be cut from a standard universal beam or column section. Designation is given by the nominal overall breadth and depth × mass per metre run.

Typical standard steel sections are shown in Fig. 10.5.1.

■■■ COLD-ROLLED SECTIONS

Cold-rolled steel sections to BS EN 10162 have become a popular alternative. They make more effective use of steel to permit selection of profiles more appropriate to the design loading. They are particularly useful in lightweight situations where standard hot-rolled sections would be excessive. These include prefabricated construction of modular units for 'slotting' into the main structural frame and for other factory-made units such as site huts.

Development of Zed and Sigma sections have proved very successful alternatives to the traditional use of steel angles for truss purlins. They have many advantages, for example:

- They are more cost-effective per unit of weight.
- Deck fixings are shorter and simpler to locate around the upper flange.
- A shear centre is contained within the section to provide greater resistance to twisting.

Several examples of standard cold-rolled sections are shown in Fig. 10.5.2, with Zed and Sigma profiles in Fig. 10.5.3.

■■■ CASTELLATED UNIVERSAL SECTIONS

These are formed by flame-cutting a standard universal beam or column section along a castellated line; the two halves so produced are welded together to form an open web beam. The resultant section is one and a half times the depth of the section from which it was cut (see Fig. 10.5.4). This increase in depth gives greater resistance to deflection without adding extra weight, but will reduce the clear headroom under the beams unless the overall height of the building is increased. Castellated sections are economical when used to support lightly loaded floor or roof slabs, and the voids in the web can be used for housing services. With this form of beam the shear stresses at the supports can be greater than the resistance provided by the web; in these cases one or two voids are filled in by welding metal blanks into the voids.

■■■ CONNECTIONS

Connections in structural steelwork are classified as either shop connections or site connections, and can be made by using bolts, rivets or by welding.

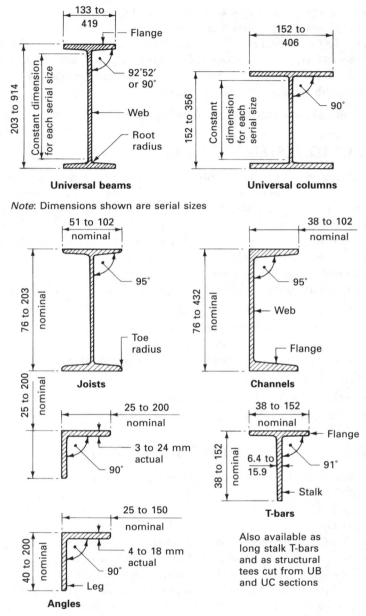

Figure 10.5.1   Typical hot–rolled steel sections.

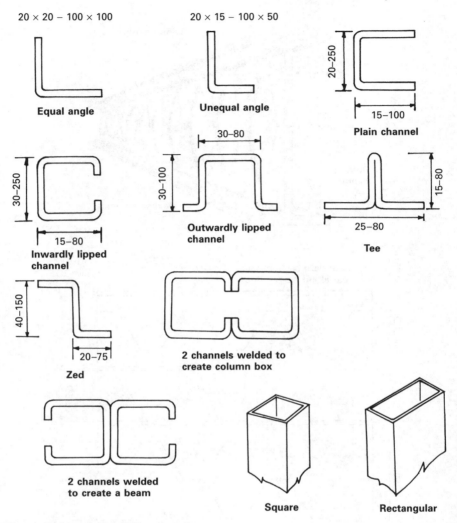

Figure 10.5.2   Cold-rolled steel sections.

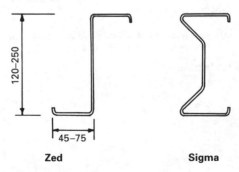

Figure 10.5.3   Cold-rolled steel purlins.

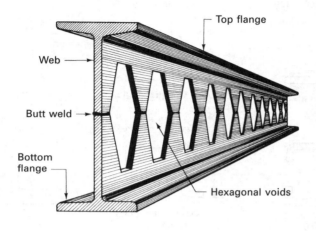

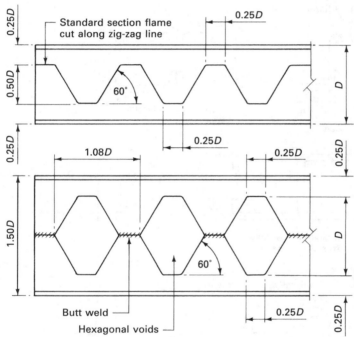

Note: Castellated joists, universal columns and zed sections also available

Figure 10.5.4   Castellated beams.

BOLTS

Black bolts

These are the cheapest form of bolt available; the black bolt can be either hot or cold forged, the thread being machined onto the shank. The allowable shear stresses for this type of bolt are low, and therefore they should be used only for end connections of secondary beams or in conjunction with a seating cleat that has been designed and fixed to resist all the shear forces involved. The clearance in the hole for this form of bolt is usually specified as 1.6 mm over the diameter of the bolt. The term 'black bolts' does not necessarily indicate the colour but is the term used to indicate the comparatively wide tolerances to which these products are usually made. Diameters range from 5 to 68 mm inclusive.

Bright bolts

These have a machined shank and are therefore of greater dimensional accuracy, fitting into a hole with a small clearance allowance. The stresses allowed are similar to those permitted for rivets. Bright bolts are sometimes called turned and fitted bolts.

High-strength friction bolts

These are manufactured from high-tensile steels and are used in conjunction with high-tensile steel nuts and tempered washers. These bolts have generally replaced rivets and black bolts for both shop and site connections as fewer bolts are needed and hence the connection size is reduced. The object of this form of bolt is to tighten it to a predetermined shank tension in order that the clamping force thus provided will transfer the loads in the connecting members by friction between the parts and not by shear in, or bearing on, the bolts. Generally a torque-controlled spanner or pneumatic impact wrench is used for tightening; other variations to ensure the correct torque are visual indicators such as a series of pips under the head or washer, which are flattened when the correct amount of shank tension has been reached. Nominal standard diameters available are from 12 to 36 mm with lengths ranging from 40 to 500 mm, as recommended in BS 4395-1 and 2.

General

The holes to receive bolts should always be drilled in a position known as the **back mark** of the section. The back mark is the position on the flange where the removal of material to form a bolt or rivet hole will have the least effect upon the section properties. Actual dimensions and recommended bolt diameters are given in the *Handbook of Structural Steelwork*.

RIVETS

Made from mild steel to the recommendations of BS 4620, rivets have been generally superseded by bolted and welded connections for structural steel frames.

Rivets are available as either cold or hot forged, with a variety of head shapes ranging from an almost semicircular or snap head to a countersunk head for use when the projection of a snap, universal or flat head would create an obstruction. Small-diameter rivets can be cold driven, but the usual practice is to drive rivets while they are hot. Rivets, like bolts, should be positioned on the back mark of the section; typical spacings are $2\frac{1}{2}$ diameters centre to centre and $1\frac{3}{4}$ diameters from the end or edge to the centre line of the first rivet.

WELDING

Welding is considered primarily as a shop connection, as the cost, together with the need for inspection – which can be difficult on site – generally makes this method uneconomic for site connections.

   The basic methods of welding are oxyacetylene and electric arc. A blowpipe is used for oxyacetylene, which allows the heat from the burning gas mixture to raise the temperature of the surfaces to be joined. A metal filler rod is held in the flame, and the molten metal from the filler rod fuses the surfaces together. For most structural steel applications, the oxyacetylene method is limited to cutting.

   In the alternative method an electric arc is struck between a metal rod connected to a suitable low-voltage electrical supply and the surface to be joined, which must be earthed or resting on an earthed surface. The heat of the arc causes the electrode

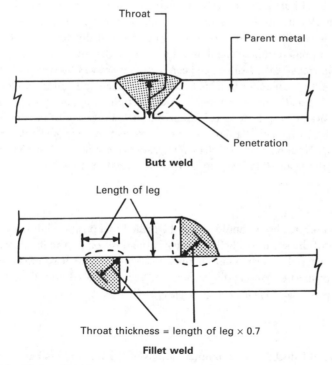

Figure 10.5.5   Types of weld.

or metal rod to melt, and the molten metal can be deposited in layers to fuse the pieces to be joined together. With electric arc welding the temperature rise is confined to the local area being welded, whereas the oxyacetylene method causes a rise in metal temperature over a general area.

Welds are classified as either **fillet** or **butt** welds. Fillet welds are used on the edges and ends of members, and form a triangular fillet of welding material. Butt welds are used on chamfered end-to-end connections (see Fig. 10.5.5).

◼◼◼ STRUCTURAL STEEL CONNECTIONS

BASE CONNECTIONS

These are of one or two forms: the slab or bloom base and the gusset base. In both methods a steel base plate is required to spread the load of the column onto the foundation. The end of the column and the upper surface of the base plate should be machined to give a good interface contact when using a bloom base. The base plate and column can be connected together by using cleats, or by fillet welding (see Fig. 10.5.6).

The gusset base is composed of a number of members that reduce the thickness of the base plate and can be used to transmit a high bending moment to the foundations. A machined interface between column and base plate will enable all the components to work in conjunction with one another, but if this method is not adopted the connections must transmit all the load to the base plate (see Fig. 10.5.6). The base is joined to the foundation by holding–down bolts, which must be designed to resist the uplift and tendency of the column to overturn. The bolt diameter, bolt length and size of plate washer are therefore important. To allow for fixing tolerances the bolts are initially housed in a void or pocket, which is filled with grout at the same time as the base is grouted onto the foundation. To level and plumb the columns, steel wedges are inserted between the underside of the base plate and the top of the foundation (see Fig. 10.5.6).

BEAM-TO-COLUMN CONNECTIONS

These can be designed as simple connections, where the whole of the load is transmitted to the column through a seating cleat. This is an expensive method, requiring heavy sections to overcome deflection problems. The usual method employed is the semi–rigid connection, where the load is transmitted from the beam to the column by means of top cleats and/or web cleats; for ease of assembly an erection cleat on the underside is also included in the connection detail (see Fig. 10.5.7). A fully rigid connection detail, which gives the greatest economy on section sizes, is made by welding the beam to the column (see Fig. 10.5.7). The uppermost beam connection to the column can be made by the methods described above; alternatively, a bearing connection can be used, which consists of a cap plate fixed to the top of the column to which the beams can be fixed, either continuously over the cap plate or with a butt joint (see Fig. 10.5.7).

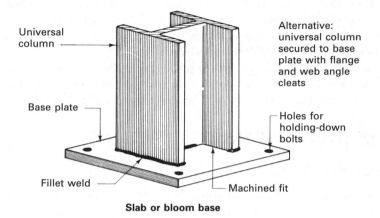

Universal
column

Base plate

Holes for
holding-down
bolts

Alternative:
universal column
secured to base
plate with flange
and web angle
cleats

Fillet weld

Machined fit

Slab or bloom base

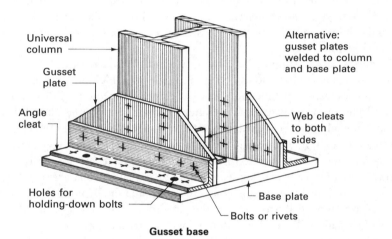

Universal
column

Gusset
plate

Angle
cleat

Alternative:
gusset plates
welded to column
and base plate

Web cleats
to both
sides

Holes for
holding-down bolts

Base plate

Bolts or rivets

Gusset base

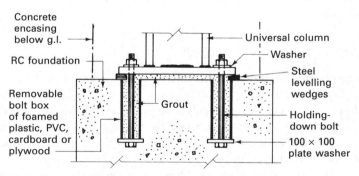

Concrete
encasing
below g.l.

RC foundation

Removable
bolt box
of foamed
plastic, PVC,
cardboard or
plywood

Grout

Universal column

Washer

Steel
levelling
wedges

Holding-
down bolt

100 × 100
plate washer

Figure 10.5.6   Structural steel column bases.

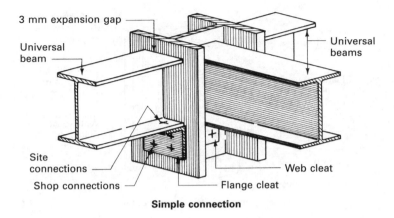

3 mm expansion gap

Universal beam

Universal beams

Site connections

Shop connections

Web cleat

Flange cleat

Simple connection

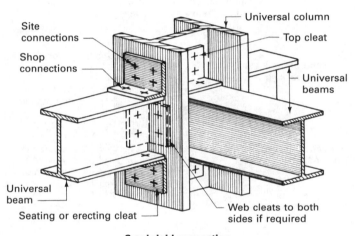

Site connections

Universal column

Top cleat

Shop connections

Universal beams

Universal beam

Seating or erecting cleat

Web cleats to both sides if required

Semi-rigid connection

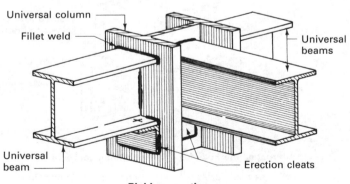

Universal column

Fillet weld

Universal beams

Universal beam

Erection cleats

Rigid connection

Figure 10.5.7   Structural steel beam to column connections.

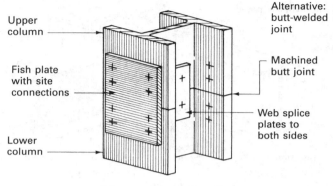

Upper column

Fish plate with site connections

Lower column

Alternative: butt-welded joint

Machined butt joint

Web splice plates to both sides

Columns with equal sections

Note: For columns of same serial size but of different sections splice is made using 4 No. fish plates fixed on the inside of flanges

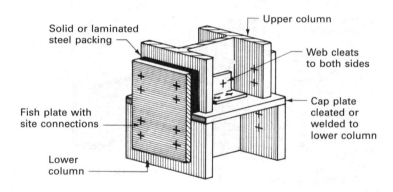

Solid or laminated steel packing

Upper column

Web cleats to both sides

Fish plate with site connections

Cap plate cleated or welded to lower column

Lower column

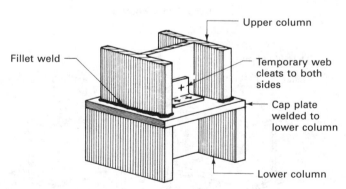

Upper column

Fillet weld

Temporary web cleats to both sides

Cap plate welded to lower column

Lower column

Alternative methods for columns of unequal sections

Note: Splices made at floor level but above beam connections

Figure 10.5.8   Structural steel column splices.

COLUMN SPLICES

These are made at floor levels but above the beam connections. The method used will depend upon the relative column sections (see Fig. 10.5.8).

BEAM-TO-BEAM CONNECTIONS

The method used will depend upon the relative depths of the beams concerned. Deep beams receiving small secondary beams can have a shelf angle connection, whereas other depths will need to be connected by web cleats (see Fig. 10.5.9).

■■■ FRAME ERECTION

This operation will not normally be commenced until all the bases have been cast and checked, because the structural steelwork contractor will need a clear site for manoeuvring the steel members into position. The usual procedure is to erect two storeys of steelwork before final plumbing and levelling take place.

The grouting of the base plates and holding-down bolts is usually left until the whole structure has been finally levelled and plumbed. The grout is a neat cement or cement/sand mixture, depending on the gap to be filled:

- 12–25 mm gap – stiff mix of neat cement;
- 25–50 mm gap – fluid mix of 1:2 cement/sand and tamped;
- Over 50 mm gap – stiff mix of 1:2 cement/sand and rammed.

With large base plates a grouting hole is sometimes included, but with smaller plates three sides of the base plate are sealed with puddle clay, bricks or formwork and the grout is introduced through the open edge on the fourth side. To protect the base from corrosion it should be encased with concrete up to the underside of the floor level, giving a minimum concrete cover of 75 mm to all the steel components.

■■■ FIRE PROTECTION OF STEELWORK

The Building Regulations, Approved Document B, gives minimum fire resistance periods for steel structures and components according to a building's purpose group and its top floor height above ground. The traditional method of fire protection is to encase the steel section with concrete. This requires formwork, and will greatly add to the structural dead load. 50 mm cover will provide up to 4 hours' fire protection, whereas the more attractive hollow protection from brickwork will require at least 100 mm thickness for equivalent protection and will occupy more space. A number of lighter-weight and less bulky sheet materials based on plasterboard can be used to 'box' steel columns and beams, but because of their porous nature are limited to interior use. They are also less practical where more than 2 hours' fire protection is required. Some basic examples of solid and hollow protection to steelwork are shown in Fig. 10.5.10. For additional details and specific treatment of the subject of fire protection consult the complementary volume, *Advanced Construction Technology*.

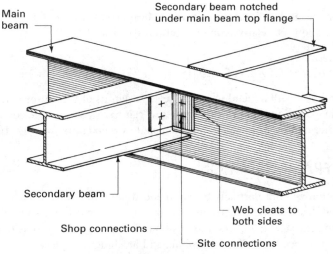

Main beam

Secondary beam notched under main beam top flange

Secondary beam

Shop connections

Web cleats to both sides

Site connections

Beam-to-beam connections

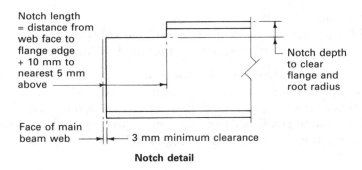

Notch length = distance from web face to flange edge + 10 mm to nearest 5 mm above

Notch depth to clear flange and root radius

Face of main beam web

3 mm minimum clearance

Notch detail

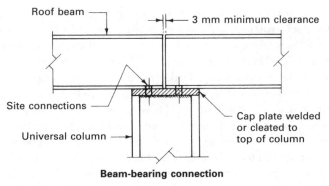

Roof beam

3 mm minimum clearance

Site connections

Universal column

Cap plate welded or cleated to top of column

Beam-bearing connection

Figure 10.5.9   Structural steel beam-to–beam connections.

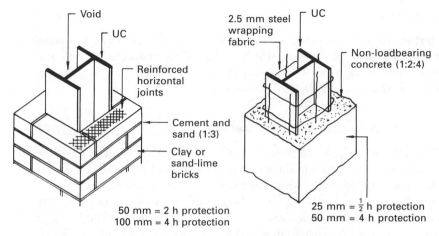

Figure 10.5.10 Fire protection to steel columns.

■■■ PRINCIPLES OF STRUCTURAL STEEL DESIGN

Design of structural steel members can be a complex and specialised exercise. For guidance, this section considers some simple applications to beams and columns.

BEAM DESIGN

Basic design formula:

$M = fZ$
where M = moment of resistance
f = fibre stress
Z = section or elastic modulus

Moment of resistance and bending moment

To appreciate the moment of resistance, it is necessary to understand the terms **couple** and **moment**. A couple occurs in beams where the upward forces (beam supports) oppose and balance the downward forces (beam loads). If they did not balance, the beam would rotate! A moment can therefore be considered as a rotational force. The moment of the balancing couple due to the resisting forces in the beam material is known as the **moment of resistance** at the specific beam section being considered. When a beam has taken up its loads and deflected to a position of equilibrium, its moment of resistance at any section along the beam has to balance the resultant moment, at that section, of all the forces acting on the beam. This resultant moment is commonly known as the **bending moment**, and it will equal the moment of resistance. Both moment of resistance and bending moment are expressed in units of kN m.

Fibre stress

The fibre stress of a material is the safe strength of the material against bending stresses. For steel this property will vary, depending on the quality, type and constituents of the material. A typical figure for rolled mild steel sections is 165 N/mm^2 and stainless steel 210 N/mm^2.

Section modulus and moment of inertia

The section modulus is otherwise known as the **elastic modulus**. The section modulus is an expression of the properties of a beam by geometric profile, derived from moment of inertia or second moment of area calculations. Steel beams are generally irregular, i.e. 'I' shaped, making calculation of the section modulus and its sources a tedious exercise. Fortunately, both section modulus and moment of inertia figures are available from steel design tables (see Table 10.5.1). Section modulus is expressed in cm^3 and moment of inertia in cm^4 about the x–x or y–y axis, as depicted in Fig. 10.5.13. Beams are normally used with the x–x axis horizontal. In simple loading situations the central axis is usually where compressive and tensile forces are at zero. It is known as the **neutral axis**.

Table 10.5.1   Universal beam (UB) design data

Serial size (mm)	Mass per m. (kg)	Section depth (mm)	Section width (mm)	Thickness Web (mm)	Thickness Flange (mm)	Root radius (mm)	Section area (cm^2)
305 × 165	54	310.9	166.8	7.7	13.7	8.9	68.4
	46	307.1	165.7	6.7	11.8	8.9	58.9
	40	303.8	165.1	6.1	10.2	8.9	51.5
305 × 127	48	310.4	125.2	8.9	14.0	8.9	60.8
	42	306.6	124.3	8.0	12.1	8.9	53.2
	37	303.8	123.5	7.2	10.7	8.9	47.5

Serial size (mm)	Moment of inertia (x–x) (cm^4)	Moment of inertia (y–y) (cm^4)	Radius of gyration (x–x) (cm)	Radius of gyration (y–y) (cm)	Section modulus (x–x) (cm^3)	Section modulus (y–y) (cm^3)
305 × 165	11 710	1061	13.09	3.94	753.3	127.3
	9 948	897	13.00	3.90	647.9	108.3
	8 523	763	12.86	3.85	561.2	92.4
305 × 127	9 504	460	12.50	2.75	612.4	73.5
	8 143	388	12.37	2.70	531.2	62.5
	7 162	337	12.28	2.67	471.5	54.6

Bending moment calculations

For simple spans between two opposing supports, the bending moment (BM) can be calculated by the standard formulae shown in Fig. 10.5.11. Fig. 10.5.12 shows an application of bending moment and moment of resistance (*M*) calculations.

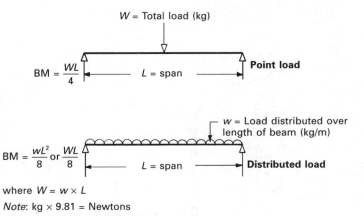

$$BM = \frac{WL}{4}$$

W = Total load (kg)

L = span

Point load

$$BM = \frac{wL^2}{8} \text{ or } \frac{WL}{8}$$

w = Load distributed over length of beam (kg/m)

L = span

Distributed load

where *W* = *w* × *L*

Note: kg × 9.81 = Newtons

Figure 10.5.11 Bending moment formulae.

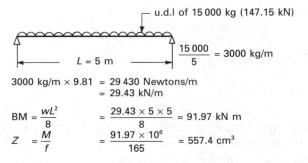

u.d.l of 15 000 kg (147.15 kN)

L = 5 m

$$\frac{15\,000}{5} = 3000 \text{ kg/m}$$

3000 kg/m × 9.81 = 29 430 Newtons/m
= 29.43 kN/m

$$BM = \frac{wL^2}{8} = \frac{29.43 \times 5 \times 5}{8} = 91.97 \text{ kN m}$$

$$Z = \frac{M}{f} = \frac{91.97 \times 10^6}{165} = 557.4 \text{ cm}^3$$

Figure 10.5.12 Bending moment calculation.

From design tables based on BS 4: *Structural steel sections* it is possible to find a suitable beam. The extract (Table 10.5.1) provides a choice of two universal beams:

305 mm × 165 mm @ 40 kg/m (561.2 cm³ section modulus), or
305 mm × 127 mm @ 48 kg/m (612.4 cm³ section modulus).

The properties of the selected 305 mm × 165 mm section @ 40 kg/m are shown in Fig. 10.5.13.

Check beam resistance to shearing

Maximum shear normally occurs towards each end of a beam. The shear stress is calculated on the web sectional area of 1853.18 mm², relative to the total load at each end of the beam:

147.15 kN (see Fig. 10.5.12)/2 = 73.58 kN

Therefore, the average shear stress = shear force/web sectional area
$$= 73.58 \times 10^3/1853.18$$
$$= 39.7 \text{ N/mm}^2$$

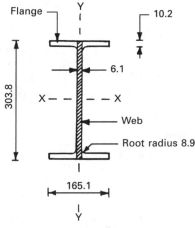

Web sectional area shown shaded
= 303.8 × 6.1 = 1853.18 mm²

Figure 10.5.13   Universal beam – 305 mm × 165 mm @ 40 kg/m.

With reference to BS 449-2: *Specification for the use of structural steel in building*, grade 43 steel (see note) has an allowable shear stress of 100 N/mm². With an actual shear stress of only 39.7 N/mm², the chosen beam section of 305 mm × 165 mm @ 40 kg/m is quite adequate.

[*Note*: See BS 7668, BS EN 10029 and BS EN 10113 for grades of steel suitable for structural use.]

Check beam for limits of deflection

The maximum limit of deflection of steel beams is expressed as 1/360 of the span. Deflection in excess of this is likely to affect the integrity of attached finishes and fixings. The formulae for calculating deflection of simply loaded and supported beams are shown in Fig. 10.5.14.

Continuing with the previous selected beam example, and substituting figures into the distributed load formula:

$$\text{Deflection} = \frac{5 \times 147.15 \times (5)^3}{384 \times 21\,000 \times 8523} = 1.33 \text{ cm or } 13.3 \text{ mm}$$

Max. deflection = $1/360 \times 5$ m span = 1.38 cm or 13.8 mm
Therefore a deflection of 13.3 mm is just acceptable.

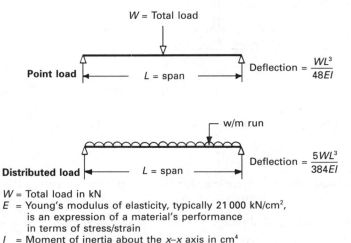

W = Total load in kN
E = Young's modulus of elasticity, typically 21 000 kN/cm², is an expression of a material's performance in terms of stress/strain
I = Moment of inertia about the *x*–*x* axis in cm⁴

Figure 10.5.14 Beam deflection formulae.

COLUMN DESIGN

Columns are designed with regard to their tendency to buckle or bend under load. Design factors to be considered relative to the loading are:

■ column effective length (see Fig. 10.2.12);
■ cross-sectional area of the steel column;
■ profile or shape of steel section.

Cross-sectional area and section shape are combined in a property known as the **radius of gyration**. It can be calculated as follows:

$r = \sqrt{I/A}$
where r = radius of gyration (cm)
        I = Moment of inertia about the *y*–*y* axis (cm⁴)
        A = Cross sectional area (cm²)

Conveniently, figures for r can be taken from design tables. The *y*–*y* axis is used for column calculations, as this is normally the weaker axis.

For example: a universal column of 254 mm × 254 mm @ 107 kg/m. Data for this size of column are shown in Table 10.5.2 and illustrated in Fig. 10.5.15.

The UC is retained top and bottom: therefore position and direction are fixed, giving an effective length of $0.7 \times L$. If L, the actual length, is 8 m, then the effective length is 0.7×8 m = 5.6 m (5600 mm). From Table 10.5.2 the radius of gyration about the *y*–*y* axis is 6.57 cm or 65.7 mm.

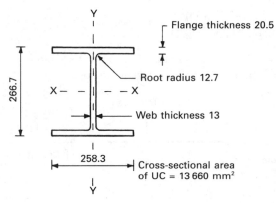

Figure 10.5.15   Universal column – 254 mm × 254 mm @ 107 kg/m.

Slenderness ratio = effective length/radius of gyration
= 5600/65.7 = 85.23

Allowable stresses for axial compression for various grades of steel are tabulated in BS 449-2. For grade 43 steel with a slenderness ratio of 85.23, the allowable stress is shown as 97 N/mm².

Cross-sectional area of UC (see Table 10.5.2) =136.6 cm² or 13 660 mm².

Total axial load = 97 × 13 660 = 1 325 000 N or 1325 kN
Approximately 135 tonnes

Table 10.5.2   Universal column (UC) design data

| Serial size (mm) | Mass per m. (kg) | Section depth (mm) | Section width (mm) | Thickness | | Root radius (mm) | Section area (cm²) |
				Web (mm)	Flange (mm)		
254 × 254	167	289.1	264.5	19.2	31.7	12.7	212.4
	132	276.4	261.0	15.6	25.1	12.7	167.7
	107	266.7	258.3	13.0	20.5	12.7	136.6
	89	260.4	255.9	10.5	17.3	12.7	114.0
	73	254.0	254.0	8.6	14.2	12.7	92.9

| Serial size (mm) | Moment of inertia | | Radius of gyration | | Section modulus | |
	x–x (cm⁴)	y–y (cm⁴)	x–x (cm)	y–y (cm)	x–x (cm³)	y–y (cm³)
254 × 254	29 914	9796	11.9	6.79	2070	740.6
	22 416	7444	11.6	6.66	1622	570.4
	17 510	5901	11.3	6.57	1313	456.9
	14 307	4849	11.2	6.52	1099	378.9
	11 360	3873	11.1	6.46	894	305.0

10.6

CLADDINGS

Claddings to buildings can be considered under two classifications:

- claddings fixed to a structural backing;
- claddings to framed structures.

▨ ■ ■ CLADDINGS FIXED TO A STRUCTURAL BACKING

Materials used in this form of cladding are generally considered to be small-unit claddings, and are applied for one of two reasons. If the structural wall is unable to provide an adequate barrier to the elements, a covering of small-unit claddings will generally raise the wall's resistance to an acceptable level. Alternatively, small-unit claddings can be used solely as a decorative feature, possibly to break up the monotony of a large plain area composed of a single material.

The materials used are tiles, slates, shingles, timber boarding, plastic boards and stone facings. The general method of fixing these small units is to secure them to timber battens fixed to the structure backing. Stone and similar facings, however, are usually secured by special mechanical fixings, as described later when considering claddings to framed structures.

TILE HANGING

The tiles used in the hanging can be ordinary plain roofing tiles, or a tile of the same dimensions but having a patterned bottom edge solely for a decorative appearance. The tiles are hung and fixed to tiling battens, although nibless tiles fixed directly to the backing wall are sometimes used (see Fig. 10.6.1). The battens should be impregnated to prevent fungal and insect attack, so that their anticipated life is comparable to that of the tiles. Each tile should be twice nailed to its support batten with corrosion-resistant nails of adequate length.

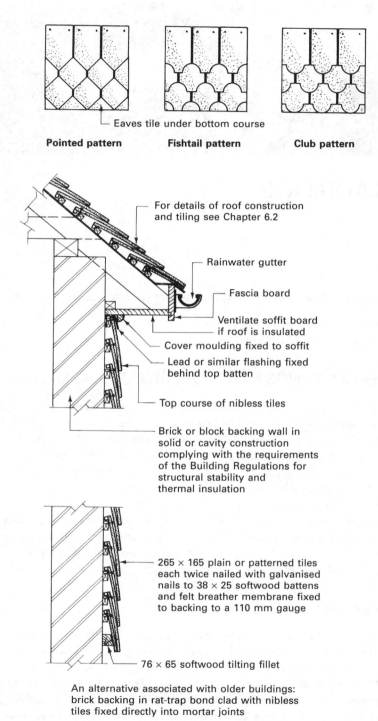

Pointed pattern   **Fishtail pattern**   **Club pattern**

Eaves tile under bottom course

For details of roof construction and tiling see Chapter 6.2

Rainwater gutter

Fascia board

Ventilate soffit board if roof is insulated

Cover moulding fixed to soffit

Lead or similar flashing fixed behind top batten

Top course of nibless tiles

Brick or block backing wall in solid or cavity construction complying with the requirements of the Building Regulations for structural stability and thermal insulation

265 × 165 plain or patterned tiles each twice nailed with galvanised nails to 38 × 25 softwood battens and felt breather membrane fixed to backing to a 110 mm gauge

76 × 65 softwood tilting fillet

An alternative associated with older buildings: brick backing in rat-trap bond clad with nibless tiles fixed directly into mortar joints

Figure 10.6.1   Vertical tile hanging – typical details 1.

The general principles of tile hanging are similar to those of double-lap roof tiling, and the gauge is calculated in the same manner. The minimum lap recommended is 40 mm, which would give a gauge of 112.5 mm using a standard 265 mm long tile.

A gauge dimension of 112.5 mm is impracticable, and therefore a gauge of 110 mm would be usual. Typical details of top edge finishes, bottom edge finishes, corners and finishes at windows are shown in Figs 10.6.1 and 10.6.2. Note that, if the structural backing is of timber framing, an impervious tiling felt breather membrane should be placed over the framing immediately underneath the battens to prevent any moisture that is blown in between the tiles from having adverse effects upon the structure. As a matter of course, it is good practice to underlay the tile battens with a breather membrane, whatever the wall structure material. In this situation building paper is not considered to be a suitable substitute. The application of slates as a small-unit hung cladding follows the principles outlined above for tile hanging.

TIMBER CLADDINGS

Timber claddings are usually in the form of moulded or shaped boards fixed to battens as either a horizontal or vertical cladding; typical examples are shown in Fig. 10.6.3. Timber claddings will require regular maintenance to preserve their resistance to the elements. Softwoods are generally painted, and will need repainting at intervals of three to five years according to the exposure. Hardwoods are sometimes treated with a preservative and left to bleach naturally; the preservative treatment needs to be carried out at two- to five-year intervals. Western red cedar is a very popular wood for timber cladding, as it has a natural immunity to insect and fungal attack under normal conditions. It also has a pleasing natural red/brown colour, which can be maintained if the timber is coated with a clear sealer such as polyurethane; however, it will bleach to a grey/white colour if exposed to the atmosphere. Plastic boards are a substitute for timber and are fixed in a similar manner.

■ ■ ■ CLADDINGS TO FRAME STRUCTURES

The methods available to clad a frame structure are extensive and include panels of masonry constructed between the columns and beams, light infill panels of metal or timber, precast concrete panels, and curtain walling that completely encloses the structure. This volume includes panel walls and the facings that can be attached to them. The other forms of cladding are considered in the complementary volume, *Advanced Construction Technology*.

BRICK PANEL WALLS

These are non-loadbearing walls, which must fulfil the following requirements:

■ adequate resistance to the elements;
■ sufficient strength to support their own weight plus any attached finishes;

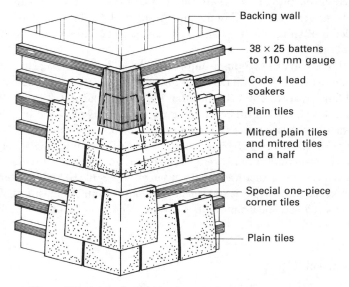

Alternative external angle treatments
(internal angles treatments similar)

- Backing wall
- 38 × 25 battens to 110 mm gauge
- Code 4 lead soakers
- Plain tiles
- Mitred plain tiles and mitred tiles and a half
- Special one-piece corner tiles
- Plain tiles

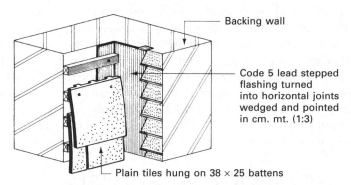

Typical abutment detail

- Backing wall
- Code 5 lead stepped flashing turned into horizontal joints wedged and pointed in cm. mt. (1:3)
- Plain tiles hung on 38 × 25 battens

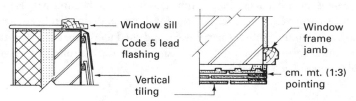

Typical opening details

- Window sill
- Code 5 lead flashing
- Vertical tiling
- Window frame jamb
- cm. mt. (1:3) pointing

Figure 10.6.2   Vertical tile hanging – typical details 2.

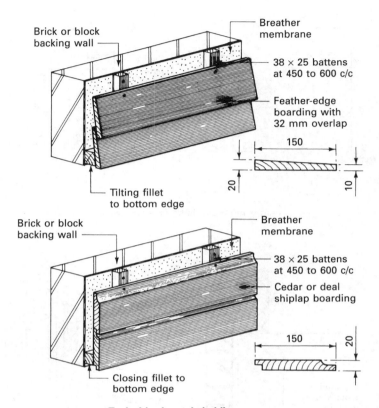

Brick or block backing wall

Breather membrane

38 × 25 battens at 450 to 600 c/c

Feather-edge boarding with 32 mm overlap

150

20

10

Tilting fillet to bottom edge

Brick or block backing wall

Breather membrane

38 × 25 battens at 450 to 600 c/c

Cedar or deal shiplap boarding

150

20

Closing fillet to bottom edge

Typical horizontal claddings

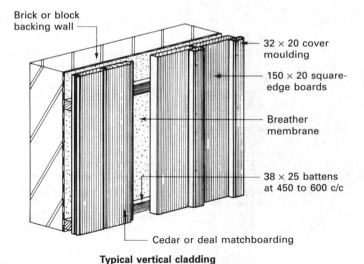

Brick or block backing wall

32 × 20 cover moulding

150 × 20 square-edge boards

Breather membrane

38 × 25 battens at 450 to 600 c/c

Cedar or deal matchboarding

Typical vertical cladding

Figure 10.6.3 Timber wall claddings.

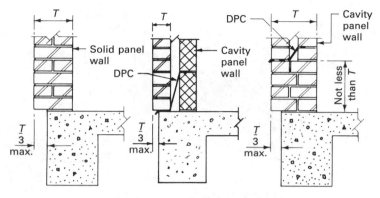

Maximum overhang for panel walls

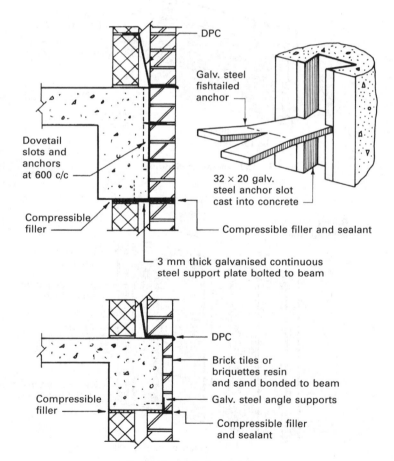

Figure 10.6.4   Brick panel walls.

- sufficient strength to resist both positive and negative wind pressures;
- provision of the required thermal and sound insulation;
- provision of the required fire resistance;
- adequate durability.

Brick panel walls are constructed in the same manner as ordinary solid or cavity walls, and any openings for windows or doors are formed by traditional methods. The panels must be supported at each structural floor level and tied to the structure at the vertical edges. Projection of the panel in front of the structural members is permissible, provided such overhangs do not impair the stability of the panel wall; acceptable limits are shown in Fig. 10.6.4. The top edge of the panels should not be pinned rigidly to the frame, because the effect of brick panel expansion together with frame shrinkage may cause cracking and failure of the brickwork. A compression joint should therefore be formed between the top edge of the panel and the underside of the framing member at each floor level (see Fig. 10.6.4).

Two methods of tying the panel to the vertical structural members are in common use:

- Butterfly wall tiles are cast into the column and built into the brick joints at four-course intervals.
- Galvanised pressed steel dovetail slots are cast into the column, and dovetail anchors are used to form the tie (see Fig. 10.6.4).

The second method gives greater flexibility with the location and insertion of adequate ties, but is higher in cost.

Location of masonry and its interaction with the structural frame may also be achieved with the support framework shown in Fig. 10.6.5. A galvanised steel dovetailed profile is cast into the beam face to create a recess for fixings. It supports a continuous steel angle, which in turn supports the outer leaf of masonry from the edge beam. A sliding anchor system is also bolted to the recess to retain both elements of the cavity wall.

Facings to brick panel walls

Any panel wall must have an acceptable and durable finish: this can be achieved by using facing bricks with a neat pointed joint, or by attaching to the face of a panel of common bricks a stone or similar cladding. Suitable materials are natural stone, artificial stone, reconstructed stone, and precast concrete of small units up to 1 m^2 and with a thickness related to the density of the material. Dense materials such as slate and marble need be only 40 mm thick, whereas the softer stones such as sandstone and limestone should be at least 75 mm thick.

Two major considerations must be taken into account when deciding on the method to be used to fix the facings to the brick backing:

- transferring the load to the structure.
- tying back the facing units.

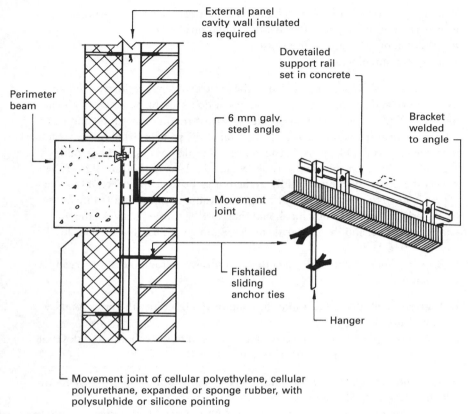

Figure 10.6.5   Masonry cladding support system.

The load of the facings can be transferred by using bonder stones or support corbels at each floor level, which should have a compression joint incorporated in the detail for the same reasons given above when considering brick panels (see Fig. 10.6.6).

The tying back of the facings is carried out by various metal fixing devices called **cramps,** which should be of a non-ferrous metal such as gunmetal, copper, phosphor bronze or stainless steel. To avoid the problem of corrosion caused by galvanic action between dissimilar metals a mixture of fixing materials should not be used. Typical examples of fixings and cramps for thick and thin facings are shown in Fig. 10.6.6.

To provide for plumbing and alignment a bedding space of 12–15 mm should be left between the face of the brick panel and the back of the facing. Dense facings such as marble are usually bedded on a series of cement mortar dabs, whereas the more porous facings are usually placed against a solid bed, which ensures that any saturation that occurs will be uniform over the entire face.

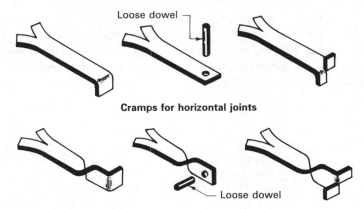

Cramps for horizontal joints

Loose dowel

Half-twist cramps for vertical joints

Loose dowel

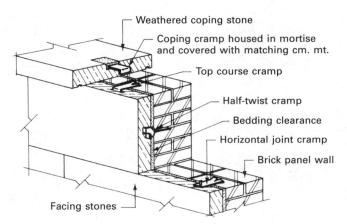

Weathered coping stone

Coping cramp housed in mortise and covered with matching cm. mt.

Top course cramp

Half-twist cramp

Bedding clearance

Horizontal joint cramp

Brick panel wall

Facing stones

Typical cladding fixing details

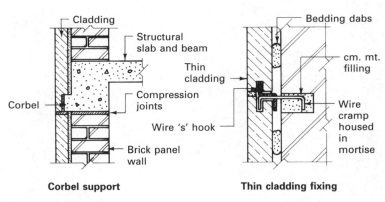

Cladding

Structural slab and beam

Thin cladding

Compression joints

Corbel

Wire 's' hook

Brick panel wall

Bedding dabs

cm. mt. filling

Wire cramp housed in mortise

Corbel support

Thin cladding fixing

Figure 10.6.6 Cladding fixings.

[*Note*: With all forms of masonry cladding to structural frames, due regard must be given to provision of adequate thermal insulation and avoidance of thermal bridging. The details shown in this chapter indicate the principles of various cladding techniques, but care must be taken to ensure that the structure and claddings selected are appropriate to the building's location and function, with regard for local building regulations.]

10.7

STEEL ROOF TRUSSES AND COVERINGS

The function of any roof is to provide a protective covering to the upper surface of the structure. By virtue of its position a roof is subjected to the elements to a greater degree than the walls: therefore the durability of the covering is of paramount importance. The roof structure must have sufficient strength to support its own weight and the load of the coverings, together with any imposed loadings such as snow and wind pressures, without collapse or excessive deflection.

Roofs can be considered as:

- short span;
- medium span;
- long span.

Short span
These are roofs up to 7.000 m, generally of traditional timber construction with a flat or pitched profile. Flat roofs are usually covered with a flexible sheet material, whereas pitched roofs are generally covered with small units such as tiles or slates.

Medium span
These are roofs of 7.000–24.000 m, except where reinforced concrete is used; the usual roof structure for a medium span is a truss or lattice of standard steel sections supporting a deformed sheeting such as corrugated asbestos cement or a structural decking system.

Long span
These are roofs over 24.000 m; they are generally designed by a specialist using girder, space deck or vaulting techniques, and are beyond the scope of a basic technology course.

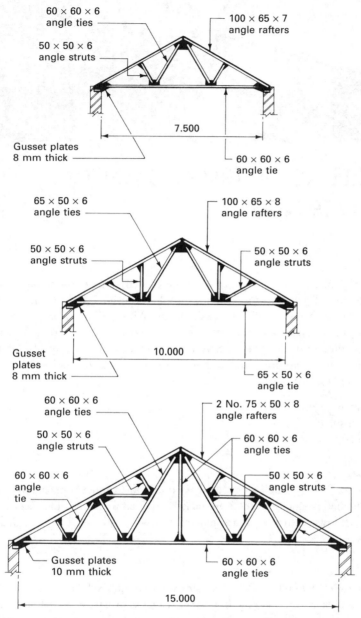

Figure 10.7.1   Typical mild steel angle roof trusses.

■■■ STEEL ROOF TRUSSES

This form of roof structure is used mainly for short- and medium-span single-storey buildings intended for industrial or recreational use. A steel roof truss is a plane frame consisting of a series of rigid triangles composed of compression and tension members. The compression members are called **rafters** and **struts**, and the tension members are termed **ties**. Standard mild steel angles are usually employed as the structural members, and these are connected together, where the centre lines converge, with flat-shaped plates called **gussets**. They can be riveted, bolted or welded together to form a rigid triangulated truss; typical arrangements are shown in Fig. 10.7.1. The internal arrangement of the struts and ties will be governed by the span. The principal or rafter is divided into a number of equal divisions, which locates the intersection point for the centreline of the internal strut or tie.

Angle purlins are used longitudinally to connect the trusses together, and to provide the fixing medium for the roof covering. It is the type of covering chosen that will determine the purlin spacing and the pitch of the truss; ideally the purlins should be positioned over the strut or tie intersection points to avoid setting up local bending stresses in the rafters. Purlins are connected to cleats attached to the backs of the rafters; alternatively a zed section can be used, thus dispensing with the need for a fixing cleat. Steel roof trusses are positioned at 3.000–7.500 m centres and supported by capped universal columns or bolted to padstones bedded onto brick walls or piers. The main disadvantage of this form of roofing is the large and virtually unusable roof space. Other disadvantages are the frequent necessity of painting the members to inhibit rust, and the fact that the flanges of the angles provide an ideal ledge on which dust can accumulate. Typical steel roof truss details are shown in Fig. 10.7.2.

Suitable truss and girder arrangements can be fabricated from welded steel tubes, which are lighter in weight, cleaner in appearance, have less surface area on which to collect dust and therefore less surface area to protect with paint.

Another variation on traditional steel angle assembly combines cold-rolled outward-lipped channels or chord sections of high-yield steel with interlacing of steel bars, welded tubes or hollow square sections. This provides considerable design opportunities for different forms of lightweight lattice truss. Figure 10.7.3 shows standard chord sections with the possible inclusion of a timber insert to provide a simple means of attaching fixtures. Figures 10.7.4 and 10.7.5 show profiles and details.

COVERINGS

The basic requirements for covering materials to steel roof trusses are:

- sufficient strength to support imposed wind and snow loadings;
- resistance to the penetration of rain, wind and snow;
- low self-weight, so that supporting members of an economic size can be used;
- acceptable standard of thermal insulation if habitable or occupational accommodation requiring space heating;
- acceptable fire resistance and resistance to spread of flame;
- durability to reduce the maintenance required during the anticipated life of the roof.

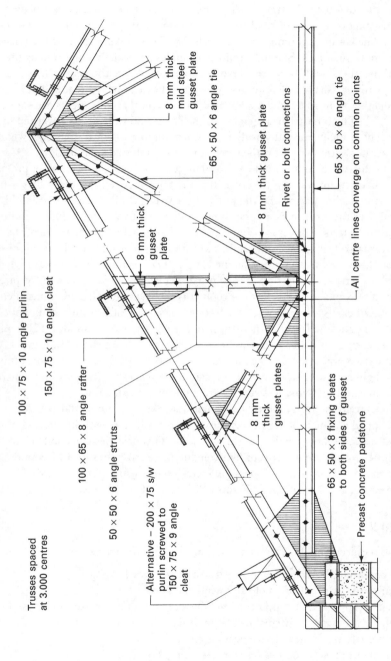

Figure 10.7.2   Typical medium span mild steel roof truss.

Trusses spaced at 3.000 centres

100 × 75 × 10 angle purlin

150 × 75 × 10 angle cleat

100 × 65 × 8 angle rafter

50 × 50 × 6 angle struts

Alternative – 200 × 75 s/w purlin screwed to 150 × 75 × 9 angle cleat

8 mm thick gusset plate

8 mm mild steel gusset plate

65 × 50 × 6 angle tie

8 mm thick gusset plate

Rivet or bolt connections

65 × 50 × 6 angle tie

All centre lines converge on common points

8 mm thick gusset plates

65 × 50 × 8 fixing cleats to both sides of gusset

Precast concrete padstone

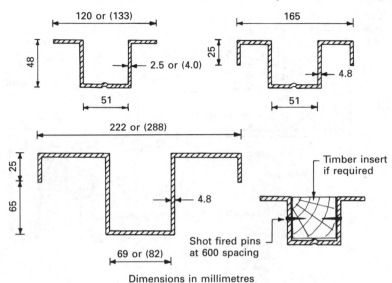

Dimensions in millimetres

Figure 10.7.3 Standard outward lipped channels or chord sections.

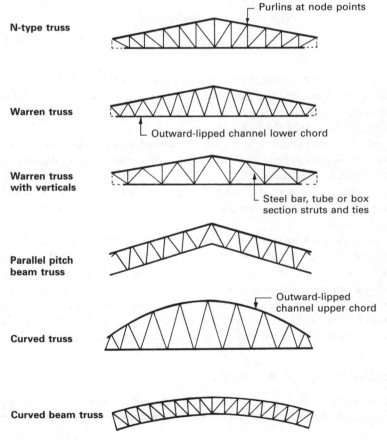

Figure 10.7.4 Chord section roof truss profiles.

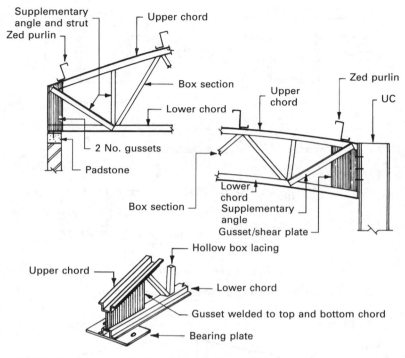

Figure 10.7.5   Further assembly details for chord section trusses.

Most of the materials used for covering a steel roof structure have poor thermal insulation properties, and therefore a combination of materials is required if heat loss, or gain, is to be controlled to satisfy legal requirements or merely to conserve the fuel required to heat the building.

Corrugated roofing materials, if correctly applied, will provide a covering that will exclude the rain and snow but will allow a small amount of wind penetration unless the end laps are sealed with two ribbons of silicon or butyl mastic sealant. These coverings are designed to support normal snow loads, but are not usually strong enough to support the weight of operatives, and therefore a crawling ladder or board should be used.

Owing to the poor thermal insulation value of these roofing materials there is a risk of condensation occurring on the internal surface of the sheets. This risk can be reduced by using a suitable lining at rafter level or by a ceiling at the lower tie level. Unless a vapour control layer is included on the warm side of the lining, water vapour may pass through the lining and condense on the underside of the covering material, which could give rise to corrosion of the steel members. An alternative method is to form a 25 mm wide ventilation cavity between the lining and the covering.

Galvanised corrugated steel sheets

These sheets are often referred to as 'corrugated iron' and have been widely used for many years for small industrial and agricultural buildings; they can also be used as a cladding to post and rail fencing. They are generally made to the recommendations of BS 3083, which specifies the sizes, the number of corrugations, and the quality of the zinc coating or galvanising.

The pitch of the corrugations, which is the distance between centres of adjacent corrugations, is 75 mm with 7, 8, 9, or 10 corrugations per sheet width with lengths ranging from 1.500 to 3.600 m. A wide range of fittings for ridge, eaves and verge treatments are available. The sheets are secured to purlins with hook bolts, drive screws or nuts and bolts in a similar manner to that detailed for fibre cement sheets in Fig. 10.7.8. The purlins are spaced at centres from 1.500 to 3.000 m according to the thickness of the sheeting being used. To form a weathertight covering the sheets are lapped at their ends and sides according to the pitch and exposure conditions:

- **End laps** Up to 20° pitch 150 mm minimum and sealed with a butyl or silicon sealant.
- **Side laps** Formed on edge away from the prevailing wind with a $1\frac{1}{2}$ corrugation lap for conditions of normal exposure and two-corrugation lap for conditions of severe exposure.

When fixed, galvanised corrugated steel sheets form a roof that is cheap to construct, strong, rigid and non-porous. On exposure the galvanised coating oxidises, forming a thin protective film that is easily broken down by acids in the atmosphere. To extend the life of the sheeting it should be regularly coated with paint containing a pigment of zinc dust, zinc oxide, calcium plumbate or zinc chromate. The use of these paints will eliminate the need for an application of mordant solution to provide a key. When laying new sheeting it is advisable to paint under the laps before fixing, because the overlap is very vulnerable to early corrosion.

The main disadvantages of this form of roof covering are:

- poor thermal insulation properties – 8.6 W/m^2 K, which can be reduced by using a 12 mm insulation fibre board in conjunction with a 25 mm cavity to 1.7 W/m^2 K;
- although a non-combustible material, galvanised corrugated steel sheets tend to buckle under typical fire conditions;
- inclined to be noisy during rain, which produces a 'drumming' sound.

Corrugated asbestos cement sheets (now obsolete)

This was the major covering material used for cladding steel roof structures, and was made from asbestos fibres and cement in the approximate proportions of 1:7 together with a controlled amount of water. Corrugated asbestos cement sheets were produced to the recommendations of BS 690: Part 3 (now withdrawn), together with a wide range of fittings for the ridge, eaves and verge, which were used in conjunction with the various profiles produced (see Figs 10.7.6 and 10.7.7).

Concern with the health risk attached to the manufacture and use of asbestos-based products has led to the development and production of alternative fibre-based materials including profiles to match the corrugated asbestos cement sheets. These now conform to BS EN 494.

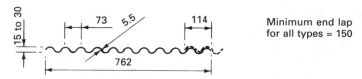

Class 1: min. pitch 10°; max. purlin spacing 900 (symmetrical)

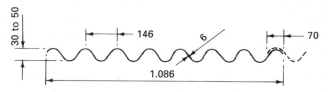

Class 2: min. pitch 10°; max. purlin spacing 1.400 (symmetrical)

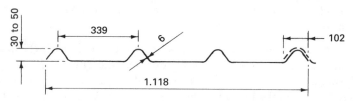

Class 2: min. pitch 10°; max. purlin spacing 1.400 (asymmetrical)

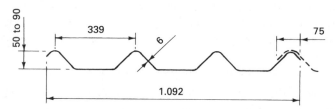

Class 3: min. pitch 4°; max. purlin spacing 1.680 (asymmetrical)

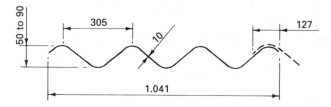

Class 4: min. pitch 4°; max. purlin spacing 1.980 (symmetrical)

Lengths	
Class 1	1.225 to 3.050
Class 2	1.525 to 3.050
Class 3	1.675 to 3.050
Class 4	1.825 to 3.050

Figure 10.7.6   Typical corrugated sheet profiles.

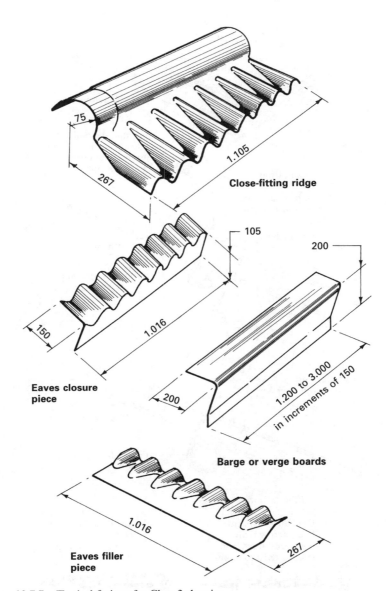

Figure 10.7.7   Typical fittings for Class 2 sheeting.

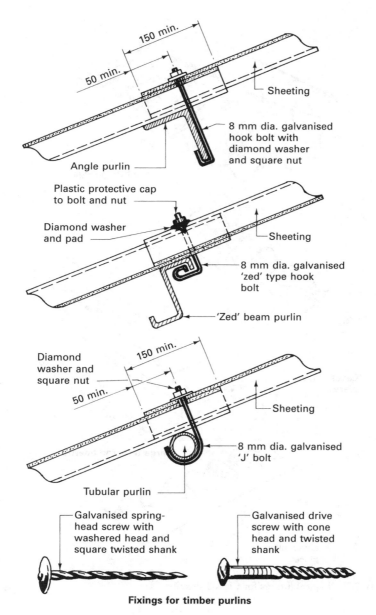

Fixings for timber purlins

Figure 10.7.8 Typical roof sheeting fixings.

Fibre cement profiled sheets

This alternative material to asbestos cement has been developed to meet the same technical specification with a similar low maintenance performance. Fibre cement sheets are made by combining natural and synthetic non-toxic fibres and fillers with Portland cement and, unlike asbestos cement sheets, which are rolled to form the required profile, these sheets are pressed over templates. The finished product has a natural grey colour, but sheets with factory-applied surface coatings are available.

The sheets and fittings are fixed through the crown of the corrugation using either shaped bolts to steel purlins or drive screws to timber purlins. At least six fixings should be used for each sheet, and to ensure that a weathertight seal is achieved at the fixing positions a suitable felt or lead pad with a diamond-shaped curved washer can be used. Alternatively a plastic sealing washer can be employed (see Figs 10.7.8 and 10.7.9). The sheets can be easily drilled for fixings, which should be 2 mm larger in diameter than the fixing and sited at least 40 mm from the edge of the sheet. Side laps should be positioned away from the prevailing wind, and end laps on low pitches should be sealed with a mastic or suitable preformed compressible strip.

The U value of fibre cement sheets is high, generally about $6.0 \text{ W/m}^2 \text{ K}$, therefore if a higher degree of thermal resistance is required it will be necessary to use a system of underlining sheets with an insulating material sandwiched between the profile and underlining sheet.

Aluminium sheeting

This form of roof covering is available in a corrugated or troughed profile usually conforming to the requirements of BS 4868. The sheets are normally made from an aluminium–manganese alloy resulting in a non-corrosive, non-combustible lightweight sheet ($2.4–5.0 \text{ kg/m}^2$). Aluminium sheeting oxidises on the surface to form a protective film upon exposure to the atmosphere, and therefore protective treatments are not normally necessary. Fixings of copper or brass should not be used, because the electrolytic action between the dissimilar materials could cause harmful corrosion, and where the sheets are in contact with steelwork the steel members should be painted with at least two coats of zinc chromate or bituminous paint.

The general application in design and construction of an aluminium covering is similar to that described and detailed for fibre cement sheeting. A wide range of fittings are available and, like the asbestos cement, sheets can be fixed with hook bolts, bolts and clips or special shot fasteners. The sheets are intended for a 15° pitched roof with purlins at 1.200 m centres for the 75 mm corrugated profile and at 2.700 m centres for the trough profiles. Laps should be 1½ corrugations for the side lap or 45–57 mm for trough sheets with a 150 mm minimum end lap for all profiles.

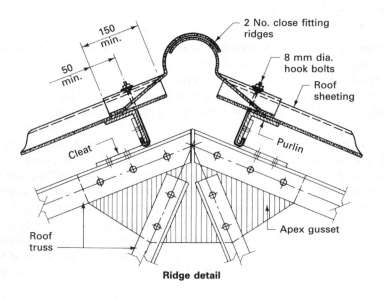

150 min.

50 min.

2 No. close fitting ridges

8 mm dia. hook bolts

Roof sheeting

Cleat

Purlin

Roof truss

Apex gusset

Ridge detail

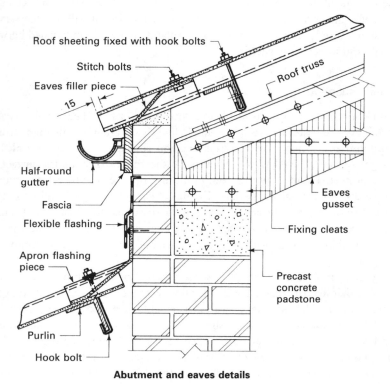

Roof sheeting fixed with hook bolts

Stitch bolts

Eaves filler piece

15

Half-round gutter

Fascia

Flexible flashing

Apron flashing piece

Purlin

Hook bolt

Roof truss

Eaves gusset

Fixing cleats

Precast concrete padstone

Abutment and eaves details

Figure 10.7.9 Typical corrugated sheet roofing details.

Aluminium (BS EN 485) and coated steel (BS EN 10326)

Profiled sheeting

These are produced in a trapezoidal profile in varying depth and pitch dimensions depending on the source of manufacture. Surface treatment may be galvanised for steel, painted or more commonly plastic coated. Various types of plastic coating are possible, including acrylic, polyester and resins. All can be colour pigmented to provide a wide range of choice to suit all types of building, including schools, hospitals, workshops, warehousing and offices. Fixings can be through the valley for direct connection to the purlin with self-drilling and tapping screws as shown in Fig. 10.7.10. Crown fixing, also shown, has self-tapping screws in pre-drilled holes with a butyl mastic (or equivalent) sealant to prevent ingress of rainwater.

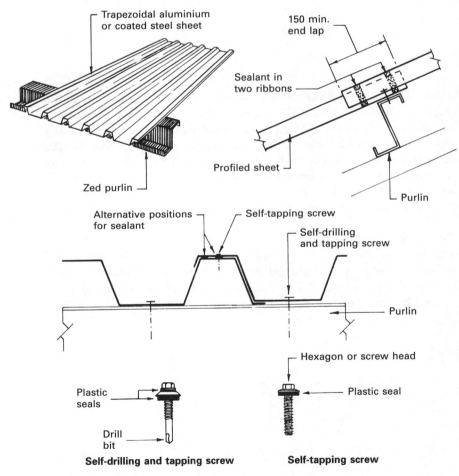

Figure 10.7.10 Profiled sheet metal roof covering.

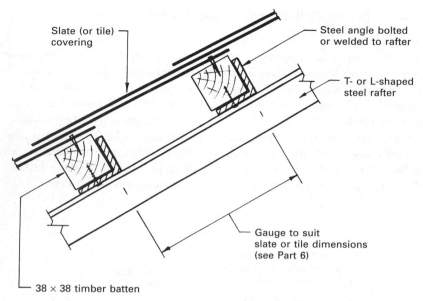

Figure 10.7.11   Slate covering.

The single skin corrugated/profiled systems of roof covering shown in this chapter are limited to unheated structures such as warehousing, workshops, garages and loading bays. Current thermal insulation legislation for industrial and commercial buildings in the UK requires a substantial thickness of insulation material integral with the roof construction. This may be sandwiched between two layers of profiled covering or incorporated above a ceiling lining. More information on this and the topic of thermal insulation to non-dwellings is to be found in the associated volume, *Advanced Construction Technology*.

Tiles or slate coverings
The traditional tile or slate roof is not limited to dwellings. Many industrial and commercial premises with medium/large-span steel roof trusses are finished in these materials. Figure 10.7.11 shows a method of fixing to standard T- or L-shaped steel rafters.

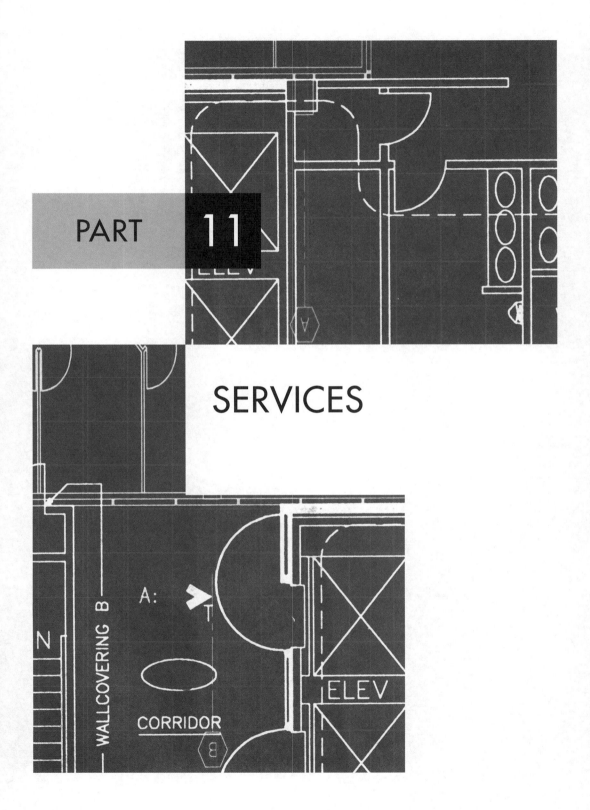

PART 11

SERVICES

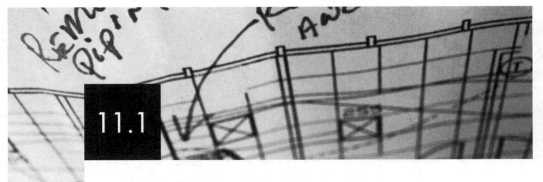

DOMESTIC WATER SUPPLY

An adequate supply of water is a basic requirement for most buildings, for reasons of personal hygiene or for activities such as cooking and manufacturing processes. In most areas a piped supply of water is available from the local utility company's water main.

Water is, in the first instance, produced by condensation in the form of clouds, and falls to the ground as rain, snow or hail; it then either becomes surface water in the form of a river, stream or lake, or percolates through the subsoil until it reaches an impervious stratum, or is held in a water-bearing subsoil.

The water authority extracts water from surface storage reservoirs, rivers or through boreholes into water-bearing subsoil. Thereafter it is filtered, chlorinated, aerated, and treated with fluoride for human consumption before distribution through a system of mains.

The water company's mains are laid underground at a depth where they will be unaffected by frost or traffic movement. The layout of the system is generally a circuit with trunk mains feeding a grid of subsidiary mains for distribution to specific areas or districts. The materials used for main pipes are cast iron and uPVC, which can be tapped while under pressure; a plug cock is inserted into the crown of the mains pipe to provide the means of connecting the communication pipe to supply an individual building.

Terminology

- **Main** A pipe for the general conveyance of water, as distinct from the conveyance to individual premises.
- **Service** A system of pipes and fittings for the supply and distribution of water in any individual premises.
- **Service pipe** A pipe in a service that is directly subject to pressure from a main, sometimes called the rising main, inside the building.

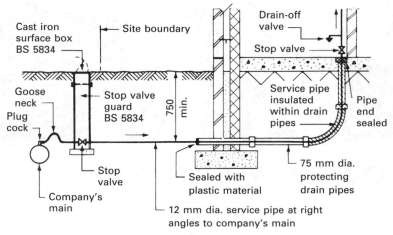

Typical cold water service layout

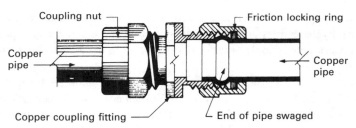

Typical manipulative compression joint

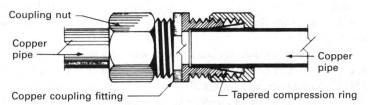

Typical non-manipulative compression fitting

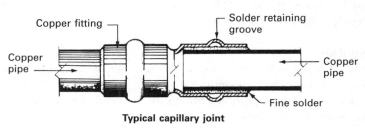

Typical capillary joint

Figure 11.1.1 Supply pipe and copper pipe joint details.

- **Communication pipe**   That part of the service pipe that is vested in the water undertaking.
- **Distribution pipe**   Any pipe in a service conveying water from a storage cistern.
- **Cistern**   A container for water in which the stored water is under atmospheric pressure.
- **Storage cistern**   Any cistern other than a flushing cistern.
- **Tank**   A rectangular vessel completely closed and used to store water.
- **Cylinder**   A closed cylindrical tank.

■ ■ ■ COLD WATER SUPPLY

The water company will provide from their mains tapping plug cock a communication pipe to a stop valve and protection chamber just outside the boundary; this is a chargeable item to the building owner. The gooseneck bend is included to relieve any stress likely to be exerted on the mains connection.

A service pipe is taken from this stop valve to an internal stop valve, preferably located just above floor level and housed under the sink unit. The stop valve should have a drain–off valve incorporated in it, or just above it, so that the service pipe or rising main can be drained.

Care must be taken when laying a service pipe that it is not placed in a position where it can be adversely affected by frost, heavy traffic or building loads. A minimum depth of 750 mm is generally recommended for supplies to domestic properties; where the pipe passes under a building it should be housed in a protective duct or pipe suitably insulated within 750 mm of the floor level (see Fig. 11.1.1).

Suitable materials for service pipes are copper, PVC, polythene and galvanised steel. Copper service pipes can be laid on and covered by a layer of sand to prevent direct contact with the earth, or wrapped with a suitable proprietary insulating material. Plastic-coated copper pipes are also available for underground pipework. Steel pipes should have a similar protection, but plastic pipes are resistant to both frost and corrosion and have largely superseded metal pipes in this situation.

WATER METER

These are a standard installation for all new buildings and conversions within buildings, as a resource conservation measure. Existing premises may have meters installed at the water authority's discretion. Where practicable, the meter is located on the service pipe, in a small compartment below ground and just inside the property boundary. If this proves impossible owing to lack of space, the meter may be positioned at the base of the rising main.

DIRECT COLD WATER SUPPLY

In this system the whole of the cold water to the sanitary fittings is supplied directly from the service pipe. The direct system is used mainly in northern districts where

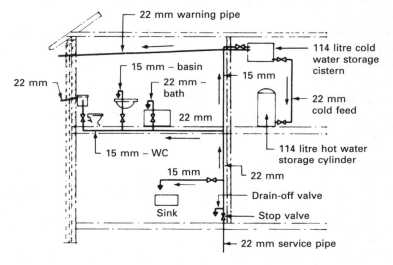

Direct system of cold water supply

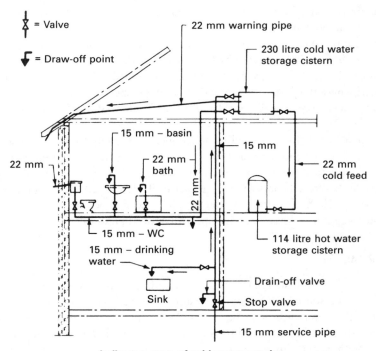

Indirect system of cold water supply

Note: All pipe sizes shown are standard copper outside diameters

Figure 11.1.2   Cold water supply systems.

large, high-level reservoirs provide a good mains supply and pressure. With this system only a small cold water storage cistern to feed the hot water tank is required; this can usually be positioned below the roof ceiling level, giving a saving on pipe runs to the roof space and eliminating the need to insulate the pipes against frost (see Fig. 11.1.2).

Another advantage of the direct system is that drinking water is available from several outlet points. The main disadvantage is the lack of reserve should the mains supply be cut off for repairs; also there can be a lowering of the supply during peak demand periods.

When sanitary fittings are connected directly to a mains supply there is always a risk of contamination of the mains water by back-siphonage. This can occur if there is a negative pressure on the mains and any of the outlets are submerged below the water level, such as a hand spray connected to the taps.

INDIRECT COLD WATER SUPPLY

In this system all the sanitary fittings, except for a drinking water outlet at the sink, are supplied indirectly from a cold water storage cistern positioned at a high level, usually in the roof space. This system requires more pipework, but it gives a reserve supply in case of mains failure, and it also reduces the risk of contamination by back-siphonage (see Fig. 11.1.2). Note that the local water authority determines the system to be used in the area.

PIPEWORK

Any of the materials that are suitable for the service pipe are equally suitable for distribution pipes, and the choice is very often based on individual preference, initial costs and possible maintenance costs.

Copper pipes

Copper pipes have a smooth bore, giving low flow resistance; they are strong, and easily jointed and bent. Joints in copper pipes can be made by one of three methods:

- **Manipulative compression joint** The end of the pipe is manipulated to fit into the coupling fitting by means of a special tool. No jointing material is required, and the joint offers great resistance to being withdrawn. It is usually a by-law requirement that this type of joint is used on service pipes below ground.
- **Non-manipulative compression joint** No manipulation is required to the cut end of the pipe; the holding power of the joint relies on the grip of a copper cone, ring or olive within the joint fitting.
- **Capillary joint** The application of heat makes the soft solder contained in a groove in the fitting flow around the end of the pipe, which has been cleaned and coated with a suitable flux to form a neat and rigid joint.

Typical examples of copper pipe joints are shown in Fig. 11.1.1.

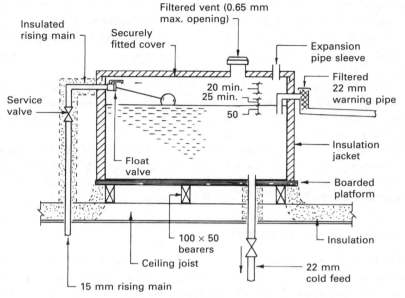

Typical cold water storage cistern

Piston float valve

Diaphragm float valve

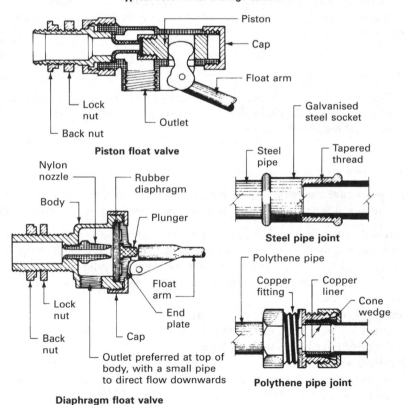

Steel pipe joint

Polythene pipe joint

Figure 11.1.3   Cisterns, float valves and joints.

Steel pipes

Steel pipes for domestic water supply can be obtained as black tube, galvanised or coated and wrapped for underground services. The joint is usually made with a tapered thread and socket fitting, and to ensure a sound joint, stranded hemp and jointing paste should be wrapped around the thread; alternatively a non-contaminating polytetrafluoroethylene (PTFE) seal tape can be used (see Fig. 11.1.3).

Polythene pipe

Polythene pipe is very light in weight, easy to joint, non-toxic, and is available in long lengths, which gives a saving on the number of joints required. Jointing of polythene pipes is generally of the compression type using a metal or plastic liner to the end of the tube (see Fig. 11.1.3). To prevent undue sagging, polythene pipes should be adequately fixed to the wall with saddle clips: recommended spacings are 14 times the outside diameter for horizontal runs and 24 times outside diameter for vertical runs.

Unplasticised PVC (uPVC)

This is plastic pipe for cold water services that is supplied in straight lengths up to 9000 mm long and in standard colours of grey, blue or black. Blue is the water authorities' preferred colour, for easy identification when buried. Jointing can be by a screw thread, but the most common method is by solvent welding. This involves cleaning and chamfering the end of the pipe, which is coated with the correct type of adhesive and pushed into a straight coupling that has been similarly coated. The solvent will set within a few minutes, but the joint does not achieve its working strength for 24 hours. Heat fusion is also used, but mainly for the larger-diameter water authority mains. Here the pipe spigot and the coupling are heated with circular dies to melt the uPVC and effect a push fit weld.

COLD WATER STORAGE CISTERNS

The size of cold water storage cisterns for dwelling houses will depend upon the reserve required and whether the cistern is intended to feed a hot water system. Minimum actual capacities recommended in model water by-laws are 115 litres for cold water storage only and 230 litres for cold and hot water services.

Cisterns should be adequately supported and installed in such a position as to give reasonable access for maintenance purposes. The cistern must be installed so that its outlets are above the highest discharge point on the sanitary fittings, as the flow is by gravity. If the cistern is housed in the roof the pipes and cistern should be insulated against possible freezing of the water; preformed casings of suitable materials are available to suit most standard cistern sizes and shapes. The inlet and outlet connections to the cistern should be on opposite sides to prevent stagnation of water. A securely fitting cover must be provided to prevent ingress of dust, dirt and insects. To prevent a vacuum occurring as water is drawn, the cover is fitted with a screened vent. It also has a moulded sleeve/boss for adapting the hot water expansion pipe. The overflow or warning pipe is also fitted with a filter. A typical cistern installation is shown in Fig. 11.1.3.

Plastic cisterns have many advantages over the traditional galvanised mild steel cisterns: they are non-corrosive, rot proof and frost resistant, and have good resistance to mechanical damage. Materials used are polythene, polypropylene and glass fibre: these cisterns are made with a wall thickness to withstand the water pressure, and have an indefinite life. Some forms of polythene cistern can be distorted to enable them to be passed through an opening of 600 mm × 600 mm, which is a great advantage when planning access to a roof space. However, it is always better to deposit the cistern within the roof structure during construction, rather than have access difficulties later!

Float valves

Every pipe supplying a cold water storage cistern must be fitted with a float valve to prevent an overflow. The float valve must be fitted at a higher level than the overflow to prevent it becoming submerged and creating the conditions where back-siphonage is possible. A float valve is designed to automatically regulate the supply of water by a floating ball closing the valve when the water reaches a predetermined level.

Two valves are in common use for domestic work, the piston valve and the diaphragm valve. The piston valve has a horizontal piston or plunger that closes over the orifice of a diameter to suit the pressure; high-, medium- and low-pressure valves are available (see Fig. 11.1.3). The diaphragm valve closes over an interchangeable nylon nozzle orifice. This type of valve is quieter in operation, easily adjustable, and less susceptible to the corrosion trouble caused by a sticking piston – this is one of the problems that can be encountered with the piston valve (see Fig. 11.1.3). Diaphragm valves now have the outlet at the top to increase the air gap between outlet and water level, thus reducing the possibility of back-siphonage if the cistern water level were to rise excessively.

■ ■ ■ HOT WATER SUPPLY

The supply of hot water to domestic sanitary fittings is usually taken from a hot water tank or cylinder. The source of heat is usually in the form of a gas-fired, oil-fired or solid-fuel boiler; alternatives are a back boiler to an open fire or an electric immersion heater fixed into the hot water storage tank. When a quantity of hot water is drawn from the storage tank it is immediately replaced by cold water from the cold water storage cistern. Two main systems are used to heat the water in the tank – these are called the **direct** and **indirect** systems. For any hot water system copper or steel pipes are generally used. These materials are satisfactory if used independently throughout the system, but they must not be mixed or interchanged within a system. This is due to the electrolytic attraction that dissimilar metals have in the presence of water, which eventually leads to corrosion. Copper and zinc are particularly renowned for this problem, as zinc is the protective plating given to steel pipes under the name of galvanising.

DIRECT HOT WATER SYSTEM

This is the simplest and cheapest system; the cold water flows through the water jacket in the boiler, where its temperature is raised and convection currents are induced, which causes the water to rise and circulate. The hot water leaving the boiler is replaced by colder water descending from the hot water cylinder or tank by gravity, thus setting up the circulation. The hot water supply is drawn off from the top of the cylinder by a horizontal pipe at least 450 mm long to prevent 'one pipe' circulation being set up in the vent or expansion pipe. This pipe is run vertically from the hot water distribution pipe to a discharge position over the cold water storage cistern (see Fig. 11.1.4).

The direct system is not suitable for supplying a central heating circuit or for hard water areas because the pipes and cylinders will become furred with lime deposits. This precipitation of lime occurs when hard water is heated to temperatures of between 50 and 70 °C, which is the ideal temperature range for domestic hot water supply.

INDIRECT HOT WATER SYSTEM

This system is designed to overcome the problems of furring, which occurs with the direct hot water system. The basic difference is in the cylinder design, which now becomes a heat exchanger. The cylinder contains a coil or annulus that is connected to the flow and return pipes from the boiler. A transfer of heat takes place within the cylinder, and therefore, after the initial precipitation of lime within the primary circuit and boiler, there is no further furring as fresh cold water is not being constantly introduced into the boiler circuit.

The supply circuit from the cylinder follows the same pattern as the direct hot water system, but a separate feed and expansion system is required for the boiler and primary circuit for initial supply, also for any necessary topping up due to evaporation. The feed cistern is similar to a cold water storage cistern but of a much smaller capacity. The water levels in the two cisterns should be equal so that equal pressures act on the indirect cylinder.

A gravity heating circuit can be taken from the boiler, its distribution being governed by the boiler capacity (see Fig. 11.1.4). Alternatively a small-bore forced system of central heating may be installed.

Hot water storage cylinders

Copper cylinders are produced to the recommendations of BS 1566. The Standard recommends sizes, capacities and positions for screwed holes for pipe connections.

To overcome the disadvantage of the extra pipework involved when using an indirect cylinder, a single-feed indirect or 'Primatic' cylinder can be used. This form of cylinder is entirely self-contained, and is installed in the same manner as a direct cylinder but functions as an indirect cylinder. It works on the principle of the buoyancy of air, which is used to form seals between the primary and secondary water systems. When the system is first filled with water the cylinder commences to

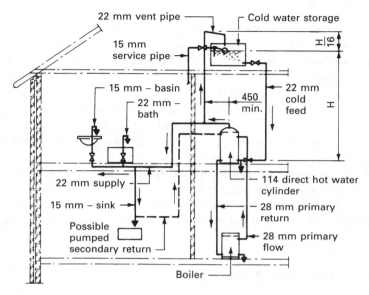

22 mm vent pipe ⌐ Cold water storage

15 mm service pipe

$\frac{H}{16}$

15 mm – basin

22 mm – bath

450 min.

22 mm cold feed

H

22 mm supply

15 mm – sink

Possible pumped secondary return

114 direct hot water cylinder

28 mm primary return

28 mm primary flow

Boiler

Note: Not suitable in hard water areas

Direct hot water system

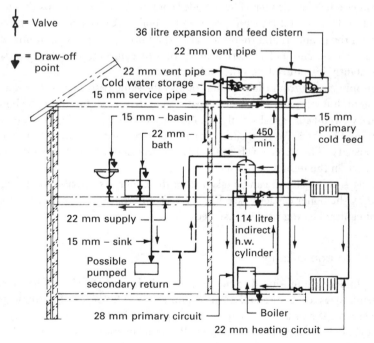

= Valve

= Draw-off point

36 litre expansion and feed cistern

22 mm vent pipe

22 mm vent pipe
Cold water storage
15 mm service pipe

15 mm – basin

22 mm – bath

450 min.

15 mm primary cold feed

22 mm supply

15 mm – sink

Possible pumped secondary return

114 litre indirect h.w. cylinder

28 mm primary circuit

Boiler

22 mm heating circuit

Indirect hot water system

Note: Pipe sizes shown are for standard copper outside diameters

Figure 11.1.4   Hot water system.

fill and fully charges the primary circuit to the boiler with water. When the cylinder water capacity has been reached two air seals will have formed, the first in the upper chamber of the primatic unit and the second in the air vent pipe. These volumes of air are used to separate the primary and secondary water. When the water is heated in the primary system expansion displaces some of the air in the upper chamber to the lower chamber. This is a reciprocating action: the seals transfer from chamber to chamber as the temperature rises and falls.

Any excess air in the primary system is vented into the secondary system, which will also automatically replenish the primary system should this be necessary. As with indirect systems careful control over the heat output of the boiler is advisable to prevent boiling and consequent furring of the pipework. Typical examples of cylinders are shown in Fig. 11.1.5.

Faults in hot water systems

Unless a hot water system is correctly designed and installed a number of faults may occur, such as airlocks and noises. Airlocks are pockets of trapped air in the system, which will stop or slow down the circulation. Air suspended in the water will be released when the water is heated, and rise to the highest point. In a good installation the pipes are designed to rise 25 mm in 3000 mm towards the vent where the air is released through the vent pipe. The most common positions for airlocks are sharp bends and the upper rail of a towel rail; the only cure for the latter position is for the towel rail to be vented.

Noises from the hot water system usually indicate a blocked pipe caused by excessive furring or corrosion. The noise is caused by the imprisoned expanded water, and the faulty pipe must be descaled or removed, or an explosion may occur.

Inadequate bracketing of pipes is another common cause of plumbing noise, particularly in the rising main, where pressures are higher than elsewhere. This is known as **water hammer**.

Mains supply hot water

This has become a popular alternative to the traditional installation. It is very economic in pipework, space requirements and installation time as conventional expansion facilities are not required. The system is mains fed and sealed, with expansion accommodated in a specially designed vessel containing an air 'cushion'. Safety facilities are essential and include both pressure and thermal relief valves. Figure 11.1.6 indicates the principles of operation.

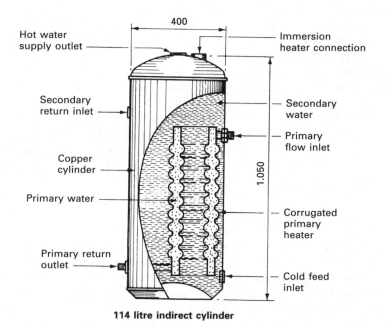

114 litre indirect cylinder

135 litre single-feed or 'Primatic' cylinder

Figure 11.1.5   Typical hot water cylinders.

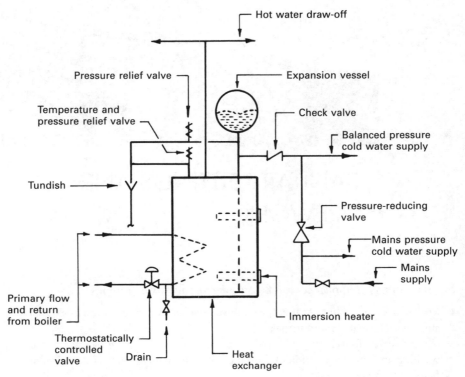

Figure 11.1.6 Schematic of mains-fed hot water supply.

SANITARY FITTINGS AND PIPEWORK

Sanitary fittings or appliances can be considered under two headings:

- **Soil fitments**   Those that are used to remove soil water and human excreta such as water closets and urinals.
- **Waste water fitments**   Those that are used to remove the waste water from washing and the preparation of food including appliances such as wash basins, baths, showers and sinks.

All sanitary appliances should be made from impervious materials, be quiet in operation, easy to clean and of a convenient shape fixed at a suitable height. A number of materials are available for most domestic sanitary fittings.

Vitreous china

This is a white clay body that is vitrified and permanently fused with a vitreous glazed surface when fired at a very high temperature, generally to the recommendations of BS 3402. Appliances made from this material are non-corrosive, hygienic, and easily cleaned with a mild detergent or soap solution.

Glazed fireclay

This consists of a porous ceramic body glazed in a similar manner to vitreous china; it is exceptionally strong and resistant to impact damage but will allow water penetration of the body if the protective glazing is damaged. Like vitreous china, these appliances are non-corrosive, hygienic and easily cleaned.

Vitreous enamel

This is a form of glass that can be melted and used to give a glazed protective coating over a steel or cast iron base. Used mainly for baths, sinks and draining boards, it produces a fitment that is lighter than those produced from a ceramic material, is hygienic, easy to clean, and has a long life. The finish, however, can be chipped and is subject to staining, especially from copper compounds from hot water systems.

Plastic materials

Acrylic plastics, glass-reinforced polyester resins and polypropylene sanitary fittings made from these plastics require no protective coatings, are very strong, light in weight, and chip resistant, but generally cost more than ceramic or vitreous enamel products. Care must be taken with fitments made of acrylic plastics as they become soft when heated: therefore they should be used for cold water fitments or have thermostatically controlled mixing taps. Plastic appliances can be easily cleaned using warm soapy water, and any dullness can be restored by polishing with ordinary domestic polishes.

Stainless steel

This is made from steel containing approximately 18% chromium and 8% nickel, which gives the metal a natural resistance to corrosion. Stainless steel appliances are very durable and relatively light in weight; for domestic situations the main application is for sinks and draining boards. Two finishes are available, polished or 'mirror' finish and 'satin' finish: the latter has a greater resistance to scratching.

General considerations

The factors to be considered when selecting or specifying sanitary fitments are as follows:

- cost – outlay, fixing and maintenance;
- hygiene – inherent and ease of cleaning;
- appearance – size, colour and shape;
- function – suitability, speed of operation and reliability;
- weight – support required from wall and/or floor;
- design – ease with which it can be included into the general services installation.

Building Regulation G1 establishes requirements for sanitary conveniences, i.e. water closets, urinals and wash basins, and Regulation G2 extends coverage to bathrooms containing bath or shower installations. The quantity of WCs and sanitary appliances in various locations is detailed in BS 6465: *Sanitary installations*. This Standard is in two parts: Part 1 is the code of practice for provision and installation, and Part 2 relates to spatial layout and design for sanitary accommodation.

The subject of water conservation with respect to the use of sanitary fitments is very topical, with manufacturers endeavouring to adapt their products and develop new ideas in line with conservation and environmental issues. Building Research Establishment Information Paper ref. IP 8/97, Parts 1 and 2, provides further reading on this topic with application to the design of low-flush WCs.

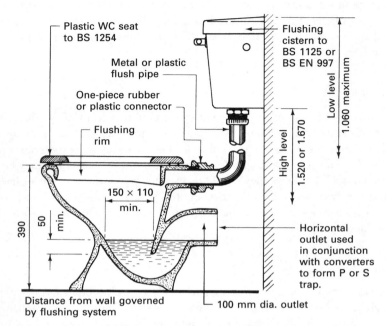

Plastic WC seat to BS 1254

Metal or plastic flush pipe

One-piece rubber or plastic connector

Flushing rim

Flushing cistern to BS 1125 or BS EN 997

Low level 1.060 maximum

High level 1.520 or 1.670

150 × 110 min.

390

50 min.

Horizontal outlet used in conjunction with converters to form P or S trap.

Distance from wall governed by flushing system

100 mm dia. outlet

BS EN 37 ceramic washdown WC pan

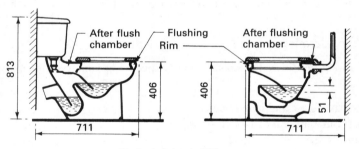

After flush chamber

Flushing Rim

After flushing chamber

813

406

406

51

711

711

Typical siphonic WC pans

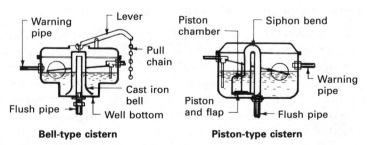

Warning pipe

Lever

Pull chain

Cast iron bell

Flush pipe

Well bottom

Bell-type cistern

Piston chamber

Siphon bend

Warning pipe

Piston and flap

Flush pipe

Piston-type cistern

Figure 11.2.1   WC pans and cisterns.

■ ■ ■ WATER CLOSETS

Most water closets are made from a ceramic base to the requirements of BS EN 33 or 37 with a horizontal outlet. The complete water closet arrangement consists of the pan, seat, flush pipe and flushing cistern. The cistern can be fixed as high level, low level (BS EN 37) or closed coupled (BS EN 33); the latter arrangement dispenses with the need for a flush pipe. A typical arrangement is shown in Fig. 11.2.1. The British Standard water closet is termed a washdown type and relies on the flush of water to wash the contents of the bowl round the trap and into the soil-pipe. An alternative form is the siphonic water closet, which is more efficient and quieter in operation but has a greater risk of blockage if misused. Two types are produced – the single-trap and the double-trap.

The single-trap siphonic water closet has a restricted outlet that serves to retard the flow, when flushed, so that the bore of the outlet connected to the bowl becomes full and sets up a siphonic flushing action, completely emptying the contents of the bowl (see Fig. 11.2.1). With the double-trap siphonic pan the air is drawn from the pocket between the two traps; when the flushing operation is started this causes atmospheric pressure to expel the complete contents of the bowl through both traps into the soil pipe (see Fig. 11.2.1).

The pan should be fixed to the floor with brass screws and bedded on a suitable compressible material; the connection to the soil-pipe socket can be made with cement mortar or preferably a mastic to allow for any differential movement between the fitment and the structure. Connections to PVC soil pipes are usually made with compression rings. The flush pipe is invariably connected to the pan with a special plastic or rubber one-piece connector.

Flushing cisterns together with the flush pipes are usually constructed to the recommendations of BS 1125 and BS EN 997, and can be made from enamelled cast iron, enamelled pressed steel, ceramic ware or of plastic materials. Two basic types have been produced, the bell or cone and the piston. The former is now obsolete, but may still be found in service. It is activated by pulling a chain, which raises and lowers the bell or cone and in so doing raises the water level above the open end of the flush pipe, thus setting up a siphonic action. These cisterns are efficient and durable but are noisy in operation (see Fig. 11.2.1). The piston-type cistern is the one in general use, and is activated by a lever or button. When activated, the disc or flap valve piston is raised and with it the water level, which commences the siphonage (see Fig. 11.2.1). The level of the water in the cistern is controlled by a float valve, and an overflow or warning pipe of a larger diameter than the inlet is fitted to discharge so that it gives a visual warning, usually in an external position. The capacity of the cistern will be determined by local water board requirements, the most common being 6, 7.5 and 9 litres.

There is a wide range of designs, colours and patterns available for water closet suites, but all can be classified as one of the types described above.

FLUSHING VALVES

These have been a popular marine plumbing installation for many years. In compact lavatory compartments they have proved effective in restricted space, and they are

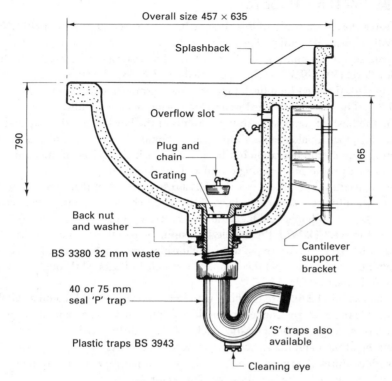

Overall size 457 × 635

Splashback

Overflow slot

Plug and
chain

Grating

790

165

Back nut
and washer

BS 3380 32 mm waste

Cantilever
support
bracket

40 or 75 mm
seal 'P' trap

'S' traps also
available

Plastic traps BS 3943

Cleaning eye

Typical lavatory basin details

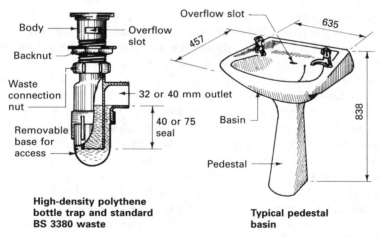

Body

Overflow
slot

Backnut

Waste
connection
nut

Removable
base for
access

32 or 40 mm outlet

40 or 75
seal

**High-density polythene
bottle trap and standard
BS 3380 waste**

Overflow slot

635

457

Basin

838

Pedestal

**Typical pedestal
basin**

Figure 11.2.2   Basins, traps and wastes.

economic with the use of limited water resources. They are becoming a credible alternative for use in buildings, particularly where usage is frequent, as they can flush at any time without delay. They comprise a large equilibrium valve, which functions on the principle of water displacement and pressure equalisation limiting the flow. Current research and recommendations from the Water Regulations Advisory Committee (WRAC) to the government are likely to succeed in establishing these valves in the Water Supply Byelaws. See BRE IP 8/97 Part 2 for more detail.

■■■ WASH BASINS

Wash basins for domestic work are usually made from a ceramic material, but metal basins complying with BS 1329 are also available. A wide variety of shapes, sizes, types and colours are available, the choice usually being one of personal preference. BS 1188 and BS 5506 provide recommendations for ceramic wash basins and pedestals, and specify two basic sizes, 635 mm × 457 mm and 559 mm × 406 mm. Small wall-hung hand-rinse basins are also available for use in limited space. These are produced to BS EN 111. All these basins are a one-piece fitment, having an integral overflow, separate waste outlet and generally pillar taps (see Fig. 11.2.2).

Wash basins can be supported on wall-mounted cantilever brackets, leg supports or pedestals. The pedestals are made from identical material to the wash basin, and are recessed at the back to receive the supply pipes to the taps and the waste pipe from the bowl. Although pedestals are designed to fully support the wash basin, most manufacturers recommend that small wall-mounted safety brackets are also used.

■■■ BATHS AND SHOWERS

Baths are available in a wide variety of designs and colours, made from porcelain-enamelled cast iron, vitreous-enamelled sheet steel, 8 mm cast acrylic sheet, or 3 mm cast acrylic sheet reinforced with a polyester resin/glass fibre laminate. Most bath designs today are rectangular in plan and made as flat bottomed as practicable with just sufficient fall to allow for gravity emptying and resealing of the trap. The British Standards for the materials quoted above recommend a coordinating plan size of 1700 mm × 700 mm with a height within the range of $n \times 50$ mm, where n is any natural number including unity. Baths are supplied with holes for pillar taps or mixer fittings, and for the waste outlets. Options include handgrips, built-in soap and sponge recesses and overflow outlets. It is advisable always to specify overflow outlets as a precautionary measure to limit the water level and to minimise splashing; most overflow pipes are designed to connect with the waste trap beneath the bath (see Fig. 11.2.3). Support for baths is usually by adjustable feet for cast iron and steel and by a strong cradle for the acrylic baths. Panels of enamelled hardboard or moulded high-impact polystyrene or glass fibre are available for enclosing the bath. These panels can be fixed by using stainless steel or aluminium angles or direct to a timber stud framework.

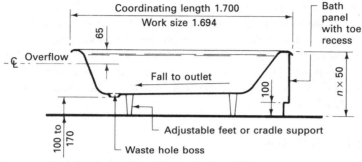

Coordinating length 1.700
Work size 1.694
65
Bath panel with toe recess
Overflow
Fall to outlet
n × 50
100
100 to 170
Adjustable feet or cradle support
Waste hole boss

Coordinating width = 700; work size = 697
Coordinating sizes can be varied within range of
$n \times 100$ mm, where n = any natural number including unity

BS 1189 (cast iron), BS 4305 (acrylic) and BS 1390 (vitreous enamelled steel) baths. Also, BS EN 60335–2–60 whirlpool baths.

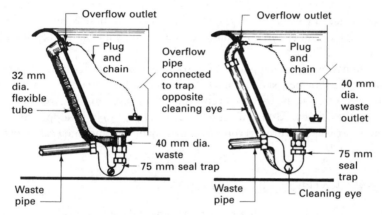

Overflow outlet

Plug and chain

Overflow pipe connected to trap opposite cleaning eye

32 mm dia. flexible tube

40 mm dia. waste
75 mm seal trap

Waste pipe

Overflow outlet

Plug and chain

40 mm dia. waste outlet

75 mm seal trap

Waste pipe

Cleaning eye

Alternative overflow connections

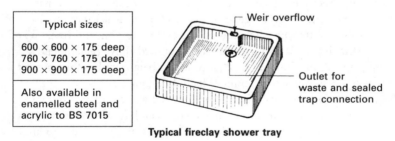

Typical sizes
600 × 600 × 175 deep 760 × 760 × 175 deep 900 × 900 × 175 deep
Also available in enamelled steel and acrylic to BS 7015

Weir overflow

Outlet for waste and sealed trap connection

Typical fireclay shower tray

Figure 11.2.3   Baths and shower trays.

Shower sprays can be used in conjunction with a bath by fitting a rigid plastic shower screen or flexible curtain to one end of the bath. A separate shower fitment, however, is considered preferable. Such fitments require less space than the conventional bath, use less hot water, and are considered to be more hygienic as the used water is being continuously discharged. A shower fitment consists of the shower tray with a waste outlet, the impervious cubicle, and a door or curtain (see Fig. 11.2.3). Materials available are similar to those described for baths. The spray outlet is normally fixed to the wall, and is connected to a mixing valve so that the water temperature can be controlled.

■■■ SINKS

Sinks are used mainly for the preparation of food, washing of dishes and clothes, and are usually positioned at the drinking water supply outlet. Their general design follows that described for basins, except that they are larger in area and deeper. Any material considered suitable for sanitary appliance construction can be used. Designs range from the simple Belfast sink with detachable draining boards of metal, plastic or timber to the combination units consisting of integral draining boards and twin bowls. Support can be wall-mounted cantilever brackets, framed legs or a purpose-made cupboard unit; typical details are shown in Fig. 11.2.4.

The layout of domestic sanitary appliances is governed by size of fitments, personal preference, pipework system being used and the space available. Building Regulations G1 and G2 detail requirements and minimum facilities to be made available in a dwelling, and the need for sanitary accommodation to be separated from kitchen areas.

■■■ PIPEWORK

Building Regulations, Approved Document H1: *Sanitary pipework and drainage* sets out in detail the recommendations for soil pipes, waste pipes and ventilating pipes. These regulations govern such things as minimum diameters of soil pipes, material requirements, provision of adequate water seals by means of an integral trap or non-integral trap, the positioning of soil pipes on the inside of a building, overflow pipework, and ventilating pipes. The only pipework that is permissible on the outside of the external wall is any waste pipe from a waste appliance situated at ground floor level, provided such a pipe discharges into a suitable trap with a grating, and the discharge is above the level of the water but below the level of the grating.

PIPEWORK SYSTEMS

Three basic pipework systems have been used for domestic work:

- one-pipe system;
- two-pipe system;
- single-stack system.

Whichever system is adopted, the functions of quick, reliable and quiet removal of the discharges to the drains remain constant.

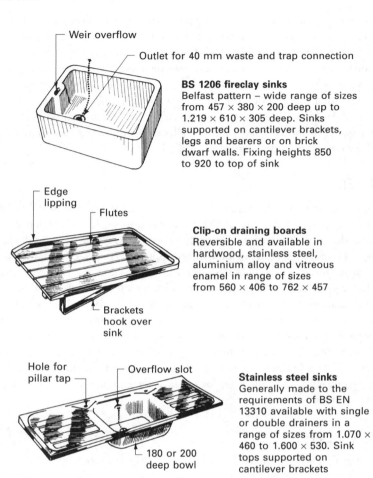

Weir overflow

Outlet for 40 mm waste and trap connection

BS 1206 fireclay sinks
Belfast pattern – wide range of sizes
from 457 × 380 × 200 deep up to
1.219 × 610 × 305 deep. Sinks
supported on cantilever brackets,
legs and bearers or on brick
dwarf walls. Fixing heights 850
to 920 to top of sink

Edge
lipping

Flutes

Clip-on draining boards
Reversible and available in
hardwood, stainless steel,
aluminium alloy and vitreous
enamel in range of sizes
from 560 × 406 to 762 × 457

Brackets
hook over
sink

Hole for
pillar tap

Overflow slot

Stainless steel sinks
Generally made to the
requirements of BS EN
13310 available with single
or double drainers in a
range of sizes from 1.070 ×
460 to 1.600 × 530. Sink
tops supported on
cantilever brackets

180 or 200
deep bowl

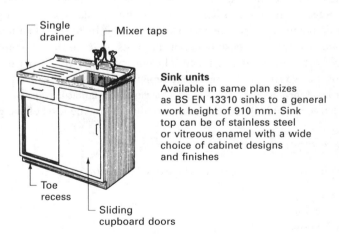

Single
drainer

Mixer taps

Sink units
Available in same plan sizes
as BS EN 13310 sinks to a general
work height of 910 mm. Sink
top can be of stainless steel
or vitreous enamel with a wide
choice of cabinet designs
and finishes

Toe
recess

Sliding
cupboard doors

Note: Waste fittings to BS EN 274-1

Figure 11.2.4   Sinks and draining boards.

One-pipe system

This consists of a single discharge pipe, which conveys both soil and waste water directly to the drain. To ensure that water seals in the traps are not broken, deep seals of 75 mm for waste pipes up to 65 mm diameter and 50 mm for pipes over 75 mm diameter are required. To allow for unrestricted layout of appliances most branch pipes will require an anti-siphon arrangement (see Fig. 11.2.5). The advantage of this system is the flexibility of appliance layout, the main disadvantage is cost, and generally the one-pipe system has been superseded by the more restricted but economic single-stack system described below.

Two-pipe system

As its name implies, this system consists of two discharge pipes: one conveys soil discharges and the other all the waste discharges. It is a simple, reliable and costly system, but has the advantages of complete flexibility in appliance layout and the fact that deep seal traps are not usually required. Like the one-pipe system, it has been largely superseded by the single-stack system. A comparison of the one- and two-pipe systems is shown in Fig. 11.2.5.

Single-stack system

This system was developed by the Building Research Establishment, and is fully described in their Digest No. 249. It is also fully detailed in BS EN 12056-2, and is now adapted into the Building Regulations, Approved Document H1. It is a simplification of the one-pipe system by using deep seal traps, relying on venting by the discharge pipe and placing certain restrictions on basin waste pipes, which have a higher risk of self-siphonage than other appliances. A diagrammatic layout is shown in Fig. 11.2.6.

Materials that can be used for domestic stack pipework include galvanised steel prefabricated stack units (BS 3868), cast iron (BS 416) and uPVC (BS 4514). The latter is standard contemporary practice, being light and easy to install with simple push-fit or solvent-weld joints. Branch waste pipes can be of a variety of plastics, including uPVC, polyethylene and polypropylene, or of copper.

Most manufacturers of soil pipes, ventilating pipes and fittings produce special components for various plumbing arrangements and appliance layouts. These fittings have the water closet socket connections, bosses for branch waste connections and access plates for cleaning and maintenance arranged as one prefabricated assembly to ease site work and ensure reliable and efficient connections to the discharge or soil pipe.

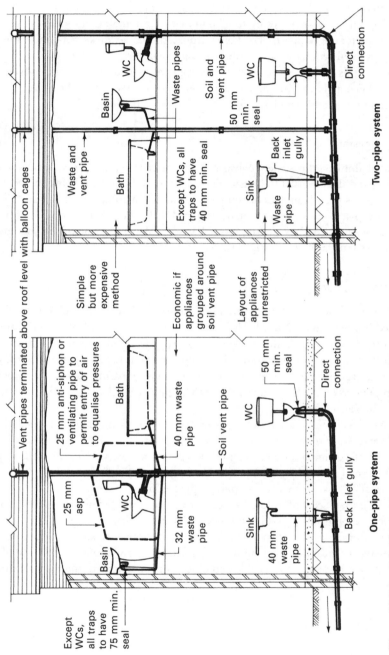

Figure 11.2.5 Comparison of one-pipe and two-pipe systems.

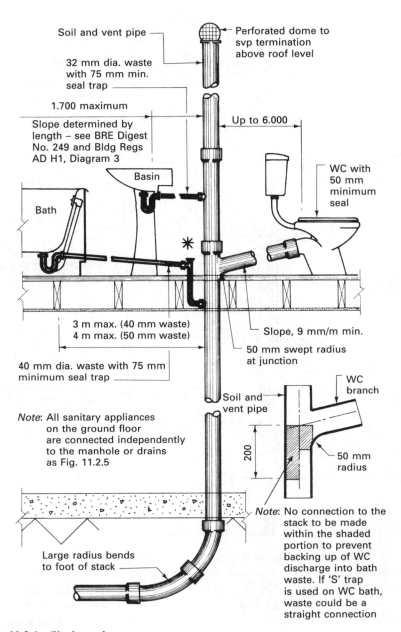

Soil and vent pipe

Perforated dome to svp termination above roof level

32 mm dia. waste with 75 mm min. seal trap

1.700 maximum

Up to 6.000

Slope determined by length – see BRE Digest No. 249 and Bldg Regs AD H1, Diagram 3

Basin

WC with 50 mm minimum seal

Bath

✳

3 m max. (40 mm waste)
4 m max. (50 mm waste)

Slope, 9 mm/m min.

50 mm swept radius at junction

40 mm dia. waste with 75 mm minimum seal trap

Soil and vent pipe

WC branch

200

50 mm radius

Note: All sanitary appliances on the ground floor are connected independently to the manhole or drains as Fig. 11.2.5

Large radius bends to foot of stack

Note: No connection to the stack to be made within the shaded portion to prevent backing up of WC discharge into bath waste. If 'S' trap is used on WC bath, waste could be a straight connection

Figure 11.2.6   Single-stack system.

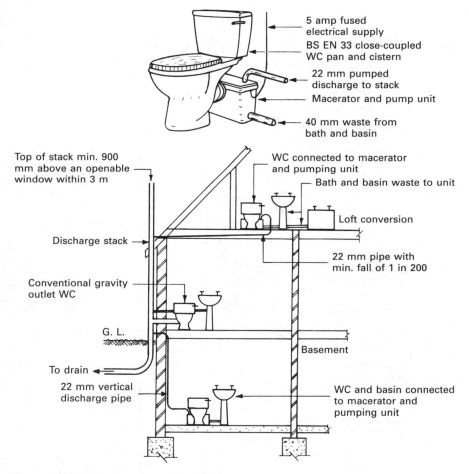

Figure 11.2.7   Macerator/pump installation.

Small-diameter pumped discharge systems

A macerator and pump unit may be connected to the WC pan outlet to collect, process and discharge soil and waste from the WC and other sanitary appliances. The discharge is conveyed in a relatively small-diameter pipe to the nearest convenient soil stack. Figure 11.2.7 shows the possible installations from the box attachment at the back of a WC pan, through vertical and near-horizontal pipes. Structural disruption is minimal, and the small discharge pipes can be used for considerable distances between WC and stack.

Macerator installation is permitted in a dwelling only as a supplement to an existing gravity-discharging WC. See Building Regulation G: Hygiene, Approved Document G1, *Sanitary conveniences and washing facilities*. This requirement is a safeguard that, in the event of an electrical power failure, the conventional gravity-discharge WC will still function.

Most applications are to loft conversions, extensions and basements, or to other situations where it is impractical to install a normal discharge branch pipe. The Care Standards Act has also produced a need for additional facilities. Care home proprietors must provide their residents with access to suitable and sufficient conveniences. This places greater emphasis for en-suite facilities, particularly in care homes for elderly people.

Installation guidelines
- By application to, and at the discretion of the local authority building control department.
- Building Regulations approval will apply only where appliance specification satisfies strict test standards, such as that set by the British Board of Agrément (BBA).
- Discharge pipework:
  - 22 or 28 mm outside diameter copper or equivalent in stainless steel or polypropylene.
  - Long radius 'pulled' bends at direction changes, not proprietary elbow fittings.
  - Minimum fall or gradient, 1 in 200.
  - Up to 50 m horizontally and 5 m vertically.
- Not suitable for use with siphonic WC pans.
- Short spigot, 100 mm diameter horizontal outlet pans to BS EN 33 or 37 required.
- Electrical requirements:
  - 230 volts single phase.
  - 5 amp fused non-switched neon light spur box.
  - 450–650 watts pump specification.

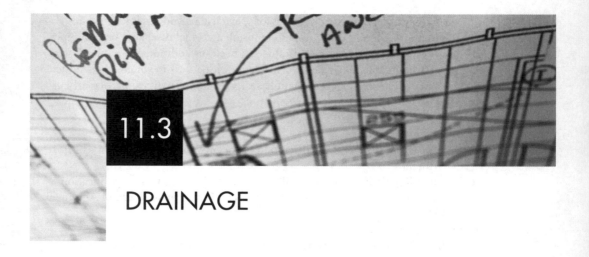

11.3

DRAINAGE

Drainage is a system of pipework, usually installed below ground level, to convey the discharge from sanitary fittings, rainwater gutters and downpipes to a suitable disposal installation. The usual method of disposal is to connect the pipework to the public sewer, which will convey the discharges to a local authority sewage treatment plant. Alternatives are a small self-contained treatment plant on site or a cesspool; the latter is a collection tank to hold the discharge until it can be collected in a special tanker lorry and taken to the local authority sewage treatment installation for disposal.

■■■ PRINCIPLES OF GOOD DRAINAGE

- Materials should have adequate strength and durability.
- Diameter of drain to be as small as practicable: for soil drains the minimum diameter allowed is 100 mm and for surface water the minimum diameter is 75 mm.
- Every part of a drain should be accessible for the purposes of inspection and cleansing.
- Drains should be laid in straight runs as far as possible.
- Drains must be laid to a gradient that will render them efficient. The fall or gradient should be calculated according to the rate of flow, the velocity required and the diameter of the drain. Individual domestic buildings have an irregular flow, and tables to accommodate this are provided in the Building Regulations, Approved Document H1. Alternatively, Maguire's rule can be used to give a gradient with reasonable velocity. This well-established rule of thumb calculates the gradient by dividing the pipe diameter (mm) by 2.5. Thus a 100 mm drain can have a fall of 1 in 40, a 150 mm drain 1 in 60, and so on. Velocity of flow should be at least 0.75 m/s to ensure self-cleansing with a drain capacity not more than 0.75 proportional depth to ensure clear flow conditions. For more detailed drainage calculations, see *Building Services, Technology and Design* by R. Greeno (Pearson Education and CIOB co-publishers).

- Every drain inlet should be trapped to prevent the entry of foul air into the building; the minimum seal required is 50 mm. The trap seal is provided in many cases by the sanitary fitting itself; rainwater drains need not be trapped unless they connect with a soil drain or sewer.
- A rodding access should be located at the head or start of each drain run.
- Inspection chambers, manholes, rodding eyes or access fittings should be placed at changes of direction and gradient if these changes would prevent the drain from being readily cleansed.
- Inspection chambers must also be placed at a junction, unless each run can be cleared from an access point.
- A change of drain pipe size will also require access.
- Junctions between drains must be arranged so that the incoming drain joints at an oblique angle in the direction of the main flow.
- Avoid drains under buildings if possible; if unavoidable they must be protected to ensure watertightness and to prevent damage. The usual protection methods employed are:
  - encase the drain with 100 mm (minimum) of granular filling;
  - use cast iron pipes under the building.
- Drains that are within 1 m of the foundations to the walls of buildings and below the foundation level must be backfilled with concrete up to the level of the underside of the foundations. Drains more than 1 m from the foundations are backfilled with concrete to a depth equal to the distance of the trench from the foundation less 150 mm.
- Where possible, the minimum invert level of a drain should be 450 mm to avoid damage by ground movement and 700 mm for traffic. The invert level is the lowest level of the bore of a drain.

▮▮▮ DRAINAGE SCHEMES

The scheme or plan layout of drains will depend upon a number of factors:

- number of discharge points;
- relative positions of discharge points;
- drainage system and location of the local authority sewers;
- internal layout of sanitary fittings;
- external positions of rainwater pipes;
- disposition of buildings;
- topography of the area to be served.

Drainage systems must be designed within the limits of the terrain, so that the discharges can flow by gravity from the point of origin to the point of discharge. The pipe sizes and gradients must be selected to provide sufficient capacity to accommodate maximum flows, while at the other extreme they must have adequate self-cleansing velocity at minimum flows to prevent debris from accumulating. Economic and construction factors control the depth to which drains can be laid. Therefore, in flat and opposingly inclined areas it may be necessary to provide pumping stations to raise the drainage discharge to higher-level sewers.

There are three drainage systems used by local authorities in this country, and the method employed by any particular authority will determine the basic scheme to be used for the drain runs from individual premises.

COMBINED SYSTEM

All the drains discharge into a common or combined sewer. It is a simple and economic method as there is no duplication of drains. This method has the advantages of easy maintenance, all drains are flushed when it rains, and it is impossible to connect to the wrong sewer. The main disadvantage is that all the discharges must pass through the sewage treatment installation, which could be costly and prove to be difficult with periods of heavy rain.

TOTALLY SEPARATE SYSTEM

The most common method employed by local authorities; two sewers are used in this method. One sewer receives the surface water discharge and conveys this direct to a suitable outfall such as a river, where it is discharged without treatment. The second sewer receives all the soil or foul discharge from baths, basins, sinks, showers and toilets; this is then conveyed to the sewage treatment installation. More drains are required, and it is often necessary to cross drains one over the other. There is a risk of connection to the wrong sewer, and the soil drains are not flushed during heavy rain, but the savings on the treatment of a smaller volume of discharge leads to an overall economy.

PARTIALLY SEPARATE SYSTEM

This is a compromise of the other two systems, and is favoured by some local authorities because of its flexibility. Two sewers are used, one to carry surface water only and the other to act as a combined sewer. The amount of surface water to be discharged into the combined sewer can be adjusted according to the capacity of the sewage treatment installation.

Soakaways, which are pits below ground level designed to receive surface water and allow it to percolate into the soil, are sometimes used to lessen the load on the surface water sewers. Typical examples of the three drainage systems are shown in Fig. 11.3.1.

PRIVATE SEWERS

A sewer can be defined as a means of conveying waste, soil or rainwater below the ground that has been collected from the drains and conveying it to the final disposal point. If the sewer is owned and maintained by the local authority it is generally called a **public sewer**, whereas one owned by a single person or a group of people and maintained by them can be classed as a **private sewer**.

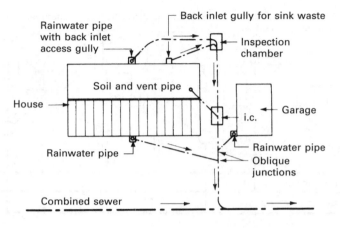

Combined system

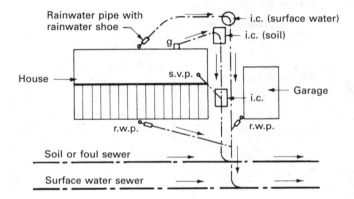

Separate system

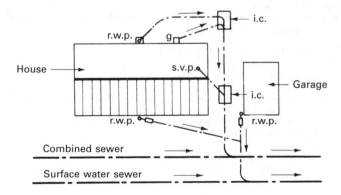

Partially separate system

Figure 11.3.1   Drainage systems.

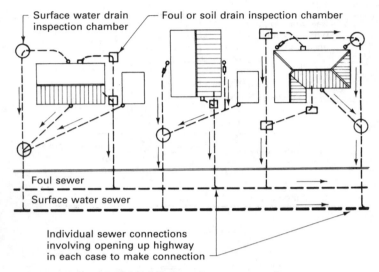

Individual drain and sewer connections

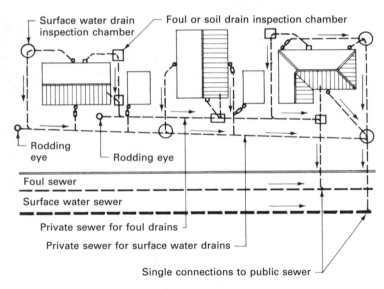

Note: Generally 20 houses can be connected to a 100 mm dia.
      private sewer at a gradient of 1:70 and 100 houses
      can be connected to a 150 mm dia. private sewer
      laid to a fall of 1:150

Figure 11.3.2 Example of a private sewer arrangement.

When planning the connection of houses to the main or public sewer one method is to consider each dwelling in isolation, but important economies in design can be achieved by the use of a private sewer. A number of houses are connected to the single sewer, which in turn is connected to the public sewer. Depending upon the number of houses connected to the private sewer, and the distance from the public sewer, the following savings are possible:

- total length of drain required;
- number of connections to public sewer;
- amount of openings in the roads;
- number of inspection chambers.

A comparative example is shown in Fig. 11.3.2.

DRAINAGE MATERIALS

Drain pipes are considered as either rigid or flexible according to the material used in their manufacture. Clay is a major material used for rigid drain pipes in domestic work, with cast iron as an expensive alternative. Flexible drain pipes are produced from unplasticised PVC.

Clay pipes

This is the traditional material used for domestic drainage, with current manufacturing standards in accordance with BS 65 and BS EN 295. These pipes were treated with a glazing of common salt, borax or boric acid during the firing process to render them impervious. Today's high standards of manufacture combined with quality dense materials no longer justify this process. Several qualities of pipes are produced, ranging from standard pipes for general use to surface water pipes and pipes of extra strength to be used where heavy loadings are likely to be encountered. The type and quality of pipes are marked on the barrel so that they can be identified after firing.

Clay pipes are produced in a range of diameters from 100 to 300 mm nominal bore for general building work. Lengths vary between manufacturers from 1300 to 1600 mm for plain-end pipes, with spigot and socketed pipes produced in shorter lengths down to 600 mm. They can be obtained with sockets and spigots prepared for rigid or flexible jointing. Rigid jointing is rarely used now, except perhaps for small repairs or alterations to an existing drain, as it is difficult to make a sound joint in wet trenches; also the system is unable to absorb movement without fracturing. Most pipes are supplied plain ended for use with push-fit polypropylene sleeve couplings. A wide variety of fittings for use with clay pipes are manufactured to give flexibility when planning drainage layouts and means of access. Typical examples of clay pipes, joints and fittings are shown in Figs 11.3.3 and 11.3.4.

Clay pipes are resistant to attack by a wide range of acids and alkalis: they are therefore suitable for all forms of domestic drainage.

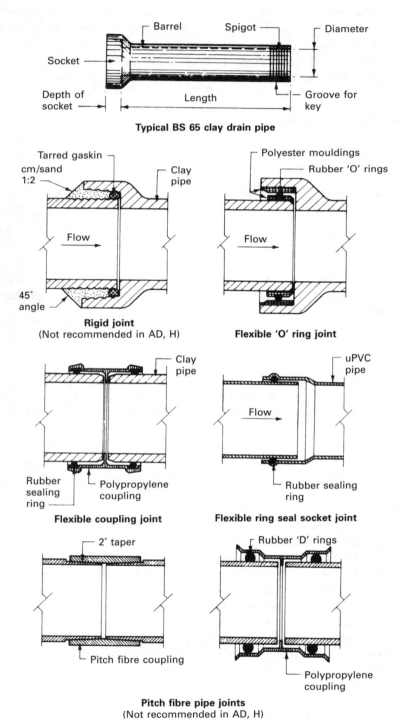

Typical BS 65 clay drain pipe

Rigid joint
(Not recommended in AD, H)

Flexible 'O' ring joint

Flexible coupling joint

Flexible ring seal socket joint

Pitch fibre pipe joints
(Not recommended in AD, H)

Figure 11.3.3   Pipes and joints.

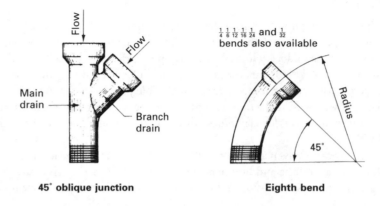

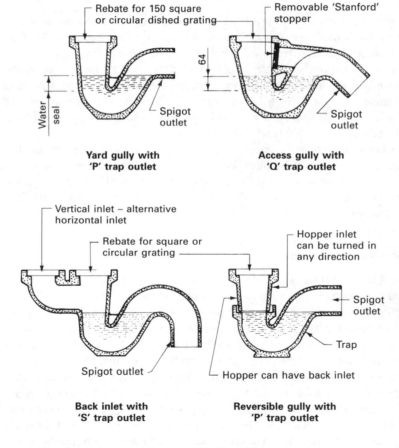

Figure 11.3.4 Clay drain pipe fittings and gullies.

Cast iron pipes

These pipes are generally considered for domestic drainage only in special circumstances such as sites with unstable ground, drains with shallow inverts, and drains that pass under buildings. Like clay pipes, cast iron pipes are made with a spigot and socket for rigid or flexible joints. The rigid joint is made with a tarred gaskin and caulked lead, whereas the flexible joint has a sealing strip in the socket, allowing a 5° deflection. Lengths, diameters and fittings available are similar to those produced for clay pipes but to the recommendations of BS 437 and BS EN 598.

Cast iron pipes are given a protective coating of a hot tar composition or a cold solution of a naphtha and bitumen composition. This coating gives the pipes good protection against corrosion and reasonable durability in average ground conditions.

Pitch fibre pipes

Pitch fibre pipes and fittings preceded uPVC as a flexible drainage material. They are rarely used now, but will be found in many existing systems. They are made from preformed felted wood cellulose fibres thoroughly impregnated under vacuum and pressure, with at least 65% by weight of coal tar pitch or bituminous compounds. They have been used for all forms of domestic drainage, and because of the smooth bore with its high flow capacity could be laid to lower gradients than most other materials of their day. Sizes range from 50 to 225 mm nominal bore, with lengths generally 2400 and 3000 mm.

The original joints had a machined 2° taper on the ends of the pipe, which made a drive fit to machined pitch fibre couplings. These joints are watertight, but do not readily accommodate axial or telescopic movement. The snap joint used a rubber 'D' ring in conjunction with a polypropylene coupling, giving a flexible joint. Both methods are shown in Fig. 11.3.3.

Unplasticised PVC pipes

These pipes and fittings are made from polyvinyl chloride plus additives that are needed to facilitate the manufacture of the polymer and produce a sound, durable pipe. BS 3506 gives the requirements for pipes intended for industrial purposes, and BS 4660 covers the pipes and fittings for domestic use. Standard outside diameters are 110 and 160 mm (100 and 150 mm nominal bore) with non-standard outside diameters of 82, 200, 250 and 315 mm being available from most manufacturers. Long lengths of 3 m and 6 m can be easily cut by hacksaw. BS 5481 and BS EN 1401-1 are also available as specifications for uPVC sewer pipes in nominal diameters of 200–630 mm. The pipes are obtainable with socket joints for either a solvent-welded joint or a ring seal joint (see Fig. 11.3.3). Like pitch fibre pipes, uPVC pipes have a smooth bore, are light and easy to handle, long lengths reducing to a minimum the number of joints required; they can be jointed and laid in all weathers.

■■■ DRAIN LAYING

Domestic drains are laid in trenches that are excavated and if necessary timbered in a similar manner to that described for foundations; the main difference is that drain trenches are excavated to the required fall or gradient. It is good practice to programme the work to enable the activities of excavation, drain laying and backfilling to be carried out in quick succession so that the excavations remain open for the shortest possible time.

The technique used in the laying and bedding of drains will depend upon two factors:

- material – rigid or flexible;
- joint – rigid or flexible.

Approved Document H recommends drains to be of sufficient strength and durability and so jointed that the drain remains watertight under all working conditions, including any differential movement between the pipe and ground.

Many examples of traditional bedding and haunching in concrete will be found in existing drains, but this is no longer practised unless the drain requires a complete surround of concrete for total protection. The cost is prohibitive, the quality of concrete is difficult to control in wet trenches, and the system is prone to failure in ground movement and settlement. Both clay and uPVC drains can accommodate axial flexibility and extensibility by combining flexible jointing with a granular flexible bedding medium. Several examples of bedding specifications are provided in the Building Regulations, Approved Document H and BS EN 752-1 to 4: *Drain and sewer systems outside buildings*. Two popular applications are shown in Fig. 11.3.5.

The selected material required for granular bedding and for tamping around pipes laid on a jointed concrete base must be of the correct quality. Pipes depend to a large extent upon the support bedding for their strength and must therefore be uniformly supported on all sides by a material that can be hard compacted. Generally a non-cohesive granular material with a particle size of 5–20 mm is suitable, and if not present on site it will have to be 'imported'.

Pipes with socket joints are laid from the bottom of the drain run with the socket end laid against the flow, each pipe being aligned and laid to the correct fall. The collar of the socket is laid in a prepared 'hollow' in the bedding, and the bore is centralised. In the case of a rigid joint a tarred gaskin is used, which also forms the seal, whereas the mechanical or flexible joints are self-aligning. Most flexible joints require a special lubricant to ease the jointing process, and those that use a coupling can be laid in any direction.

MEANS OF ACCESS

Drains require access for testing, maintenance and clearance of blockages. Four possibilities exist:

- rodding eye;
- shallow access fitting;
- inspection chamber;
- manhole.

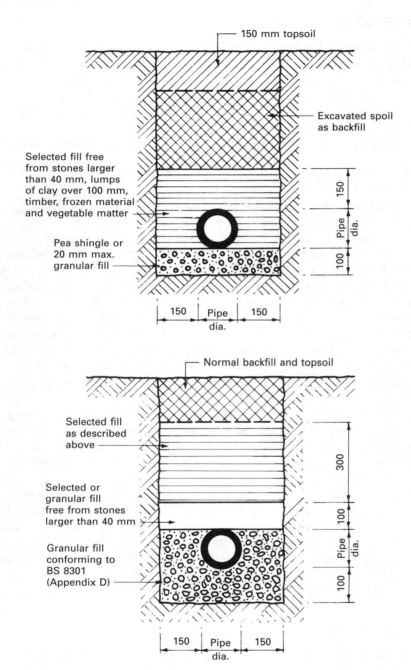

Figure 11.3.5   Typical pipe bedding details.

When fitted with lightweight covers these should be secured by screws to deter unauthorised interference.

Rodding eyes

These generally locate at the head of a drain and are effectively a swept extension of the drain to surface level, with a sealed access plate. With local authority approval they can replace a more expensive inspection chamber to provide means for clearance in one direction only.

Shallow access fittings

These provide vertical access to both directions of a drain run that is not over 600 mm deep.

Inspection chambers

These are an enlarged version of the shallow access fitting, used to invert depths of up to 1 m. They provide for limited access, and contain facilities for a few junctions and branch connections. Materials include plastics, precast concrete and traditional brick masonry. Figure 11.3.6 shows examples of means of drain access, with the plastic inspection chamber containing optional adaptors, which can be cut to suit 100 or 150 mm pipes at various approach angles. The depth of these units is cut to suit drain trench levels, and a cast iron or aluminium frame and cover is fitted.

Manholes

These are inspection chambers over 1 m to invert. They are a compartment containing half or three-quarter section round channels to enable the flow to be observed, and to provide a drain access point for cleansing and testing. Both inspection chambers and manholes are positioned to comply with the access recommendations of Approved Document H (see Fig. 11.3.1).

Simple domestic drainage is normally concerned only with shallow manholes up to an invert depth of 1800 mm. The internal sizing is governed by the depth to invert, the number of branch drains, the diameter of branch drains, and the space required for a person to work within the manhole. A general guide to the internal sizing of inspection chambers and manholes is given in Approved Document H, Table 9.

Manholes can be constructed of brickwork or of rectangular or circular precast concrete units (see Fig. 11.3.7). The access covers used in domestic work are generally of cast iron or pressed steel, and are light duty as defined in BS EN 124. They have a single seal, which should be bedded in grease to form an airtight joint; double seal covers would be required if the access was situated inside the building. Concrete access covers are available for use with surface water manholes.

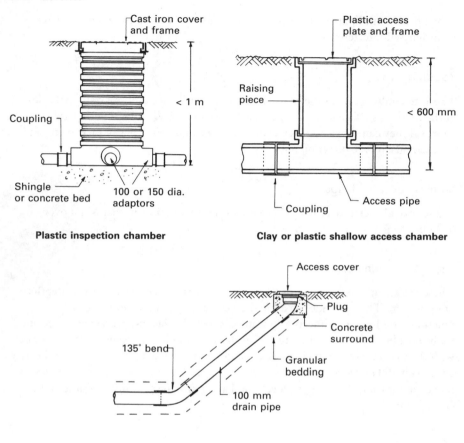

Plastic inspection chamber

Clay or plastic shallow access chamber

Rodding eye

Figure 11.3.6   Drains – means of access up to 1 m deep.

LOCATION OF DRAIN ACCESS POINTS

Access to drainage systems is required at various situations. These are outlined on page 579. Long straight drain runs must also be avoided, otherwise parts of the drain would be very difficult to access by cleaning rods in event of a blockage. For domestic drainage (up to 300 mm diameter) the maximum spacing between different means of access provision are as listed in Table 11.3.1.

VENTILATION OF DRAINS

To prevent foul air from soil and combined drains escaping and causing a nuisance, all drains should be vented by a flow of air. A ventilating pipe should be provided at or near the head of each main drain and any branch drain exceeding 10.000 in length. The ventilating pipe can be a separate pipe, or the soil discharge stack pipe

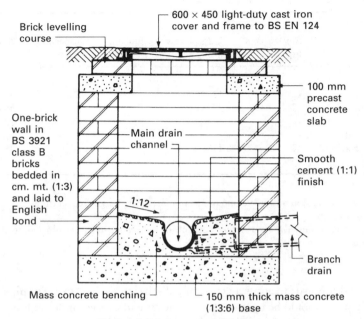

Brick levelling course

600 × 450 light-duty cast iron cover and frame to BS EN 124

100 mm precast concrete slab

One-brick wall in BS 3921 class B bricks bedded in cm. mt. (1:3) and laid to English bond

Main drain channel

Smooth cement (1:1) finish

1:12

Branch drain

Mass concrete benching

150 mm thick mass concrete (1:3:6) base

Shallow brick inspection chamber

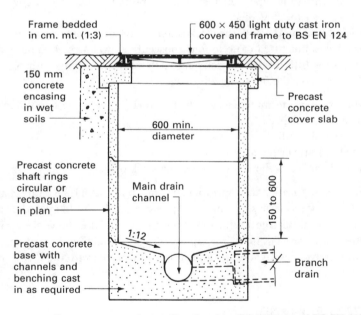

Frame bedded in cm. mt. (1:3)

600 × 450 light duty cast iron cover and frame to BS EN 124

150 mm concrete encasing in wet soils

600 min. diameter

Precast concrete cover slab

Precast concrete shaft rings circular or rectangular in plan

Main drain channel

150 to 600

Precast concrete base with channels and benching cast in as required

1:12

Branch drain

Precast concrete inspection chamber (BS 5911-4 and BS EN 1917)

Figure 11.3.7 Typical shallow inspection chambers.

Table 11.3.1   Maximum spacing of drainage access (m)

	To: Small access fitting	Large access fitting	Junction	Inspection chamber	Manhole
From:					
Start of drain	12	12	–	22	45
Rodding eye	22	22	22	45	45
Small access fitting	–	–	12	22	22
Large access fitting	–	–	12	45	45
Inspection chamber	22	45	22	45	45
Manhole	–	–	–	45	90 (note 2)

Notes:
1. Small access fitting: 150 mm × 100 mm or 150 mm diameter. Large access fitting: 225 mm × 100 mm.
2. Up to 200 m is permitted if the manhole is of substantial proportions, i.e. sufficient to allow a person entry space to work.

can be carried upwards to act as a ventilating discharge stack or soil vent pipe. Ventilating pipes should be open to the outside air and carried up for at least 900 mm above the head of any window opening within a horizontal distance of 3.000 from the ventilating pipe, which should be finished with a cage or cover that does not restrict the flow of air.

Conventional venting through the stack is not always necessary. An air admittance valve can be fitted in up to four consecutive dwellings of no more than three storeys, if the fifth dwelling is conventionally vented. Systems with air admittance valves have the following advantages:

■ Ventilating stack can terminate inside the building (above highest spillover level), typically in the roof space.
■ Greater design flexibility.
■ Adaptable to plastic or metal pipework.
■ Visually unobtrusive as there is no projecting stack pipe.

The principles of application and operation are shown in Fig. 11.3.8. A discharge of water in the stack creates a slight negative pressure, sufficient to open the valve and admit air. After discharge a return to atmospheric pressure allows the spring to re-seal the unit to prevent foul air escaping. To satisfy building regulations, the device must have received design approval in accordance with the Board of Agrément.

RAINWATER DRAINAGE

A rainwater drainage installation is required to collect the discharge from roofs and paved areas, and convey it to a suitable drainage system. Paved areas, such as garage forecourts or hardstands, are laid to fall so as to direct the rainwater into a yard gully, which is connected to the surface water drainage system. A rainwater

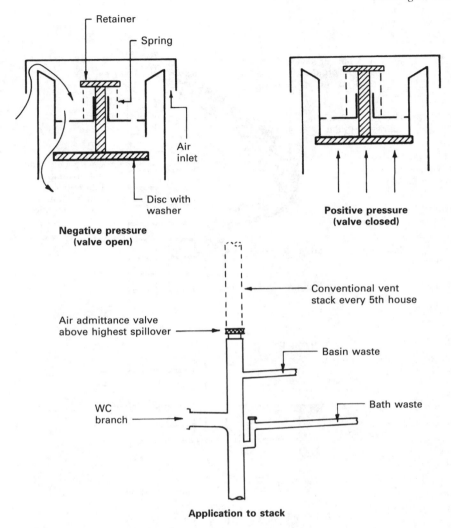

Figure 11.3.8 Air admittance valve.

installation for a roof consists of a collection channel called a **gutter,** which is connected to vertical rainwater pipes. The rainwater pipe is terminated at its lowest point by means of a rainwater shoe for discharge to a surface water drain, or a trapped gully if the discharge is to a combined drain (see Fig. 11.3.9). If a separate system of drainage or soakaways is used it may be possible to connect the rainwater pipe direct to the drains, provided there is an alternative means of access for cleansing.

The materials available for domestic rainwater installations are galvanised pressed steel, cast iron and uPVC. The usual materials for domestic work are cast iron and uPVC, the latter being the usual specification for new work.

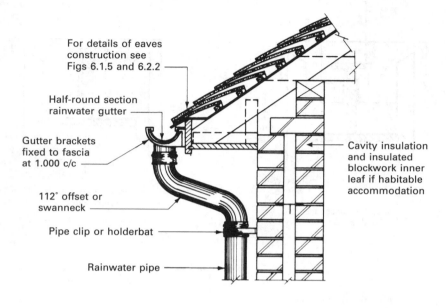

For details of eaves
construction see
Figs 6.1.5 and 6.2.2

Half-round section
rainwater gutter

Gutter brackets
fixed to fascia
at 1.000 c/c

112° offset or
swanneck

Pipe clip or holderbat

Rainwater pipe

Cavity insulation
and insulated
blockwork inner
leaf if habitable
accommodation

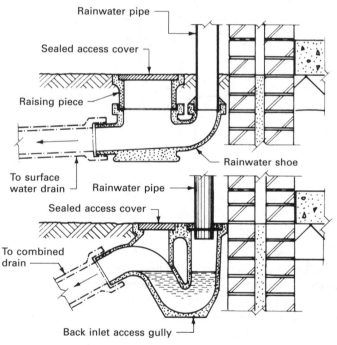

Rainwater pipe

Sealed access cover

Raising piece

To surface
water drain

Rainwater shoe

Rainwater pipe

Sealed access cover

To combined
drain

Back inlet access gully

Figure 11.3.9   Rainwater pipework and drainage.

Cast iron rainwater goods

Cast iron rainwater pipes, gutters and fittings are generally made to the requirements of BS 460, which specifies a half-round section gutter with a socket joint in diameters from 75 to 150 mm and an effective length of 1800 mm. The gutter socket joint should be lapped in the direction of the flow and sealed with either putty or an approved sealing compound before being bolted together. The gutter is supported at 1000–1800 mm centres by means of mild steel gutter brackets screwed to the feet of rafters for an open eaves or to the fascia board with a closed eaves.

Cast iron rainwater pipes are also produced to a standard effective length of 1800 mm with a socket joint that is sealed with putty, run lead or in many cases dry jointed – the pipe diameters range from 50 to 150 mm. The downpipes are fixed to the wall by means of pipe nails and spacers when the pipes are supplied with ears, or with split ring hinged holderbats when the pipes are supplied without ears cast on. A full range of fittings such as outlets, stopped ends and internal and external angles are available for cast iron half-round gutters, and, for the downpipes, fittings such as bends, offsets and rainwater heads are produced.

Unplasticised PVC rainwater goods

The advantages of uPVC rainwater goods over cast iron are:

- Easier jointing, gutter bolts are not required, and the joint is self-sealing, generally by means of a butyl or similar strip.
- Corrosion is eliminated.
- Decoration is not required; several standard colours are available including brown, black, white and grey.
- Breakages are reduced.
- Better flow properties usually enable smaller sections and lower falls.

Half-round gutters are supplied in standard effective lengths up to 6 m, with diameter ranging from 75 to 150 mm. Downpipes are supplied in two standard lengths of 2 and 4 m with diameters of 50, 63, 75 and 89 mm. The gutters, pipes and fittings are generally produced to the requirements of BS 4576 and BS 12200-1. Typical details of domestic rainwater gutter and pipework are shown in Fig. 11.3.9.

Sizing of pipes and gutters

The sizing of the gutters and downpipes to effectively cater for the discharge from a roof will depend upon:

- the area of roof to be drained;
- anticipated intensity of rainfall;
- material of gutter and downpipe;
- fall within gutter, usually in the range 1/150 to 1/600;
- number, size and position of outlets.

The requirements for Building Regulation H3 concerning rainwater drainage can be satisfied by using the design guide tables contained in Approved Document H, which gives guidance to sizing gutters, downpipes and selection of suitable materials.

███ CONNECTIONS TO SEWERS

It is generally recommended that all connections to sewers shall be made so that the incoming drain or private sewer is joined to the main sewer obliquely in the direction of flow and that the connection will remain watertight and satisfactory under all working conditions. Normally sewer connections are made by the local authority or under their direction and supervision.

The method of connection will depend upon a number of factors:

- relative sizes of sewer and connecting drain or private sewer;
- relative invert levels;
- position of nearest inspection chamber on the sewer run;
- whether the sewer is existing or being laid concurrently with the drains or private sewers;
- whether stopped or joinder junctions have been built into the existing sewer;
- the shortest and most practicable route.

If the public sewer is of a small diameter, less than 225 mm, the practical method is to remove two or three pipes and replace with new pipes and an oblique junction to receive the incoming drain. If three pipes are removed it is usually possible to 'spring in' two new pipes and the oblique junction and joint in the usual manner, but if only two pipes are removed a collar connection will be necessary (see Fig. 11.3.10).

If new connections have been anticipated stopped junctions or joinder junctions may have been included in the sewer design. A stopped junction has a disc temporarily secured in the socket of the branch arm, whereas the joinder has a cover cap as an integral part of the branch arm. In both cases careful removal of the disc or cap is essential to ensure that a clean undamaged socket is available to make the connection (see Fig. 11.3.10).

Connections to inspection chambers or manholes, whether new or existing, can take several forms, depending mainly upon the differences in invert levels. If the invert levels of the sewer and incoming drain are similar, the connection can be made in the conventional way using an oblique branch channel. Where there is a difference in invert levels the following can be considered:

- a ramp formed in the benching within the inspection chamber;
- a backdrop manhole or inspection chamber;
- an increase in the gradient of the branch drain.

The maximum difference between invert levels that can be successfully overcome by the use of a ramp is a matter of conjecture. The old CP 2005 entitled *Sewerage*, since replaced by BS 8005: Part 1 and subsequently BS EN 752, gave a maximum difference of 1.800 m, whereas BS 8301, covering building drainage, gave 1.000 for external ramps at 45°. The generally accepted limit of invert level difference

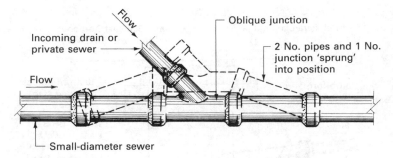

Removing 3 No. pipes and inserting oblique junction

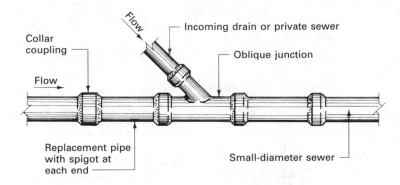

Removing 2 No. pipes and inserting oblique junction

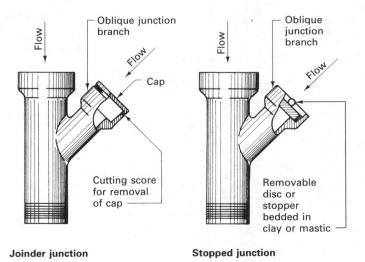

Joinder junction   **Stopped junction**

Figure 11.3.10   Connections to small-diameter sewers (traditional rigid-jointed spigot and socket clay pipes).

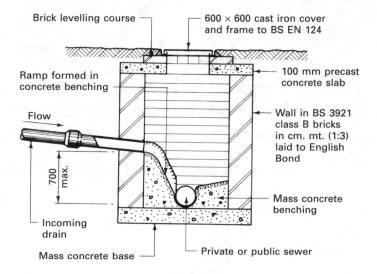

Brick levelling course

600 × 600 cast iron cover
and frame to BS EN 124

Ramp formed in
concrete benching

100 mm precast
concrete slab

Flow

Wall in BS 3921
class B bricks
in cm. mt. (1:3)
laid to English
Bond

700
max.

Mass concrete
benching

Incoming
drain

Mass concrete base

Private or public sewer

Ramp connection

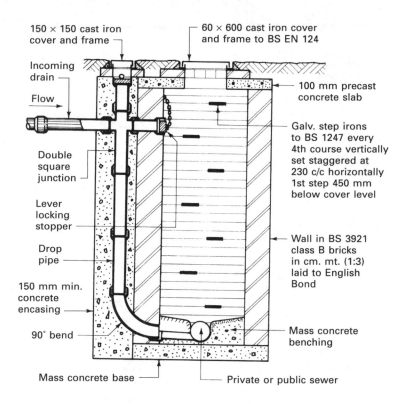

150 × 150 cast iron
cover and frame

60 × 600 cast iron cover
and frame to BS EN 124

Incoming
drain

Flow

100 mm precast
concrete slab

Double
square
junction

Galv. step irons
to BS 1247 every
4th course vertically
set staggered at
230 c/c horizontally
1st step 450 mm
below cover level

Lever
locking
stopper

Drop
pipe

Wall in BS 3921
class B bricks
in cm. mt. (1:3)
laid to English
Bond

150 mm min.
concrete
encasing

90° bend

Mass concrete
benching

Mass concrete base

Private or public sewer

Backdrop connection

Figure 11.3.11 Manhole and inspection chamber sewer connections.

for the use of internal ramps is 700 mm, which approximates to the figure quoted in the first edition of CP 301, later superseded by BS 8301. Typical constructional details are shown in Fig. 11.3.11.

Where the limit for ramps is exceeded, a backdrop manhole construction can be considered. This consists of a vertical 'drop' pipe with access for both horizontal and vertical rodding. If the pipework is of clay or concrete the vertical pipe should be positioned as close to the outside face of the manhole as possible and encased in not less than 150 mm of mass concrete. Cast iron pipework is usually sited inside the chamber and fixed to the walls with holderbolts. Whichever material is used the basic principles are constant (see Fig. 11.3.11).

Changing the gradient of the incoming drain to bring its invert level in line with that of the sewer requires careful consideration and design. Although simple in conception, the gradient must be such that a self-cleansing velocity is maintained and the requirements of Building Regulation Part H are not contravened.

Connections of small-diameter drains to large-diameter sewers can be made by any of the methods described above, or by using a saddle connection. A saddle is a short socketed pipe with a flange or saddle curved to suit the outer profile of the sewer pipe. To make the connection a hole must be cut in the upper part of the sewer to receive the saddle, ensuring that little or no debris is allowed to fall into the sewer. A small pilot hole is usually cut first, and this is enlarged to the required diameter by careful cutting and removing the debris outwards. The saddle connection is bedded onto the sewer pipe with a cement mortar, and the whole connection is surrounded with a minimum of 150 mm of mass concrete (see Fig. 11.3.12).

■ ■ ■ DRAIN TESTING

To satisfy the Building Regulations requirements for watertightness of drains, reference to Approved Document H1 permits application of either a water or an air test. A smoke test could also be applied to determine the position of any apparent leakage revealed by the other tests. The Approved Document also makes reference to the recommendations contained in BS EN 752-3, 4 and 6.

The local authority will carry out drain testing after the backfilling of the drain trench has taken place: therefore it is in the contractor's interest to test drains and private sewers before the backfilling is carried out, as the detection and repair of any failure discovered after backfilling can be time-consuming and costly.

TYPES OF TEST

There are three methods available for the testing of drains:

- **Water test**   The usual method employed; carried out by filling the drain run being tested with water under pressure and observing whether there is any escape of water (see Fig. 11.3.13).
- **Smoke test**   This is a dated practice, but if required smoke can be pumped into the sealed drain run under test and any fall in pressure, as indicated by the fall of the float on the smoke machine, observed (see Fig. 11.3.14).

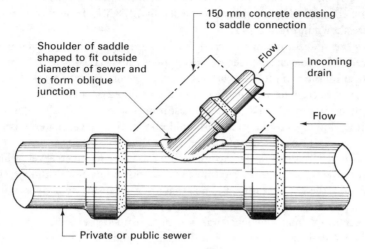

150 mm concrete encasing to saddle connection

Shoulder of saddle shaped to fit outside diameter of sewer and to form oblique junction

Flow

Incoming drain

Flow

Private or public sewer

Connection arrangement

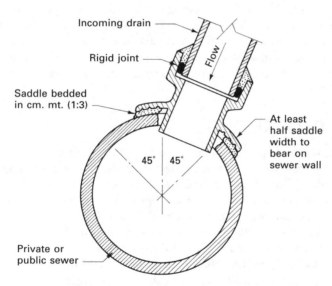

Incoming drain

Rigid joint

Flow

Saddle bedded in cm. mt. (1:3)

At least half saddle width to bear on sewer wall

45°   45°

Private or public sewer

Note: Saddle connection should be made in the crown of the sewer pipe within 45° on either side of the vertical axis

Typical section

Figure 11.3.12   Saddle connections to sewers.

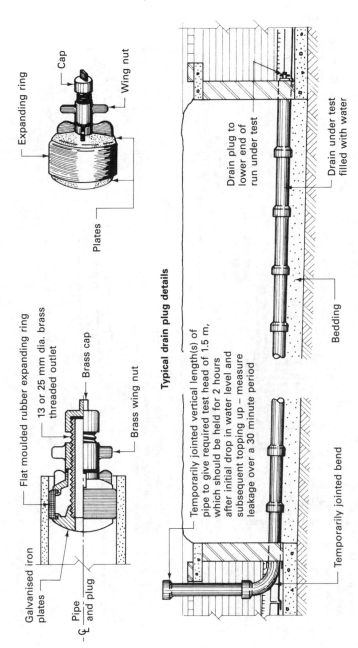

Expanding ring

Cap

Wing nut

Plates

Flat moulded rubber expanding ring

13 or 25 mm dia. brass threaded outlet

Brass cap

Brass wing nut

Galvanised iron plates

Pipe and plug

Typical drain plug details

Drain plug to lower end of run under test

Drain under test filled with water

Bedding

Temporarily jointed vertical length(s) of pipe to give required test head of 1.5 m, which should be held for 2 hours after initial drop in water level and subsequent topping up – measure leakage over a 30 minute period

Temporarily jointed bend

Figure 11.3.13 Water testing of drains.

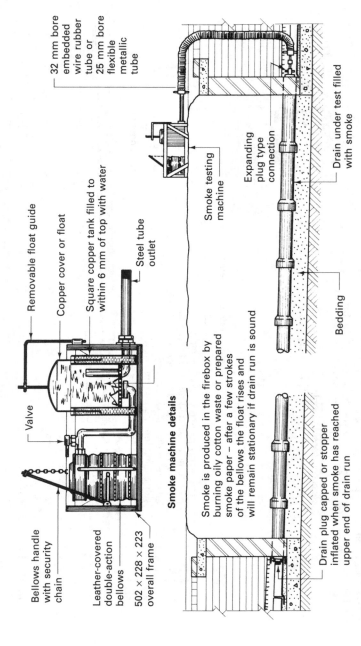

32 mm bore embedded wire rubber tube or 25 mm bore flexible metallic tube

Removable float guide

Copper cover or float

Square copper tank filled to within 6 mm of top with water

Steel tube outlet

Valve

Bellows handle with security chain

Leather-covered double-action bellows

502 × 228 × 223 overall frame

Smoke testing machine

Expanding plug type connection

Drain under test filled with smoke

Bedding

Smoke machine details

Smoke is produced in the firebox by burning oily cotton waste or prepared smoke paper – after a few strokes of the bellows the float rises and will remain stationary if drain run is sound

Drain plug capped or stopper inflated when smoke has reached upper end of drain run

Figure 11.3.14   Smoke testing of drains.

■ **Air test**   This is not a particularly conclusive test, but it is sometimes used in special circumstances such as large-diameter pipes where a large quantity of water would be required. If a failure is indicated by an air test the drain should be retested using the more reliable water test (see Fig. 11.3.15).

The illustrations of drain testing have been prepared on the assumption that the test is being carried out by the contractor before backfilling has taken place.

In general the testing of drains should be carried out between manholes. Manholes should be tested separately, and short branches of less than 6.000 m should be tested with the main drain to which they are connected; long branches would be tested in the same manner as a main drain.

■■■ SOAKAWAYS

A soakaway is a pit dug in permeable ground that receives the rainwater discharge from the roof and paved areas of a small building, and is so constructed that the water collected can percolate into the surrounding subsoil. To function correctly and efficiently a soakaway must be designed to take the following factors into account:

■ permeability or rate of dispersion of the subsoil;
■ area to be drained;
■ storage capacity required to accept sudden inflow of water, such as that encountered during a storm;
■ local authority requirements as to method of construction and siting in relation to buildings;
■ depth of water table.

Before any soakaway is designed or constructed the local authority should be contacted to obtain permission and ascertain its specific requirements. Some authorities will permit the use of soakaways only as an outfall to a subsoil drainage scheme.

The rate at which water will percolate into the ground depends mainly on the permeability of the soil. Generally, clay soils are unacceptable for soakaway construction, whereas sands and gravels are usually satisfactory. An indication of the permeability of a soil can be ascertained by observing the rate of percolation. A borehole 150 mm in diameter should be drilled to a depth of 1.000 m. Water to a depth of 300 mm is poured into the hole, and the time taken for the water to disperse is noted. Several tests should be made to obtain an average figure, and the whole procedure should be repeated at 1.000 m stages until the proposed depth of the soakaway has been reached. A suitable diameter and effective depth for the soakaway can be obtained from design guidance in Building Research Establishment Digest 365.

An alternative method, using a simple empirical formula, calculates the volume of a soakaway by allowing for a storage capacity equal to one-third of the hourly rain falling onto the area to be drained:

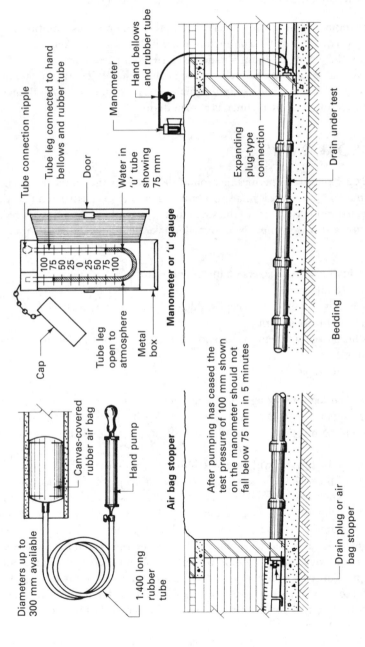

Manometer or 'u' gauge

Tube connection nipple

Tube leg connected to hand bellows and rubber tube

Door

Water in 'u' tube showing 75 mm

Cap

Tube leg open to atmosphere

Metal box

Manometer

Hand bellows and rubber tube

Expanding plug-type connection

Drain under test

Bedding

Air bag stopper

Diameters up to 300 mm available

Canvas-covered rubber air bag

Hand pump

1.400 long rubber tube

After pumping has ceased the test pressure of 100 mm shown on the manometer should not fall below 75 mm in 5 minutes

Drain plug or air bag stopper

Figure 11.3.15   Air testing of drains.

$$C = \frac{AR}{3}$$

where C = capacity (m^3)
A = area to be drained (m^2)
R = rainfall (m/h)

For example:

Area to be drained = 150 m^2
stormwater allowance = 75 mm/h

therefore

$$C = \frac{150 \times 0.075}{3} = 3.75 \text{ m}^3$$

TYPES OF SOAKAWAY

A soakaway is constructed by excavating a pit of the appropriate size and either filling the void with selected coarse granular material or lining the sides of the excavation with brickwork or precast concrete rings (see Figs 11.3.16 and 11.3.17).

Filled soakaways are usually employed only for small capacities, because it is difficult to estimate the storage capacity, and the life of the soakaway may be limited by the silting-up of the voids between the filling material. Lined soakaways are generally more efficient, have a longer life, and if access is provided can be inspected and maintained at regular intervals.

Soakaways should be sited away from buildings so that foundations are unaffected by the percolation of water from the soakaway. The minimum 'safe' distance is often quoted as 5.000 m, but local authority advice should always be sought. The number of soakaways required can be determined only by having the facts concerning total drain runs, areas to be drained, and the rate of percolation for any particular site.

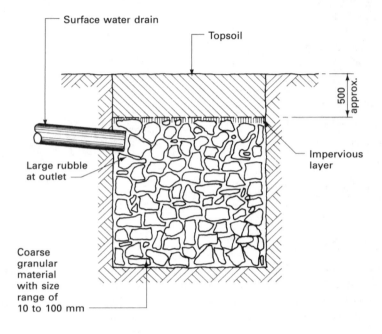

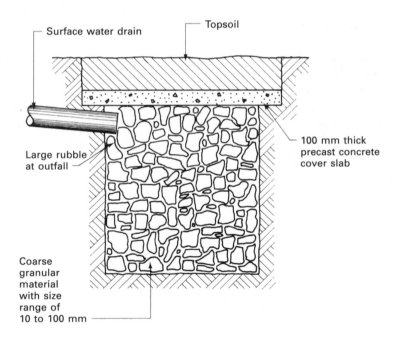

Suitable coarse granular materials include broken bricks, crushed sound rock and hard clinker

Figure 11.3.16   Filled soakaways.

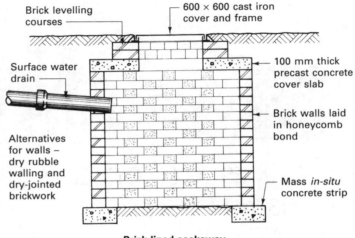

Brick-lined soakaway

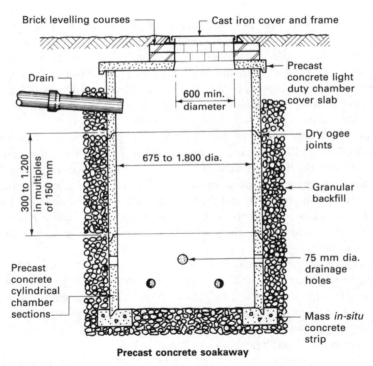

Precast concrete soakaway

Figure 11.3.17 Lined soakaways.

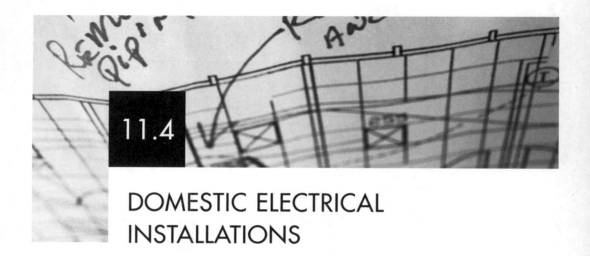

11.4

DOMESTIC ELECTRICAL INSTALLATIONS

A simple definition of the term **electricity** is not possible, but it can be considered as a form of energy due to the free movement of tiny particles called **electrons**. If sufficient of these free or loose electrons move, an electric current is produced in the material in which they are moving. Materials such as most metals and water which allow an electric current to flow readily are called **conductors** and are said to have a low resistance. Materials that resist the flow of an electric current such as rubber, glass and most plastics are called **insulators**.

For an electric current to flow there must be a complete path or circuit from the source of energy through a conductor back to the source. Any interruption of the path will stop the flow of electricity. The pressure that forces or pushes the current around the circuit is called the **voltage**. The rate at which the current flows is measured in **amperes**, and the resistance offered by the circuit to passage of electrons is measured in **ohms**. A **watt** is the unit of power, and is equal to the product of volts × amperes; similarly it can be shown that voltage is equal to the product of amperes × ohms.

Another effect of an electric current flowing through a conductor is that it will dissipate wasted energy in the form of heat according to the resistance of the conductor. If a wire of the correct resistance is chosen it will become very hot, and this heating effect can be used in appliances such as cookers, irons and fires. The conductor in a filament bulb is a very thin wire of high resistance and becomes white hot, thus giving out light as well as heat.

Most domestic premises receive a single-phase supply of electricity from an area electricity authority at a rating of 230 volts, and a frequency of 50 hertz. The area authority's cable, from which the domestic supply is taken, consists of four lines: three lines each carry a 230 volt supply, and the fourth is the common return line or neutral, which is connected to earth at the transformer or substation as a safety precaution should a fault occur on the electrical appliance. Each line or phase is tapped in turn together with the neutral to provide the single-phase 230 V supply. Electricity is generated and supplied as an **alternating current**, which means that the current flows first one way then the other; the direction change is so rapid that it is hardly discernible in fittings such as lights. The cycle of this reversal of flow is termed **frequency**.

Figure 11.4.1 shows a typical underground domestic electricity supply to an external meter box. The consumer unit should be located as close as possible to the intake on the inside of the cavity wall as shown, or on an adjacent partition.

The conductors used in domestic installations are called **cables**, and consist of a conductor of low resistance such as copper or aluminium surrounded by an insulator of high resistance such as rubber or plastic. Cable sizes are known by the nominal cross-sectional area of the conductor, and up to 2.5 mm^2 are usually of one strand. Larger cables consist of a number of strands to give them flexibility. All cables are assigned a rating in amperes, which is the maximum load the cable can carry without becoming overheated.

For domestic work, wiring drawings are not usually required; instead the positions of outlets, switches and lighting points are shown by symbols on the plans. Specification of fittings, fixing heights and cables is given either in a schedule or in a written document (see Fig. 11.4.2).

■■■ CIRCUITS

Domestic buildings are wired using a **ring circuit** as opposed to the older method of having a separate fused subcircuit to each socket outlet. Lighting circuits are carried out by using the **loop in** method.

The supply or intake cable will enter the building through ducts and be terminated in the area authority's fused sealing chamber, which should be sited in a dry accessible position. From the sealing chamber the supply passes through the meter, which records the electricity consumed in units of kilowatt/hours, to the consumer unit, which has a switch controlling the supply to the circuit fuses or miniature circuit-breakers. These fuses or circuit-breakers are a protection against excess current or overload of the circuit because, should overloading occur, the fuse or circuit-breaker will isolate the circuit from the supply.

The number of fuseways or miniature circuit-breakers contained in the consumer unit will depend upon the size of the building and the equipment to be installed. A separate ring circuit of 32 amp loading should be allowed for every 100 m^2 of floor area, and as far as practicable the number of outlets should be evenly distributed over the circuits. A typical domestic installation would have the following circuits from the consumer unit:

1. 6 amp: ground floor lighting up to 10 fittings or a total load of 6 amps;
2. 6 amp: upper floor lighting as above;
3. 16 amp: immersion heater;
4. 32 amp: ring circuit 1;
5. 32 amp: ring circuit 2;
6. 32 amp: ring circuit, kitchen;
7. 45 amp: cooker circuit.

[*Note*: A further 40 or 45 amp circuit-breaker or fuse may be installed for an electric shower unit.]

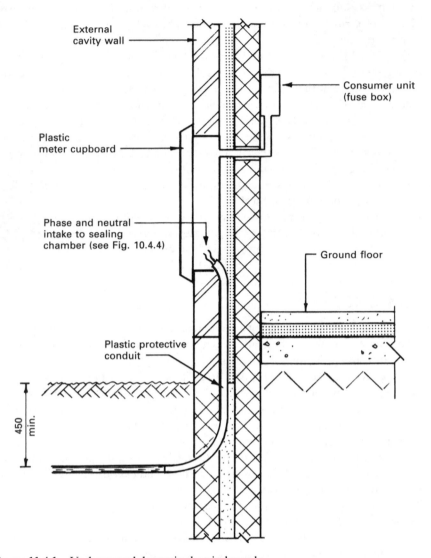

External
cavity wall

Consumer unit
(fuse box)

Plastic
meter cupboard

Phase and neutral
intake to sealing
chamber (see Fig. 10.4.4)

Ground floor

Plastic protective
conduit

450
min.

Figure 11.4.1   Underground domestic electrical supply.

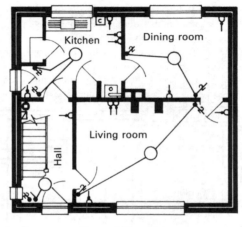

Ground floor plan

Upper floor plan

Symbols		
✓ One-way switch	⚡ Pendant switch	☑ Consumer unit
✗ Two-way switch	◯ Ceiling outlet	◉ Meter
▷— Switch socket outlet	▭— Immersion heater	⊡ Cooker control

Figure 11.4.2 Typical domestic electrical layout.

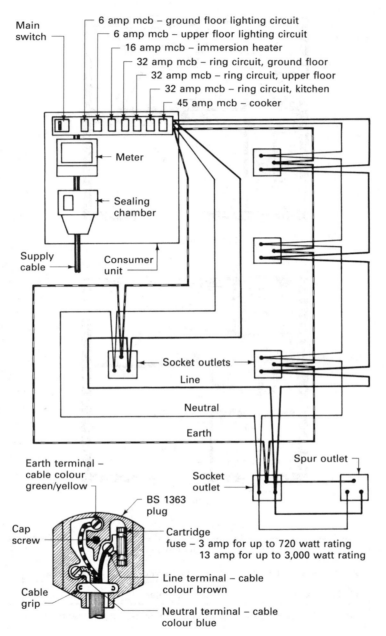

Figure 11.4.3 Ring circuits and plug wiring.

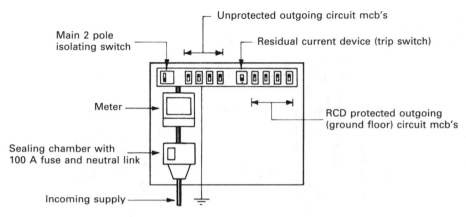

Figure 11.4.4   Split-load consumer unit.

The complete installation is earthed by connecting the metal consumer unit casing to the sheath of the supply cable, or by connection to a separate earth electrode. Figure 11.4.3 shows a standard consumer unit with disposition of fuseways or miniature circuit-breakers. A preferable alternative is the split-load consumer unit shown in Fig. 11.4.4. This contains specific protection to ground floor sockets that could have extension leads attached for use with portable garden equipment.

For lighting circuits using sheathed wiring a 1.5 mm^2 conductor is required, and therefore a twin with earth cable is used. The loop-in method of wiring is shown diagrammatically in Fig. 11.4.5. It is essential that lighting circuits are properly earthed, as most domestic fittings and switches contain metal parts or fixings, which could become live should a fault occur. Lighting circuits using a conduit installation with single-core cables can be looped from switch to switch as shown in Fig. 11.4.5. Conduit installation consists of metal or plastic tubing together with various boxes for forming junctions and housing switches, which gives a protected rewireable system. If steel-screwed conduit is used it will also serve as the earth leakage path, but if plastic conduit is used, as it is non-conductive, the circuit must have an insulated earth conductor throughout.

A ring circuit for socket outlets consists of a twin 2.5 mm^2 earthed cable starting from and returning to the consumer unit. The cables are looped into the outlet boxes, making sure that the correct cable is connected to the correct terminals (see Fig. 11.4.4). The number of outlets is unlimited if the requirement of one ring circuit per 100 m^2 of floor area has been adopted. Spur outlets leading off the main ring circuit are permissible provided the limitations that not more than two outlet sockets are on any one spur, and that not more than half the socket outlets on the circuit are on spurs, are not exceeded. Socket outlets can be switched controlled and of single or double outlet; the double outlet is considered the best arrangement as it discourages the use of multiple adaptors. Fixed appliances such as wall heaters should be connected directly to a fused spur outlet to reduce the number of trailing leads. Movable appliances such as irons, radios and standard lamps should have a fused plug for connection to the switched outlet, conforming to the requirements of BS 1363. The rating of the cartridge fuse should be in accordance with rating

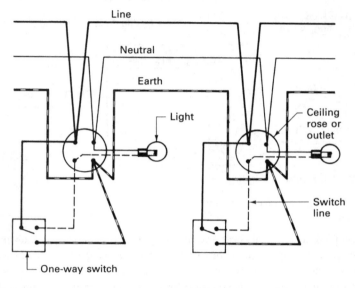

'Loop-in' method of wiring using sheathed cable

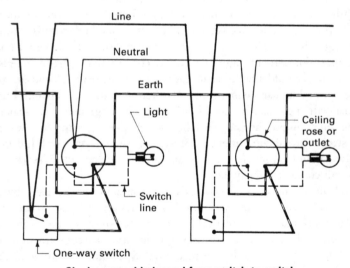

Single-core cable looped from switch to switch

Figure 11.4.5 Typical lighting circuits.

of the appliance. Appliances with a rating of not more than 720 watts should be protected by a 3 amp fuse and appliances over this rating up to 3000 watts should have a 13 amp fuse. As with the circuit, correct wiring of the plug is essential (see Fig. 11.4.3).

The number of outlets is not mandatory, but the minimum numbers recommended for various types of accommodation are:

- kitchens: 6 plus cooker control unit with one outlet socket;
- living rooms: 8;
- dining rooms: 4;
- bedrooms: 4;
- halls: 1;
- landings: 1;
- garages: 1;
- stores and workshops: 1.

The outlets should be installed around the perimeter of the rooms in the most convenient and economic positions to give maximum coverage with minimum amount of trailing leads.

Cables sheathed with PVC can be run under suspended floors by drilling small holes on the neutral axis of the joists; where the cables are to be covered by wall finishes or floor screed they should be protected either by oval conduit or by means of small metal cover channels fixed to the wall or floor. Systems using mineral-insulated covered cables follow the same principles. This form of cable consists of single strands of copper or aluminium, all encased in a sheath of the same metal, which is densely packed with fine magnesium oxide insulation around the strands. This insulating material is unaffected by heat or age and is therefore very durable, but it can absorb moisture. The sealing of the ends of this type of cable with special sealing 'pots' is therefore of paramount importance. Cables used in a conduit installation have adequate protection, but it is generally necessary to chase the walls of the building to accommodate the conduit, outlet socket boxes and switch boxes below the wall finish level. Surface run conduit is normally secured to the backing by using screwed shaped clips called **saddles**.

Guidance for the installation of electrical circuits and electrical equipment is provided in the Building Regulations, Approved Document P: Electrical safety. Further details are produced in the Institution of Electrical Engineers Regulations in accordance with BS 7671: *Requirements for electrical installations*.

With the exception of minor work such as replacement of socket outlets, switches, etc., all other proposed electrical work in dwellings requires notification to the local building control authority. Certification of a satisfactory installation is by:

- inspection by a qualified electrician appointed by the local building control office; or
- self-certification by a prescribed suitably qualified competent person.

DOMESTIC GAS INSTALLATIONS

Gas is a combustible fuel that burns with a luminous flame; it is used mainly in domestic installations as a source of heat energy in appliances such as room heaters, cookers and boilers. Historically, gas preceded electricity as a domestic energy utility, and it has been used to power washing machines, refrigerators and lighting.

BG Group plc supplies and processes gas for general distribution in the UK. It is stored in a liquid state in huge underground storage chambers, from where it is conveyed as a gas through a network of underground mains installed and maintained by the gas transporter, National Grid Transco Plc. Some gas is still stored in above-ground cylindrical holders, but these are gradually being phased out. Until the late 1960s a gas manufactured from coal known as **town gas** served domestic premises. The original pipework for conveying **town gas** was retained when supplies were changed to offshore gas supplies, which became known as **North Sea gas**. Offshore supply is a natural gas comprising about 90% methane, whereas coal gas had a hydrogen content of about 50%. Therefore, although pipework was unaffected, the different burning characteristics of the gases required major changes or replacements to domestic appliances. Table 11.5.1 shows a comparison of the calorific values of both gases and for bottled gases for use where mains gas is unavailable. Most domestic appliances manufactured for use with mains gas can be converted for bottled gas by changing the burners.

The installation of gas supplies can be considered in three stages:

1. **Main**   This is the part of the gas distribution system that is connected to form a grid. The grid is an interconnection of main pipes, normally located beneath roads or footways. The arrangement of pipes enables maintenance and repairs to be undertaken in isolation, without a major disruption of supply to a large community. For ease of identification, all new underground gas installation pipework is undertaken in a yellow-coloured plastic, not to be confused with water (blue), electricity (red) or drainage (orange).

2. **Service pipe**   This is the connection between the main and the consumer control valve positioned just before the meter. For domestic installations the service pipe diameter is between 25 and 50 mm according to the number and type of appliances being installed. The pipeline should be as short as possible, at right angles to the main, and laid at least 375 mm below the surface to avoid damage. Figure 11.5.1 shows a typical domestic installation with meter details and accessories.
3. **Internal distribution**   This commences at the consumer control valve and consists of a governor to stabilise gas pressure and volume, a meter to record the amount of gas consumed, and pipework to convey the gas supply to appliances. The meter and associated equipment are the property of the gas provider, but thereafter the installation is the responsibility of the building owner. A typical internal distribution to a house is shown in Fig. 11.5.2.

Table 11.5.1   Calorific values of gases for domestic appliances

Type of gas	Approx. calorific value (MJ/m^3)
Town gas	18
North sea gas	38
Bottled gas (Propane)	96
Bottled gas (Butane)	122

Notes:
1. Approximate values are given, as values will vary slightly depending on the source and quality.
2. Propane and butane are otherwise known as liquid petroleum gas (LPG). They are found under the North Sea bed and in other oil resources around the world. For transportation they are pressurised and liquefied to about 1/200 their gas volume.

■■■ PIPEWORK

Pipes for internal distribution can be of mild steel (not galvanised) or solid drawn copper. Flexible tubing of metal-reinforced rubber may be used for connection to cookers and portable appliances. The size of pipes will depend on such factors as gas consumption of appliances, frictional losses due to pipe length, and number of bends. A typical domestic installation is in 22 mm copper from the meter, with a 15 mm branch to a boiler and 10 mm to individual low-rated appliances. Pipes must not be accommodated in the cavity of a wall, but may pass through a wall provided a sleeve of non-corrodible material protects the pipe against any differential movement or settlement.

■■■ APPLIANCES

Gas appliances fall into two categories:

■ gas supply only: cookers, radiant convector fires, decorative fuel effect fires;
■ gas supply plus other services: central heating boilers, water heaters.

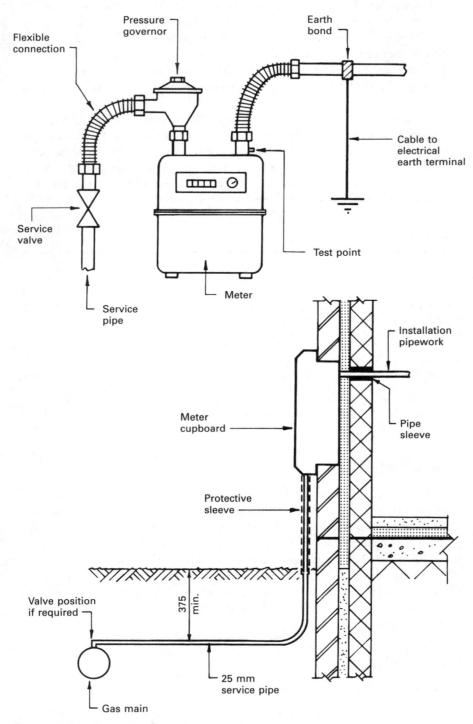

Figure 11.5.1   Domestic gas supply.

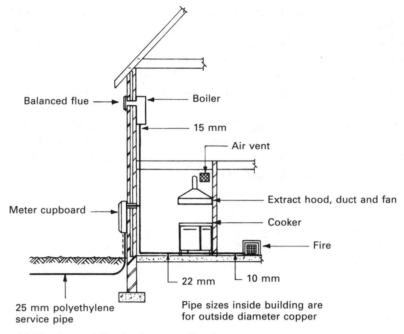

Figure 11.5.2   Internal distribution to appliances.

Gas radiant convector fires have a relatively low energy rating (generally < 3 kW). If less than 7 kW input rating, air for gas combustion can be taken from the room. They have a visible heat source as the gas burns through the fragile fireclay radiant. Gas fires can be set on a fireplace hearth with a closure back plate over the fire opening. A void in the plate allows for extraction of burnt gases directly up the chimney. Precast concrete flue blocks can also be used; see section on flue materials in Chapter 5.2.

Decorative fuel effect (DFE) fires are a popular adaptation within traditional fireplace openings. Log or coal effects are designed to resemble a real fire by burning gas and discharging the combusted products indirectly into the open flue. For efficient combustion, most DFE fires will require a room air vent of at least 10 000 mm^2 free area. The manufacturer's details should be consulted.

Gas cookers and instantaneous water heaters are an exception to the requirements for flues, but they should be installed and used only in well-ventilated rooms. The Building Regulations, Approved Document J, Section 3 should be consulted for minimum areas of air vents relative to room volume.

Room-sealed or balanced flue heaters are appliances that have the heat exchanger sealed from the room. Air for combustion of gas is drawn in from an external duct at the back of the unit, and burnt gases discharge through an adjacent flue: see Fig. 5.2.6. Acceptable locations for balanced flues are shown in Fig. 5.2.7 and Table 5.2.1.

Gas central heating consists of a wall-mounted or free-standing boiler connected to a flue (Chapter 5.2). This is the heat source for either a ducted fan-assisted warm air circulation system or a pump-circulated hot water system.

The Health and Safety Executive, through the Gas Safety (Installation and Use) Regulations, requires the installation and maintenance of gas fittings, appliances and storage vessels to be undertaken only by a competent and qualified person. A facility for assessing and accrediting individuals to undertake the various aspects of gas fitting and maintenance has been established by the Council for Registered Gas Installers (CORGI). Ref. Building Regulations, Approved Document J, Section 3.2.

The installation of gas services, like that of electricity, is a specialist subject not normally developed within the framework of construction technology. For a greater appreciation, see *Building Services, Technology and Design* by R. Greeno (Pearson Education and CIOB, co-publishers).

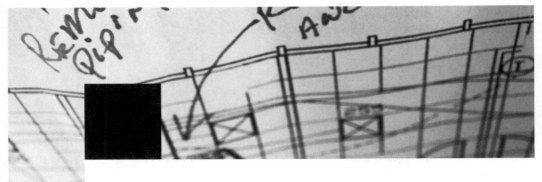

BIBLIOGRAPHY

Barry, R. *The Construction of Buildings*. Crosby Lockwood & Sons Ltd.

Elder, A. J. *A. J. Guide to the Building Regulations*. The Architectural Press.

Elder, A. J. *Relevant A. J. Handbooks*. The Architectural Press.

Fisher Cassie, W. and Napper, J. H. *Structure in Building*. The Architectural Press.

Greeno, R. *Building Services, Technology and Design*. Addison Wesley Longman and CIOB.

Greeno, R. *Principles of Construction*. Addison Wesley Longman.

Hall, F. *Plumbing*. Macmillan.

Handisyde, Cecil C. *Building Materials*. The Architectural Press.

Leech, L. V. *Structural Steelwork for Students*. Butterworth.

Llewelyn Davies, R. and Petty, D. J. *Building Elements*. The Architectural Press.

McKay, W. B. *Building Construction*, Vols 1 to 4. Longman.

Ragsdale, L. A. and Raynham E. A. *Building Materials Technology*. Edward Arnold Ltd.

Wooley, L. *Drainage Details*. Chapman & Hall.

GENERAL

Application of Mastic Asphalt. Mastic Asphalt Council and Employers Federation.

ODPM. *The Building Regulations, Approved Documents*. HMSO.

Copper Roofing. Copper Development Association.

Gas Handbook for Architects and Builders. The Gas Council.

Handbook on Structural Steelwork. The British Constructional Steelwork Association Ltd and The Steel Construction Institute.

Mitchells Building Series. Addison Wesley Longman.

'Sanitation details by Aquarius', *Building Trades Journal*. Chapman & Hall.

The Blue Book of Plasterboard. British Gypsum Ltd.

The Green Book of Plasterboard. British Gypsum Ltd.
The Plumber's Handbook. Lead Development Association.

OTHER READING

Relevant advisory leaflets. DoE.
Relevant BRE Digests. Construction Research Communications Ltd.
Relevant British Standards Codes of Practice. British Standards Institution.
Relevant British Standards. British Standards Institution.
Relevant manufacturers' catalogues contained in the Barbour Index and Building
    Products Index Libraries.

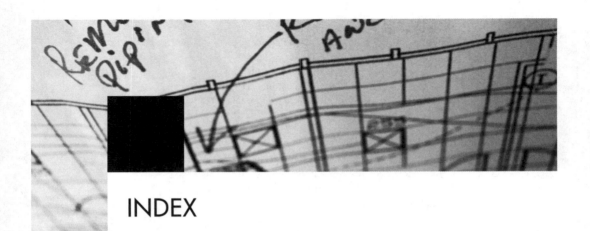

INDEX